Space Radio Science

Earth Space Institute Book Series
Editor in Chief: Dr Peter Kleber, Chairman, Earth Space Institute

The Earth Space Institute Book Series covers different aspects of the future of space. These books aim to show how space has become a new tool in further developments for the benefit of mankind. In this context, space can now be considered as part of the human experience. Each volume in the series will be devoted to selected topics in engineering, legal and financial issues, and space-based sciences (extraterrestrial, earth observation, life sciences, materials science) intended for students, researchers and space professionals. The series will also feature historical and general overviews of interest to the public.

Volume 1
Space Science in China
edited by *Wen-Rui Hu*

Volume 2
Der Mensch im Kosmos
edited by *Peter R. Sahm* and *Gerhard Thiele*

Volume 3
Multiple Gravity Assist Interplanetary Trajectories
by *Alexei V. Lubunsky, Oleg V. Papkov* and *Kostantin G. Sukhanov*

Volume 4
Large Space Structures Formed by Centrifugal Forces
by *V.M. Melnikov* and *V.A. Koshelev*

Volume 5
East Asian Archaeoastronomy: Historical Records of Astronomical Observations of China, Japan and Korea
by *Zhentao Xu, Yaotiao Jiang* and *David W. Pankenier*

Volume 6
Space Debris: Hazard Evaluation and Mitigation
edited by *Nickolay N. Smirnov*

Volume 7
Physics of Fluids in Microgravity
edited by *Rodolfo Monti*

Volume 8
Space Radio Science
by *Oleg I. Yakovlev*

Space Radio Science

Oleg I. Yakolev
*Institute of Radioengineering and Electronics, Russian
Academy of Sciences, Moscow, Russia*

Translated from the Russian by Nikolai and Olga
Golovchenko

TAYLOR & FRANCIS
ALERE FLAMMAM
Founded 1798

Front cover: 70-m parabolic antenna used in Evpatoria, Crimea, the base instrument for both space mission control and space radio science experiments. This antenna was used to broadcast the interstellar messages "Cosmic Call" in 1999 and "Teen-Age Message" in 2001 to 10 nearby Sun-like stars. Courtesy of E.V. Kazakov

Back cover: Antenna of the Centre of Deep Space Communication consisting of eight parabolic dishes, each 16m in diameter. This instrument was used to provide communication with the first artificial planet satellites Mars-2 in 1971 and Venera-9 in 1975. Courtesy of E.V. Kazakov

First published 2002 by Taylor & Francis
11 New Fetter Lane, London EC4P 4EE

Simultaneously published in the USA and Canada by Taylor & Francis Inc
29 West 35th Street, New York, NY 1001

Taylor & Francis is an imprint of the Taylor & Francis Group

Publisher's note: this book has been prepared from camera-ready-copy provided by the authors.

© 2002 Taylor & Francis

Printed and bound in Great Britain by TJ International Ltd, Padstow, Cornwall

British Cataloguing in Publication Data
A catalogue record for this book is available from the British Library

Library of Congress Cataloging in Publication Data
A catalog record for this book has been requested

ISBN 0-415-27350-1

Contents

Preface

Interest in the conditions under which radio waves propagate in space arose with the origination of radio astronomy as a science. The interpretation of data on the radio-frequency emissions of planets and the Sun required that the effects related to the propagation of radio waves through the atmospheres of planets and the solar corona be taken into account. The radio astronomic observations of the Sun and Venus provided the first convincing evidence for the efficiency of the remote sensing of distant bodies. Radio astronomy deals with the natural noiselike electromagnetic radiation of celestial bodies, characterized by power flux density. The employment of radars operating on monochromatic and modulated signals for the remote sensing of planets required the elaboration of a technique for the analysis of frequency fluctuations, energy spectrum, and the time delay of reflected radio waves. Spacecraft launchings into orbits around the Earth and other planets called for the solution of many problems, perhaps one of the most basic being the problem of the effect of planetary atmospheres, as well as the circumsolar and interplanetary plasmas, on the characteristics of radio waves. The spacecraft-based experimental studies of medium-induced changes in the parameters of radio waves contributed to the development of efficient radiophysical techniques for the exploration of remote objects. Voluminous experimental data that had accumulated since 1957 gave rise to a new branch of radiophysics, space radiophysics. This science deals with the remote sensing of the Earth and other bodies of the solar system and treats the problem of the propagation of radio waves in relation to space communications.

The outstanding space achievements of Russia, including the launchings of the first man-made satellites of the Earth, the Moon, Mars, and Venus, and the historical flight of Yuri Gagarin, was an important impetus for the development of space radiophysics. The range of the radiophysical exploration of space has greatly extended after the historical flights of American astronauts to the Moon and unmanned spacecraft to Jupiter, Saturn, Uranus, and Neptune.

The objects of investigation by space radiophysicists include the surface, atmosphere and ionosphere of the Earth, the Moon, and other planets, space and circumsolar plasma, etc. One of the objectives of space radiophysics is to elucidate the conditions of the propagation of radio waves in various media and to develop efficient methods for their remote sensing. The aim of this book is to give a comprehensive analysis of radiophysical problems and methods, as well as the benefits from the use of various spacecraft and radio signals of meter, decimeter, centimeter, and millimeter wavelengths.

Each chapter of this book is devoted to three large topics. These are (1) the treatment of the direct problem of the propagation of radio waves through various media and the effect of encountered media on the characteristics of radio waves with regard to space communications; (2) the description of methods used for the remote sensing of particular media, as well as for solving the inverse problem, that is, the retrieval of the medium parameters from the characteristics of probing radio waves; and (3) analysis of the results of radiophysical experiments in space and the comparison of experimental dependences with those derived theoretically.

Chapter I considers the conditions of ground-to-satellite communications, the effects of Earth's atmosphere and ionosphere on the propagation of radio waves (or, in other words, the properties of these media as a "radio window" to space), as well as the phenomena of absorption and amplitude fluctuations of radio waves, their refraction and time delay, and variations in their phase, frequency, and polarization. In Chapter II, the potentialities of the radio occultation technique for studying planetary atmospheres and ionospheres are evaluated with reference to those of Mars, Venus, Jupiter, Saturn and other planets of the solar system. The feasibility of the radio occultation technique for a global monitoring of the Earth's atmosphere and ionosphere is also considered. Chapter III is devoted to radio occultation observations of circumsolar and interplanetary plasmas. The data presented show the effect of these plasmas on radio waves and how the characteristics of the solar wind can be determined from fluctuations in the frequency and amplitude of radio signals and from power spectrum broadening. Chapter IV describes the mechanisms of reflection of radio waves. The appropriateness of the remote sensing technique for studying planetary surfaces is demonstrated by the results of ground-based radar observations of the Moon, asteroids, and other celestial objects and their bistatic radar studies from lunar, Venusian, and Martian satellites. The features of side-looking radar observations of the Earth and Venus from their satellites are analyzed. And finally, Chapter V is devoted to the conditions of communications to interplanetary stations and the propagation of radio waves in the Galaxy. This chapter also gives a brief account of the effect of gravitational fields on the propagation of radio waves. Each chapter is written in an independent manner, so that the reader can understand any chapter without looking through the previous ones.

Since the book deals with a greater number of media encountered by radio waves than the relevant monographs concentrating primarily on "Earth's conditions", many of the theoretical considerations are presented without deriving the inferences; in this case the reader is merely referred to the appropriate monographs. It should also be noted that the author of this book is an experiment researcher and, therefore, gives preference to experimental material. The theory of particular phenomena is given only when it can provide some valuable formulas to explain relevant experimental data.

The list of references was given a particular care. The bibliography covers all the papers on space radiophysics published over the period 1960–1999. Russian publications are indicated by the parenthesized "in Russian". It should be noted that, since 1974, the English versions of the main Russian academic journals have been available in the USA. The Russian journal *Radiofizika* is published as *Radiophysics and Quantum Electronics*, *Kosmicheskie Issledovaniya* as *Cosmic Research*, *Astronomicheskii Zhurnal* as *Russian Astronomical Reports* (until 1992, *Soviet Astronomy Journal*), *Radiotekhnika i Elektronika* as *Radio Engineering and Electronic Physics* (since 1984, as *Soviet Journal of Communications Technology and Electronics* and, since 1992, as *Journal of Communications Technology and Electronics*). In the case when both Russian and English versions of a publication are available, the reference is marked by an asterisk.

References in the text are given without mentioning the names of their authors. This is because one cannot often be entirely certain as to who was the first to express a particular idea or who was the first to accomplish a particular experiment. The references to each chapter are chosen to indicate the authors who made the main contribution to a particular research area.

The author believes that this book will be useful to radiophysicists and radio engineers who are interested in methods for the remote sensing of media, space communications, and satellite tracking. Teachers and students can use this book as a treatise in such disciplines as space radiophysics, remote sensing, propagation of radio waves, and space communications.

Chapter 1

Radio wave propagation in communications to Earth-orbiting satellites

1.1 Propagation of free radio waves

Propagation of radio waves in space is free, since they do not experience a strong influence of a medium. The propagation of free radio waves can be described by simple equations. Let us assume that a transmitting omnidirectional antenna of power W_1 uniformly irradiates a sphere of area $4\pi L^2$. In this case the radio wave energy flux P at distance L from the transmitter is given by

$$P = \frac{W_1}{4\pi L^2}. \tag{1.1}$$

Real antennae are directional: they increase G_1 times the density of energy flux in the direction of a radio beam when compared with omnidirectional antennae. Therefore, the wave energy density in the direction of the beam is equal to PG_1. A receiving antenna is characterized by the effective area A_2, so that the power of the input signal, W_2, is given by PG_1A_2. The relation $W_2 = PG_1A_2$ defines the effective area of receiving antennae. Based on eq. (1.1) and the definition of the antenna gain I_1, the relationship between the effective antenna area A_2, input signal power W_2, and transmitter power W_1 can be given by

$$W_2 = \frac{W_1 G_1 A_2}{4\pi L^2}. \tag{1.2}$$

The gain G and the effective area A of antennae are related as

$$G = 4\pi A \lambda^{-2}, \tag{1.3}$$

where λ is the wavelength. By expressing the gain G_1 of the transmitting antenna through its effective area A_1 and taking into account (1.2) and (1.3), we get another formula for the propagation of free radio waves:

$$W_2 = \frac{W_1 A_1 A_2}{L^2 \lambda^2}. \tag{1.4}$$

1

Formula (1.4) gives an idea of the significance of the antenna dimensions and the dependence of input signal on its wavelength. For aperture antennae, the effective area A is proportional to the antenna aperture; in particular for parabolic and horn antennae, $A \approx S/2$, where S is the aperture area. The term λ^{-2} in formula (1.4) takes into account the focusing properties of the transmitting antenna. The half-power beamwidth is $\Delta\varphi \sim \lambda^{-1}$; therefore, the transmitting antenna has a gain $G \sim \lambda^{-2}$, and the input signal power is $W_2 \sim \lambda^{-2}$.

If a transmitter or receiver of radio waves moves in a vacuum so that the transmitter–receiver line-of-sight moves at a rate of V_2, there appears the Doppler shift in frequency given by

$$\Delta f_0 = V_2 \lambda^{-1}. \tag{1.5}$$

Here V_2 is the projection of the spacecraft velocity vector onto the transmitter–receiver line-of-sight.

For free radio waves, the equiphase surface is a sphere of radius L, and the travel time, given by $\Delta t = Lc^{-1}$, depends only on the distance L and the speed of radio waves in a vacuum, c. In this case the rays of radio waves, which, by definition, are perpendicular to the equiphase surface, represent straight lines passing through the satellite-borne transmitters and Earth-based receivers. The polarization of radio waves is constant at any distance from the transmitting antenna, as it is determined only by the polarization characteristics of the antenna.

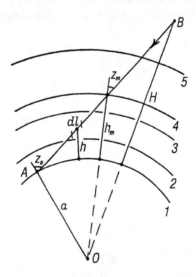

Fig. 1.1. A diagram illustrating radio communications to Earth-orbiting satellites.

The aforementioned simple relationships are only applicable to free radio waves. However, in Earth-to-satellite communications, radio waves propagate through the Earth's troposphere and ionosphere and, hence, experience the influence of these media. In this case the power of the input signal can change because of absorption and fluctuations of radio waves. Furthermore, radio waves undergo refraction, as a result of which the ray path deviates from a straight line. Since the speed of radio waves in the troposphere and ionosphere differs from that in a vacuum, the travel time of radio waves increases in these media. The troposphere and ionosphere also affect the phase, frequency, and polarization of radio waves to an extent dependent on their frequency; therefore, the range of frequencies suitable for communications to satellites and for radiophysical space investigations is determined by the properties of the troposphere and ionosphere. In particular, the ionosphere and troposphere restrict the use of meter and millimeter radio waves, respectively. The encountered medium and relative positions of a satellite and receiver are schematically shown in Fig. 1.1. The line *AB* between the satellite located at point *B* and the ground-based receiving antenna positioned at point *A* represents the path along which radio waves propagate. Circles *3* and *5* show the arbitrary boundaries of the ionosphere, and line *4* denotes the main ionospheric layer. The troposphere extends between circles *1* and *2*.

1.2 Absorption and fluctuations of meter radio waves in the ionosphere

Vertically incident radio waves can either reflect off the ionosphere or pass through it, depending on the relationship between the radio frequency f and the electron density in the principal ionospheric maximum, N_m. If the following inequality,

$$f < f_c = (80.8 N_m)^{1/2},\qquad(1.6)$$

is satisfied, radio waves reflect from the ionosphere. When $f > f_c$, radio waves pass through it [1]. In expression (1.6), the dimensions of frequency and electron density are Hz and m^{-3}, respectively. The critical frequency f_c of obliquely incident radio waves is multiplied by a factor sec z_m, so that the condition to be satisfied for the transmission of such radio waves through the ionosphere takes the form

$$f > f_c \sec z_m,\qquad(1.7)$$

where z_m is the zenith angle of radio waves propagating near the main ionization maximum (Fig. 1.1).

The critical frequency f_c is determined by the peak electron number density N_m in region F of the ionosphere. This frequency can be measured by a network of ionospheric stations that collect voluminous experimental data concerning the diurnal and seasonal dependences of critical frequency and the effect of solar activity on it. In summer, the diurnal dependence of f_c in mid-latitude regions is weak: f_c equals 6–8 MHz day and night. In winter, however, the diurnal dependence of f_c is clear-cut: f_c shows a maximum (10–12 MHz) at noon and a minimum (4–5 MHz) at night and in the early morning. The critical frequency depends on the solar activity, increasing by about a factor of 1.8 when the solar activity changes from a minimum to a maximum. Based on these data and taking into account inequality (1.7), one can estimate that, at zenith angles $z_m < 70°$, radio waves transmittable through the ionosphere should have a frequency of more than 14 MHz in summer and more than 35 MHz in winter.

Radio frequencies exceeding critical values are weakly absorbed in the ionosphere, so that their absorption during space communications can be neglected. At the same time, the absorption of meter radio waves, which are extensively employed in radiophysics, can reach noticeable values. The decrease in the power flux of radio waves propagating in a plasma is given by the expression [1]

$$Y = \exp\left(-\frac{80.8}{cf^2 \cos z_m} \int\limits_0^H N_e v \, dh \right). \tag{1.8}$$

Here N_e is electron density, v is the effective frequency of electron collisions, h is the height above the Earth's surface, and H is the satellite's altitude. Formula (1.8), which is valid for high frequencies ($f > 20$ MHz), such that $v^2 \ll f^2$, allows the dependence of the absorption coefficient on altitude to be found. Calculations in [1, 4] showed that absorption increases with altitude until the latter reaches a value of 250–300 km. At these altitudes, the absorption of radio waves becomes constant, since the upper ionosphere contributes little to absorption. Inasmuch as we deal with transmitter's altitudes higher than 250 km, the upper limit of the integral in formula (1.8) can be extended to infinity. It should be noted that the application of this formula suggests that the dependences $N_e(h)$ and $v(h)$ are known. These functions depend on the zenith angle of the Sun, its activity, and the season. The height profile of the effective frequency of collisions, $v(h)$, can only be determined experimentally. In such a case, direct measurements of absorption of radio waves in the ionosphere are preferable.

The absorption of radio waves propagating vertically through the ionosphere was investigated in detail with a riometric method. With this method, absorption is measured at ground-based stations by recording variations in the space radiation that have frequencies ranging from 20 to 35 MHz [5–11]. The absorption of these frequencies in the mid-latitude regions has a clear-cut diurnal dependence with a maximum between 10:00 and 16:00 and minimum between 2:00 and 4:00 local time. For instance, in Alma-Ata (Kazakhstan), the absorption of 24.6-MHz radio waves is 1–1.2 dB in the day and 0.3–0.5 at night [6]. In Moscow, the absorption of 28.6-MHz radio waves is 1.6 dB in the day and 0.4–0.6 dB at night [7]. In polar regions, the diurnal dependence of the absorption of radio waves is not profound; instead, there are regular periods with an anomalously high absorption. For instance, in the northern auroral oval, the absorption of 25-MHz radio waves has a value of 2–3 dB for 10% of the time [8]. Measurements performed in the Arctic auroral oval at Dikson, a small Kara Sea port in Russia, showed that the median absorption of 32-MHz radio waves is 1.9 dB [9]. At the same time, four-month observations in 1966 revealed some abrupt rises in the absorption of radio waves, amounting to 6–8 dB. Similar riometric observations in Thule (Greenland) showed that the mean absorption of 30-MHz radio waves there is 1–2 dB, whereas in September, 1971, an absorption of 4–6 dB had been observed over a few days. At the South pole, several-hour periods of an abrupt increase in the absorption of 30-MHz were observed [11]. Thus, the ionospheric absorption of vertically propagating radio waves is usually weak, even if their frequency is about 30 MHz.

As follows from (1.8), for a radio beam incident obliquely at a zenith angle of z_m, absorption expressed in dB should be multiplied by a factor of $\sec z_m$. In the mid-latitude regions, for the zenith angle $z_m = 70°$ and radio frequencies of 30–35 MHz, absorption is 4–5 dB in the daytime and 1.5 dB at night. In polar regions, the absorption of radio waves during the periods of abrupt rise may reach 9–18 dB. According to (1.8), the absorption of radio waves decreases with frequency as f^{-2}; therefore, the absorption of 60-MHz radio waves must be four times as low as that of 30-MHz waves. In other words, except for short periods of an abrupt rise in absorption in the auroral ovals, the ionosphere can be considered as transparent for radio frequencies higher than 80 MHz.

Wind-induced irregularities in electron density can affect the amplitude of radio waves. In particular, such irregularities can cause an irregular fading of radio waves, although the mean value of the wave energy flux remains constant. Radioastronomical observations of discrete sources and analysis of signals received from the first satellites indicated the occurrence of significant irregular fluctuations in the amplitude of meter radio waves. Soon it became clear that these fluctuations are caused by electron density irregularities in the vicinity of the main ionospheric maximum, i.e., at altitudes of 200–300 km. In view of this, extensive investigations of the ionospheric fading of satellite signals were undertaken.

It should be noted that the results of these studies are difficult to compare, since various teams of investigators use different characteristics to describe these fluctuations as a random process. Common fade characteristics are given by the following formulae:

$$s_1 = \langle E \rangle^{-1} \langle |E - \langle E \rangle| \rangle, \tag{1.9}$$

$$s_2 = \langle E \rangle^{-1} \left[\langle (E - \langle E \rangle)^2 \rangle \right]^{\frac{1}{2}}, \tag{1.10}$$

$$s_3 = \langle E^2 \rangle^{-1} \langle |E^2 - \langle E^2 \rangle| \rangle, \tag{1.11}$$

$$s_4 = \langle E^2 \rangle^{-1} \left[\langle (E^2 - \langle E^2 \rangle)^2 \rangle \right]^{\frac{1}{2}}. \tag{1.12}$$

Here the symbols $\langle\ \rangle$ denote averaging over time, and E represents the instantaneous strength of a radio-frequency field. Expressions (1.9) and (1.10) define the magnitude of field fluctuations, while expressions (1.11) and (1.12) define the magnitude of the fluctuations in the radio wave energy flux. Various methods for describing ionospheric fading were analyzed in [12–14]. Generally, the relationship between s_1, s_2, s_3, and s_4 depends on the distribution law of amplitude fluctuations. However, to an accuracy that is sufficient for practice, the following approximate relationships are valid [12]:

$$s_1 = 0.42 s_4,$$

$$s_2 = 0.52 s_4, \tag{1.13}$$

$$s_3 = 0.78 s_4.$$

These expressions require that the recorded signals from satellites be statistically processed. For simplicity, in early works, the fading depth was estimated through the scintillation index s_5 defined as

$$s_5 = \frac{E_{max}^2 - E_{min}^2}{E_{max}^2 + E_{min}^2}. \tag{1.14}$$

Here E_{max}^2 and E_{min}^2 are the powers of the signal maximum and signal minimum ranking third in magnitude among, respectively, signal maxima and signal minima [14]. It should be noted that the scintillation index s_5 is not stringent, and its relationship with the indices s_1, s_2, s_3, and s_4 is equivocal [15]. Nevertheless, an approximate relationship between the scintillation index s_5 and r.m.s. field fluctuations, s_2, can be established [15], as illustrated by Table 1.1.

Table 1.1. Relationship between indices s_5 and s_2

s_2, dB	0.5	1	2	3	4	5
s_5	0.21	0.38	0.67	0.85	0.92	0.96

Let us consider some examples of ionospheric fading. Figure 1.2a represents a 15-min record of the satellite signal at the Sagamer-Hill station (the United States). This record illustrates the relatively slow fadeouts of 254-MHz radio waves with a scintillation index s_5 equal to 0.72 [16]. The ordinates in Fig. 1.2 show the signal strength in relative units. Ionospheric fades are variable, being either slow or rapid. Figure 1.2 exemplifies the rapid fades of the 150-MHz (record a) and 400-MHz (record c) radio waves transmitted from a navigation satellite and received in Milston (the United States) [15]. Unlike the record a, records b and c have 1-min time scales. The r.m.s. amplitude fluctuations in Fig. 1.2b have an intensity of 5.5 dB; this corresponds to the rapid fadeout of meter radio waves. A considerably smaller value of s_2 (1.25 dB) for decimeter radio waves indicates that their fading is much lower in depth (Fig. 1.2c).

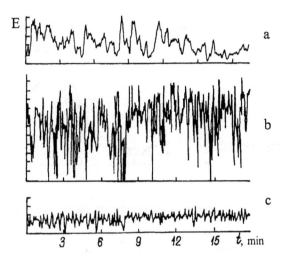

Fig. 1.2. Three records illustrating the fading of meter radio wave in the ionosphere [15, 16].

Analysis of a substantial body of ionospheric fading data made it possible to derive the diurnal and seasonal dependences of fluctuation depth, as well as its dependence upon the wavelength of radio waves and the latitude of the receiving station. Regions with deep and shallow fades were revealed. The results of the ionospheric fading studies were reviewed in [12–45]. Fadeouts were most frequently observed in the northern auroral oval [19] at geomagnetic latitudes higher than 55°. In the mid-latitude regions, ionospheric fading was weak. In equatorial regions, deep fluctuations can be observed at night.

The equatorial zone of scintillations corresponds to geomagnetic latitudes lower than 20° [20]. The boundaries of both polar and equatorial fading zones are variable and depend on the solar and magnetic activities. The latitude dependence of the fading depth was found from the data provided by a network of ground-based stations, which received 40-MHz radio signals tramsmitted from satellites [16, 21]. Figure 1.3 shows the dependence of the fading index s_5 on the geomagnetic latitude of the site where the ray path crosses the region of ionospheric inhomogeneities (the two curves in this figure correspond to the minimum and maximum fading). This figure fails to adequately describe fluctuations in polar areas, for which a strong variability of the fading depth and pattern is typical. Generally, magnetic storms in these areas are accompanied by fadeouts. A clear-cut diurnal dependence of fading was observed in equatorial areas, where fading was minimum in the day and maximum at midnight. Equatorial fades are sporadic and abrupt [22–24]. The fading depth strongly depends on the solar activity. The frequency of fade-outs during the periods of maximum solar activity is several times greater than that during the periods of minimum solar activity.

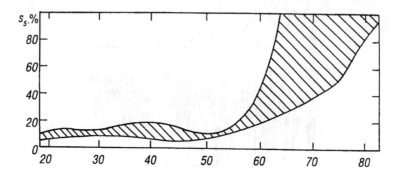

Fig. 1.3. Latitude dependence of the ionospheric fading index s_5 [16, 21].

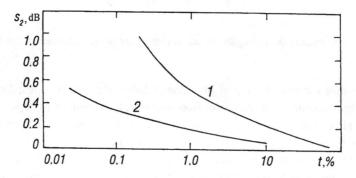

Fig. 1.4. Empirical distribution functions of the ionospheric fading of 400-MHz radio waves [15].

As was noted above, ionospheric fades are highly variable; therefore, their expected values can be estimated using the empirical distribution functions. Figure 1.4 presents the results of the three-year observations of the fading of 400-MHz radio waves [15]. Curves *1* and *2* in this figure correspond to high latitudes of 60–70° and mid latitudes of 42–54°, respectively. The data shown in Fig. 1.4 allow one to estimate the fraction of the time during which the root mean square amplitude fluctuations are greater than a specified value.

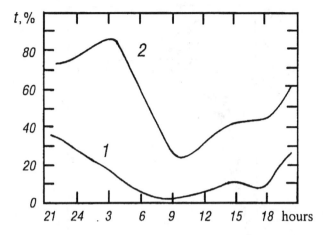

Fig. 1.5. Diurnal dependence of the relative duration of ionospheric fadeouts in the polar regions of the Earth [16].

The polar regions of the Earth are characterized by a clear-cut diurnal dependence of the ionospheric fading depth. Figure 1.5 shows how the time of day or night affects the fraction of the time during which the fading index s_5 is greater than 0.6 and, consequently, the variance s_2 is greater than 1.8 dB. Fading in the polar regions of the Earth strongly depends on magnetic activity. Curve *1* in Fig. 1.5 corresponds to magnetically quiet conditions, while curve *2* corresponds to the conditions of magnetic disturbance. It can be seen that fading is maximum before midnight under magnetically quiet conditions and after midnight under the conditions of magnetic disturbance (the data discussed were obtained during investigations in Greenland for a frequency of 254 MHz [16]). Similar dependences were derived for Antarctica, where fading was the highest at about 19:00 local time [25].

In an attempt to elucidate the dependence of ionospheric fluctuations on the wavelength of radio signals, the authors of [26, 27] performed their measurements at two different frequencies and found that the index of ionospheric fading $s_5 \sim \lambda^m$, where $m_1 = 2$. Similar measurements of the fading of 22- and 34-MHz radio waves yielded a value of $m_1 = 1.6$ provided that fluctuations are small [28]. With the accumulation of sufficient experimental data it became clear that the magnitude of fluctuations depends on the wavelength of radio signals if their fluctuations are small. For instance, the comprehensive statistical analysis of variance s_2 performed for two frequencies of 150 and 400 MHz (the results of this analysis are presented in Fig. 1.6) showed that $s_2 \sim \lambda^{1.6}$ when $s_2 > 5$ dB. However, if $s_2 > 5.6$ dB, no frequency dependence can be derived, since fading becomes saturated.

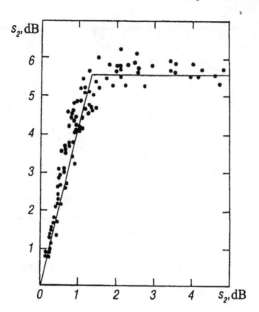

Fig. 1.6. The variance s_2 of simultaneously measured fluctuations in the amplitude of 150-MHz (ordinate) and 400-MHz (abscissa) radio waves [29].

Earlier investigations of ionospheric fluctuations provided some information about their properties. Later, analysis of these fluctuations was used to study the ionospheric plasma inhomogeneities. As follows from the theory of radio wave propagation in statistically irregular media, the temporal spectrum of amplitude fluctuations, $\Phi_E(F)$, and the wavelength dependence of amplitude fluctuations, $s(\lambda)$, are determined by the spatial spectrum of the refractive index of radio waves provided that the fluctuations are small and, consequently, unsaturated [2]. If the spatial spectrum of the refractive index fluctuations or, as in the given case, the spectrum of fluctuations in the electron density can be given by the power function [2]

$$\Phi_N \sim \ae^{-\alpha}, \tag{1.15}$$

the temporal spectrum of small fluctuations in the amplitude can be described in terms of the similar power function

$$\Phi_E \sim F^{-m_2}. \tag{1.16}$$

In (1.15), $\ae$ is the spatial wavenumber; in (1.16), F is the frequency of fluctuations. The power indices m_2 and α are related as

$$m_2 = \alpha - 1. \tag{1.17}$$

The wavelength dependence of small r.m.s. fluctuations is determined by spectrum (1.15):

$$s \sim \lambda^{m_1},$$
$$m_1 = \frac{\alpha + 2}{4}. \tag{1.18}$$

The derivation and analysis of these formulae are described in more detail in the third chapter. It should be emphasized that expressions (1.16)–(1.18) are valid only if the spatial spectrum of fluctuations in the electron density can be given by a simple power function, such as (1.15), and if the amplitude fluctuations of radio waves are small. Analysis of the temporal spectrum of amplitude fluctuations performed by the authors of the publications [30–34] showed that the dependence (1.16) satisfactorily conforms to experimental data if $m_2 \approx 3$; from whence $\alpha \approx 4$.

The spatial spectrum of electron density fluctuations thus derived can account for the above wavelength dependence of unsaturated amplitude fluctuations. Setting $\alpha = 4$ in (1.18), we get $m_1 = 1.5$; this value agrees well with the experiment. Data on fluctuations in the amplitude of radio waves of certain wavelengths presented in Figs. 1.3, 1.4, and 1.5 can be converted to other wavelengths, using the dependence $s_2 \sim \lambda^{1.5}$. If this conversion yields values for s_2 greater than 5.6 dB or values for s_5 greater than 0.9, this implies that fluctuations are saturated.

The aforementioned data refer to the radio waves that propagate vertically in the ionosphere. As follows from the theory, the magnitude of unsaturated fluctuations is proportional to the thickness of the nonuniform region of ionospheric plasma; therefore, the root mean square amplitude fluctuations are proportional to the secant of the zenith angle z_m in the ionosphere (Fig. 1.1).

Ionospheric plasma inhomogeneities were studied through the long-term observations of ionospheric plasma inhomogeneities in polar, mid-latitude, and equatorial regions [39–45]. The fluctuations in the radio signals transmitted from geostationary satellites were recorded at four frequencies. The authors of the publications [41, 42] analyzed experimental data for radio signals with $f = 3.94$ and 1.54 GHz obtained in the equatorial region and found that the dependence of ionospheric fluctuations on the wavelength of radio signals can be well approximated by formula (1.18) with $m_1 = 1.68$. This quantity can be taken as a mean experimental value for the index m_1 of ionospheric fluctuations in the equatorial regions. The temporal variations in m_1 may range from 1.5 to 1.9, thus indicating a variability of the index α. Analysis of the temporal spectra of amplitude fluctuations showed that the index m (see formula (1.16)) varies from 4.7 to 5.9 [41, 42]. These data were interpreted in terms of the complex spectrum of electron density fluctuations, with a regular molecular absorption. This phenomenon is discussed in the next section. involving two different spectral indices (α_1 and α_2) and two external scales (more and less extended) for large and small plasma irregularities. In the opinion of the authors of the publications [41, 42], their experimental data confirmed that plasma irregularities have a complex spectrum. It should be noted that the simple power-law spectrum of plasma irregularities (1.15) is, undoubtedly, an approximation, but it describes the amplitude fluctuations of radio waves well provided that the large- and small-scale irregularities are given by two different indices α . Such an approach allows researchers to avoid the consideration of the complex spectra of the electron density fluctuations.

In mid-latitude regions, the depth of the irregular fadeouts of 1.5-GHz radio waves may reach 15–18 dB [43]. These fadeouts, which are more frequent at night than in the day, look like an overshoot followed by oscillations and, hence, appreciably differ from random field oscillations, such as those shown in Fig. 1.2. The cause of these fadeouts is unknown.

Thus, the ionospheric fluctuations of radio waves have been studied well [12–45]. Theory yields the well-established dependence of the magnitude of fluctuation on the wavelength of radio waves and the geometrical parameters of the ray path but does not allow the calculation of the expected fluctuation depth. This is because of a high variability of fluctuations in the electron density and the anisotropy of ionospheric plasma irregularities. Ionospheric fluctuations can interfere appreciably with communications to satellites. The technical aspects of the interfering effect of ionospheric fluctuations on radio communications were discussed in [3, 36].

When summing up this section devoted to the effect of the ionosphere on meter radio waves, we must emphasize that ionospheric fading greatly diminishes with the increasing frequency of radio waves, so that centimeter radio waves fade negligibly. When considering millimeter radio waves, we have to deal

1.3 Attenuation and fluctuations of centi- and millimeter radio waves in the troposphere

Millimeter radio waves undergo a strong attenuation in the troposphere because of their absorption by water vapor and oxygen; therefore, the short-wave boundary of atmospheric radiotransparency is determined by the absorption of radio waves by these gases. Formal analysis of the attenuation of radio waves uses the coefficient of absorption, γ_1, which is defined based on the assumption that the decrease in the radio wave energy flux, dP, is proportional to the length element dl and energy flux P :

$$dP = -\gamma_1 P dL . \tag{1.19}$$

The attenuation of radio waves defined as the ratio of the initial radio wave energy flux P_0 to its current value P is given by the expression

$$Y = \frac{P}{P_0} = \exp\left(-\int \gamma_1 dl\right).$$

(1.20)

Here integration is performed along the ray AB within the limits of the troposphere (Fig. 1.1) allowing for the fact that the absorption coefficient γ_1 depends on the height h above the Earth's surface. If a radio signal propagates along the limited path section ΔL in a homogeneous medium at a constant height, then, according to (1.20), the attenuation of this radio signal is given by the simple formula

$$Y = \exp(-\gamma_1 \Delta L).$$

(1.21)

Since the troposphere contains two absorbing gases, the total absorption coefficient can be presented as the sum $\gamma_1 = \gamma_w + \gamma_o$, where γ_w and γ_o are the absorption coefficients of water vapor and oxygen, respectively.

The molecule of water vapor has a constant electrical torque, whose interaction with an electromagnetic field determines the absorption of radio waves. There are three absorption lines of water vapor that have resonant frequencies equal to 22.3, 183.4, and 323.8 GHz. The theory of the radio wave absorption by water vapor was formulated in [46, 47]. The experimental verification of this theory revealed a discrepancy. In order to avoid it, the authors of [48, 49] took into account the absorption of radio waves caused by the far spectral lines of water vapor. Recent measurements of the absorption of millimeter radio waves by water vapor showed that, in accordance with the theory, it is proportional to the absolute atmospheric humidity and the square of the pressure. The temperature dependence of absorption is more complex; it can be taken to be approximately proportional to T^{-1}. Then the absorptivity of radio waves by water vapor can be given by the approximate formula

$$\gamma_w = eP^2 T^{-1} F_1(\lambda).$$

(1.22)

Formula (1.22) is convenient, since the height-dependent meteorological parameters – humidity e, pressure P, and temperature T – enter into this formula as separate factors, while the effects related to the shape of absorption lines and requiring quantum-mechanical analysis are taken into account by the function $F_1(\lambda)$.

It should be noted that the account of meteorological parameters as independent factors is an approximation, since actually the shape of spectral lines depends on pressure and temperature. Near the Earth's surface, $P = 1$ atm and $T = 15–20°C$; therefore, the dependence of the absorptivity of radio signals with frequencies $f < 200$ GHz by water vapor can be approximated by the formula

$$\gamma_w = \left[0.067 + \frac{\alpha_1}{(f - 22.3)^2 + \beta_1} + \frac{\alpha_2}{(f - 183.4)^2 + \beta_2} \right] f^2 e \cdot 10^{-4}, \qquad (1.23)$$

where coefficient γ_w, frequency f, and humidity e are expressed in dB km^{-1}, GHz, and g m^{-3}, respectively. At 15 and 20°C, the parameters in expression (1.23) take the following values: α_1, 3 and 2.4; α_2, 10 and 7.3; β_1, 7.3 and 6.6; β_2, 9 and 5.

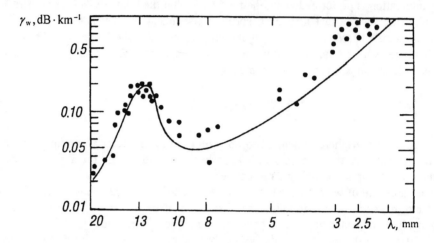

Fig. 1.7. Wavelength dependence of the absorptivity of radio waves by water vapor [69].

Figure 1.7 shows the results of calculations (solid line) and measurements (data points) of the absorptivity of radio waves by water vapor under normal conditions: $P = 1$ atm, $T = 293$ K, and $e = 7.5$ g m^{-3}. The detailed calculation of the absorption of millimeter radio waves at different pressures and temperatures can be found in the publications [50–52].

Oxygen molecules possess a magnetic dipole moment responsible for a single 119-GHz absorption line and a group of lines at 60 GHz. At a pressure of about 1 atm, the lines of this group are unresolved and look like a diffuse band. At heights greater than 15 km, these lines are resolved; the total attenuation in this case is, however, small. In terms of the theory of the absorption of radio waves by oxygen, which was formulated in [53, 54], the absorption coefficient is expressed through the line width. Therefore, high-accuracy estimations of the absorption coefficient require a knowledge of the width of the spectral lines, their relative intensity, and the dependence on pressure and altitude [55–60]. The dependence of the absorption coefficient of oxygen on pressure and temperature can be approximated by the formula

$$\gamma_0 = P^2 T^{-5/2} F_2(\lambda). \qquad (1.24)$$

The function $F_2(\lambda)$ near the resonance lines of absorption is also dependent on the pressure and temperature. At frequencies lower than the resonant frequency, i.e., at $f < 56$ GHz, the absorption coefficient for oxygen is given by the formula

$$\gamma_0 = \left[\frac{6.6}{f^2 + 0.33} + \frac{9}{(f - 57)^2 + 1.96} \right] f^2 \cdot 10^{-3}. \qquad (1.25)$$

This formula corresponds to $P = 1$ atm and $T = 20°C$ and includes frequency expressed in GHz and γ_0 given in dB km^{-1}. Figure 1.8 shows the dependence of the absorption coefficient of oxygen upon wavelength λ for the above values of P and T. The line in this figure was obtained by calculations, whereas the points indicate the experimental values of the attenuation of radio waves.

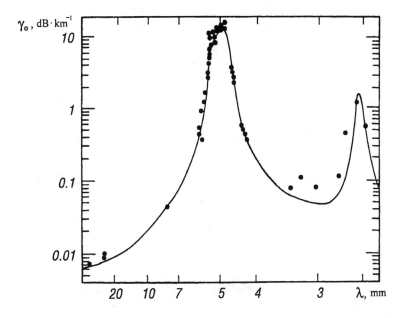

Fig. 1.8. Wavelength dependence of the radio wave absorptivity by oxygen [69].

The abridged description of the behavior of the absorption coefficients γ_0 and γ_w given above corresponds to normal conditions, i.e., to the mean values of pressure, temperature, and humidity near the Earth's surface. Analysis of the propagation of radio signal during space communications requires the determination of the attenuation of radio waves propagating either vertically or obliquely through the entire troposphere. In this case the dependence of pressure, temperature, and humidity on altitude should be taken into account. In a number of publications [61–75], analysis of the total attenuation of radio waves in the troposphere is given in full detail. If the attenuation of radio waves is small, the solution of the problem is simple. Allowing for the fact that the air humidity and temperature depend on height, the coefficients of the absorption of radio waves by water vapor and oxygen decrease with the height h almost exponentially [61, 62]. If so, one can write the following exponential dependence

$$\gamma_1 = \gamma_w \exp\left(-hH_w^{-1}\right) + \gamma_o \exp\left(-hH_o^{-1}\right),$$ (1.26)

where $H_{w,o}$ are parameters expressing the height-dependent decrease in the absorption coefficient, and γ_w and γ_o are the coefficients of the absorption of radio waves by water vapor and oxygen near the Earth's surface (they are given by formulae (1.23) and (1.25) and their dependence on the wavelength is shown in Figs. 1.7 and 1.8). The total absorption of the radio waves propagating obliquely is given by the expression

$$Y = \exp\left(-\frac{\gamma_w}{\cos z_0}\int_0^\infty e^{-hH_w^{-1}}\,dh - \frac{\gamma_o}{\cos z_0}\int_0^\infty e^{-hH_o^{-1}}\,dh\right).$$ (1.27)

In deriving this formula, we used the following approximate expression

$$dh = dl \cos z_0,$$ (1.28)

where z_0 is the zenith angle of a ray near the Earth's surface (Fig. 1.1). From expression (1.27), it follows that

$$Y = \exp\left(-\frac{\gamma_w H_w + \gamma_o H_o}{\cos z_0}\right).$$ (1.29)

A comparison of (1.21) and (1.29) suggests that parameters H_w and H_o have the meaning of the length of the path that runs along the Earth's surface and displays the same attenuation of radio waves as a vertical path. Expression (1.29) is approximate, since it is based on the approximate formula (1.28), which does not account for the sphericity of the Earth and the refraction of radio waves. Instead of the simple formula (1.29), the authors of [67] used a more complex formula, accounting for the sphericity of the Earth and the refractive bending of the ray path. Analysis of these formulae shows that, at zenith angles lower than 85°, both give comparable results. However, at $z_0 > 85°$, formula (1.29) overestimates the attenuation of radio waves.

The above expressions show the role of the parameters $H_{w,o}$. The behavior of these parameters, measured by radioastronomical methods, was analyzed in [61–67, 71]. The absorption of radio waves by oxygen does not significantly depend on the season: the annual value of parameter H_0 is $4.3 \div 5.4$ km. At the same time, the absorption of radio waves by water vapor is determined by the air humidity and, hence, must depend on the seasonal and local meteorological conditions. In [52, 71], the authors performed a detailed analysis of the integral absorption of radio waves by water vapor and obtained H_w values for various conditions. In the mid-latitude regions, the winter value of H_w is 2.5 km and its summer value is 2.1 km for both wavelengths, $\lambda = 30$ and 8 mm [71]. The values of H_w for various signal wavelengths, seasons, and latitudes, taken from [52], are summarized in Table 1.2. For vertically propagating radio waves, the parameter H_w can be taken to be proportional to the total content of water vapor in the atmosphere.

Table 1.2. Parameter H_w (km) for various signal wavelengths, seasons, and latitudes [52]

λ, mm	75° North		60° North		45° North	
	January	July	January	July	January	July
20	0.24	1.69	0.72	2.95	1.01	3.38
13.5	0.37	2.48	0.96	4.3	1.45	4.8
5÷2	0.32	1.83	0.82	3.1	1.1	3.34

Relation (1.29) and data presented in Table 1.2 and Figs. 1.7 and 1.8 make it possible to determine the absorption of radio frequencies propagating obliquely through the troposphere, except for the radio frequencies close to those of resonant absorption lines.

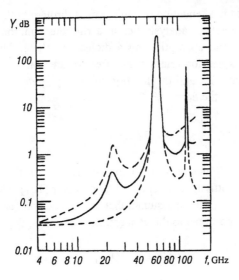

Fig. 1.9. Frequency dependence of the integral absorption of radio waves propagating vertically in the troposphere [65].

In Fig. 1.9, the solid line shows the total absorption of radio waves propagating vertically at a latitude of 45° under the meteorological conditions typical of summer. The upper dashed line corresponds to a relative humidity of 100%, while the lower dashed line
corresponds to a zero-humidity atmosphere. These dashed lines indicate the greatest hu-
midity-related variations in the absorption of radio waves in the atmosphere. As is obvious from Fig. 1.9, in a frequency range of 10–100 GHz, there are two spectral regions
(10–19 GHz and 28–35 GHz) within which the absorption of radio waves must be small.
It should be noted that the data in Fig. 1.9 refer to a vertical ray. At high zenith angles, the
absorption of radio waves is considerably greater: thus, at $z_0 = 84°$, the attenuation of radio waves is 10 times greater than that shown in this figure. It can also be seen from Fig.
1.9 that centimeter radio waves undergo a slight attenuation in the atmosphere. For instance, the attenuation of 8-GHz radio waves even at a zenith angle of 80° is as small as
0.3 dB. The absorption of radio waves in a pure atmosphere, when the effect of precipitations can be neglected, was extensively investigated in [46–75]. A simple engineering
approach for the calculation of the absorption of radio waves was proposed in [3].

The discussion above concerned the regular absorption of radio waves in the troposphere. Along with this, precipitations can bring about the irregular attenuation of radio
waves in the troposphere. It should be noted that dry snows, clouds, and fogs only affect
the absorption of radio waves, while rains can cause their irregular fadeouts. Beyond the
molecular absorption bands, the main factor that affects radio links at frequencies higher
than 10 GHz is the rain-induced fading. This fact stimulated extensive investigations into
the effect of rains on the propagation of radio waves along the Earth–satellite path [76–
100].

Rains bring about the attenuation of radio waves through both their scattering and absorption. Relevant theoretical analysis lies in a rigorous solution of the problem of the
radio wave scattering and absorption by a dielectric sphere. Theoretical consideration
yields the following expression relating the effective area of a water drop, Q, the radiation energy flux density P, and the fraction of energy spent for the scattering of radio
waves and the heating of the drop, δP:

$$\delta P = PQ(a_d, \lambda). \tag{1.30}$$

The effective area Q depends in a complex manner on the drop radius a_d, wavelength λ, and the permittivity and conductivity of water [101]. Since the spatial distribution of drops is random, it can be assumed that the decrease in energy over the unit length
the wave travels is proportional to the concentration of drops, N_d:

$$\delta P = N_d PQ(a_d, \lambda). \tag{1.31}$$

This relationship is valid for equally sized drops. Actually, drops have different radii;
this can be taken into account by introducing the size distribution of drops, $I(a_d)$. The

function $N_d I(a_d)$ represents the number of drops that have radii between a_d and $a_d + da_d$ per unit volume. Taking into account that the distribution function $I(a_d)$ is normalized to unity, we get, instead of (1.31), the following expression for the decrease in the radio wave energy:

$$\delta P = N_d P \int\limits_0^\infty I(a_d)Q(a_d,\lambda)da_d ,\qquad(1.32)$$

from whence it follows that the decrease in the energy flux by a value dP over a distance dl covered by the radio wave can be given by

$$dP = -\left[N_d \int\limits_0^\infty I(a_d)Q(a_d,\lambda)da_d \right]Pdl .\qquad(1.33)$$

By comparing this formula with (1.19) and (1.21), we can derive the following formula for calculating the rain-induced attenuation of radio waves:

$$Y = \exp(-\gamma_r \Delta L),\qquad(1.34)$$

where the coefficient of the rain-induced attenuation of radio signals is given by the formula

$$\gamma_r = N_d \int\limits_0^\infty I(a_d)Q(a_d,\lambda)da_d .\qquad(1.35)$$

Here ΔL is the length of the radio wave path in the rain.

The theoretical calculation of the rain-induced attenuation of radio waves is sophisticated. This is primarily because one has to know the distribution function $I(a_d)$, which can only be determined experimentally. Moreover, the function $I(a_d)$, which depends on the rainfall rate, may differ even for rains of equal intensity. Theoretically, γ_r is proportional to the drop concentration N_d. Since this parameter is difficult to determine, rains are commonly characterized by their rate I (i.e., the amount of the water precipitated per one hour) rather than by the concentration of raindrops. Generally, the drop concentration N_d and the distribution function I are related nonlinearly, since the speed of a falling drop depends on its diameter and, hence, on the size distribution of drops. According to formula (1.35), the absorption coefficient can be calculated if the function $Q(a_d,\lambda)$ is known. In turn, this function can be derived through a stringent electrodynamic solution [101]. The analysis of the theory and relevant experimental data for various wavelengths and rainfall rates can be found in [85, 86].

Figure 1.10 illustrates the theoretical dependences of the attenuation of radio waves on the rainfall rate for five frequencies. Numerals along the curves indicate radio frequencies in GHz; the attenuation of radio waves, Y, is given in dB per one km of the path; the rainfall rate I has a dimension of mm per hour. Since rains are usually spatially nonuniform, calculations for long beam paths can lead to considerable errors. In general, the theoretical determination of the rain attenuation of radio waves is unreliable and yields the results that poorly agree with the experiment.

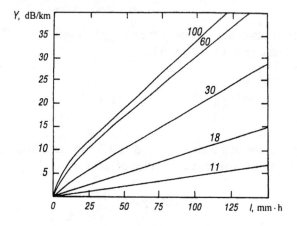

Fig. 1.10. Theoretical dependences of the absorption coefficient on the rainfall rate for five radio frequencies [86].

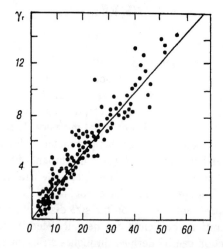

Fig. 1.11. Experimental dependence of the attenuation coefficient of 35-GHz radio waves on the rainfall rate [76].

In view of this, direct experimental data on the rain-induced attenuation of radio waves under different climatic conditions are of great practical importance. In [76], the authors presented the results of the measurements of the attenuation of 35-GHz radio waves along short near-surface paths. The data in Fig. 1.11 refer to ten heavy summer rains in the Great Britain. The experimental data in this figure are represented by points, and the line shows the theoretical dependence of the attenuation on the rainfall rate (γ_r and I have the dimensions of dB km^{-1} and mm h^{-1}, respectively). The apparent good agreement of the theoretical and experimental data is due to two factors: the properly chosen size distribution of drops, $I(a_d)$, and the shortness of the radio path, so that the rainfall rate changed little along the path. For longer radio paths, the agreement between the theory and the experiment is worse. The effective length of the radio path in the rain, ΔL, is of great practical importance. This parameter can be determined by comparing the attenuations obtained experimentally and through calculations by formulae (1.34) and (1.35). The effective length of a radio path in a 'typical' rain is 3–4 km and is virtually independent of the zenith angle, since the vertical and horizontal dimensions of rains are approximately equal.

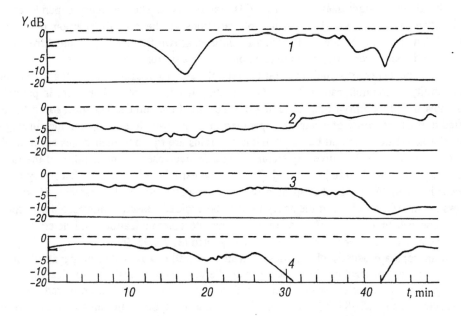

Fig. 1.12. The rain-induced attenuation of radio waves in satellite communications [68].

The authors of [68] investigated the rain attenuation of 15.3- and 31.6-GHz radio waves along the geostationary satellite–Earth path in North Carolina (the United States). Figure 1.12 shows the attenuation of 15.3-GHz radio waves by different rains. The horizontal time axes in this figure correspond to about 50-min recordings. Trace *1* shows the attenuation of radio waves about −3 dB in magnitude caused by a moderate rain. When the rainfall rate increased to 25–40 mm h⁻¹ (it took place two times over the period of measurements), the magnitude of signal fading rose to −10 and −14 dB. Trace *2* was recorded during a heavy (about 20 mm h⁻¹) long-duration rain. It is obvious from this record that the rain brought about a strong fading of the signal of up to −9 dB in magnitude. Hail, occasionally occurring during this rain, caused no noticeable increase in signal fading. Trace *3* was obtained during the light rain with an intensity of about 10 mm h⁻¹. The initial attenuation of about −3 dB in magnitude increased by almost 15 dB when the rainfall rate rose to 30–40 mm h⁻¹. Trace *4* illustrates the effect of the heavy short-duration shower that had a rate of more than 70 mm h⁻¹. In this case, the signal attenuation exceeded 20 dB, so that the receiving station lost the signal. In the author's opinion [68], the weak correlation between the degree of attenuation of radio waves and the rainfall rate measured near the receiving antenna could be explained by a strong spatial irregularity of rain. On average, the attenuation of 31.6-GHz radio waves was 2.5–3.4 times greater than that of 15.3-GHz radio waves [68].

Similarly, the authors of [76] found that the attenuation of 30.9-GHz radio waves was 2.0–2.5 times stronger than that of 18.5-GHz radio waves. The frequency dependence of rain attenuation was variable because of variations in the size distribution of drops, $I(a_d)$. The ratio of the attenuation values measured at two frequencies changed from day to day and could even change within one hour during a uniform rain.

The dependence of signal fades on the rainfall rate provides no information about the probability of communication failure. On the other hand, it is more important in practice to know the probability of signal fading of a specified magnitude. This probability is related to the probability of a rain of a specified intensity, which is a regional meteorological characteristic. It should be noted that heavy rains are rare and their zones are limited. The use of the so-called diversity technique greatly decreases the probability of the rain-induced radio link failure [3, 76, 96]. This can be explained by the fact that heavy rains occur in small spaces, so that spacing two receiving stations by, for instance, 10 km allows the interfering effect of the heavy rains to be avoided. Based on meteorological forecast, one can predict the effect of rains on satellite communications and determine the regions where strong or weak fading can be expected [3, 77, 95, 99].

Raindrops are usually oblate, due to which the radio waves that propagate through rains undergo polarization. Inasmuch as satellite communications use polarization channeling, this may interfere with communications. The rain depolarization of radio waves was studied in [3, 82–84, 88–90]. The effect of rains on space radio links was extensively investigated in [76–100].

Apart from the long-duration fades caused by rains, there may be, even during periods of fair weather, the irregular short-duration scintillations of centi- and millimeter radio waves because of their refraction from migrating tropospheric inhomogeneities. When radio waves propagate in the lower troposphere at small angles to the horizon, milli- and centimeter radio waves undergo intense fluctuations. During space communications, radio signals usually travel at some zenith angle z_0, so that the most irregular near-surface region of the troposphere affects the fluctuations of radio waves negligibly. Only at relatively high zenith angles do tropospheric fluctuations become noticeable.

According to [2], the tropospheric fluctuations of radio waves can be given by the equation

$$\left\langle \ln \frac{E}{E_0} \right\rangle^2 = 0.56 \left(\frac{2\pi}{\lambda} \right)^{7/6} \int_0^{H_3} c_n^2(h) \left[\frac{l(H_3 - l)}{H_3} \right]^{5/3} dl. \tag{1.36}$$

Here $\left\langle \ln \frac{E}{E_0} \right\rangle^2$ is the mean square of the log amplitude fluctuations; this quantity is close to the squared magnitude of field fluctuations, s_2^2, if they are small. Furthermore, $c_n^2(h)$ is the structural parameter of the refractive index irregularities, h is the altitude of the ray element dl, and H_3 is an arbitrary height of the troposphere. Integration in (1.36) is performed along the ray AB (Fig. 1.1). As follows from expression (1.36), to determine the magnitude of tropospheric fluctuations, one has to know the dependence of c_n on the altitude h. The authors of [102–104] established that the altitude profile $c_n(h)$ depends on meteorological conditions, since the troposphere is statistically inhomogeneous to an altitude of 10 km. It can be taken that below the altitude H_3, which is conventionally equal to 10 km, c_n is a constant value, while it becomes equal to zero at altitudes greater than H_3. Neglecting the refraction of radio waves and taking into account (1.28) and the conventional altitude dependence of $c_n(h)$, we can derive from (1.36) the following expression

$$\left\langle \ln \frac{E}{E_0} \right\rangle^2 = 2.6 c_n^2 H_3^{\frac{11}{6}} \lambda^{-\frac{7}{6}} \sec^{\frac{11}{6}} z_0, \tag{1.37}$$

where the term $c_n^2 H_3^{\frac{11}{6}}$ is an integral characteristic of the tropospheric inhomogeneity of the refractive index.

Let us now consider the effect of tropospheric fluctuations on space communications. The author of [105] investigated the fluctuations of 32-cm radio waves by observing slow low-magnitude fades at high zenith angles. At $z_0 = 86°$, s_2^2 averaged $1.8 \cdot 10^{-3}$; this value corresponded to 4.2% variations in the field strength. The authors of [65, 106] studied the amplitude fluctuations of 7.3-GHz radio waves emitted from a high-orbit satellite whose low-speed movement relative to the Earth's surface allowed comprehensive observations of amplitude fluctuations at various zenith angles. In these experiments, fluctuations were also small: indeed, even at $z_0 = 86°$, the quantity s_2^2 was typically as low as $(6-8) \cdot 10^{-4}$. In [107], the magnitude of the tropospheric fluctuations of 2- and 30-GHz radio waves was studied as a function of the zenith angle by means of a geostationary satellite. In full agreement with (1.37), s_2^2 was found to increase with decreasing zenith angle as $\sec^{11/6} z_0$.

The authors of [108] studied the tropospheric fluctuations of 30-GHz radio waves emitted from a geostationary satellite flying low above the horizon at a slowly changing zenith angle (these experiments were performed in Texas, the United States). Figure 1.13 shows, in relative units, typical traces of the tropospheric fluctuations of radio waves that were recorded at different zenith angles indicated by numerals along the curves. As is evident from this figure, the magnitude of the fluctuations drastically decreases with the zenith angle, so that, at $z_0 < 75°$, the tropospheric fluctuations in the amplitude of radio waves are insignificant even at a frequency of 30 GHz.

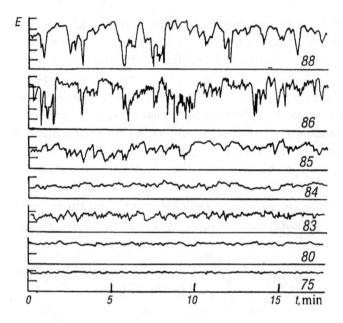

Fig. 1.13. Typical amplitude fluctuations of radio waves in the troposphere at high zenith angles [108].

For decimeter radio waves, the parameter c_n is greater than for centi- or millimeter radio waves, thus indicating that ionospheric fluctuations in the shorter-wavelength region of the decimeter wavelength band must be strong. It can be assumed that purely tropospheric fluctuations are significant only at $\lambda < 10$ cm. When evaluating the expected values of the fluctuations by formula (1.37), the term $c_n^2 H_3^{11/6}$ averages $5 \cdot 10^{-5}$ cm$^{7/6}$, from which it follows that the parameter c_n is equal to $(2 \div 6) \cdot 10^{-6}$ cm$^{-1/3}$. This value of c_n is of the same order of magnitude as the earlier estimates of this parameter reported in [102]. More information on the tropospheric fluctuations in the amplitude of radio waves and on c_n estimates can be found in [109–111]. In the above consideration, we restricted ourselves to the case of $z_0 < 86°$, since the propagation of radio waves at lower angles to the horizon leads to considerably higher fluctuations caused by the effect of stratified tropospheric inhomogeneities. Relation (1.37) and the aforementioned experimental data suggest that the magnitude of tropospheric fluctuations increases with the decreasing wavelength of radio signals; therefore, the tropospheric fluctuations of millimeter radio waves must be relatively high.

1.4 Refraction of radio waves

Let us consider the effect of the altitude dependence of the refraction index, $n(h)$, on the arrival angle of radio waves. In the atmosphere, the refraction index is close to unity. Therefore, it would be convenient to introduce a new parameter, refractivity N, defined by the equation

$$n = 1 + N.$$
(1.38)

In the troposphere, the refractivity N depends on pressure P, temperature T, and humidity e as

$$N = \frac{77.8}{T}\left(P + \frac{4810e}{T}\right) \cdot 10^{-6}.$$
(1.39)

Here pressure and humidity are expressed in millibars, and temperature is given in degrees Kelvin. In the troposphere, pressure and humidity exponentially decrease with altitude, while temperature decreases linearly; therefore, the average altitude profile of refractivity can be approximated by the exponential function

$$N = N_0 \exp(-b_1 h).$$
(1.40)

Refractivity near the Earth's surface, N_0, can be determined, according to formula (1.39), from the near-surface values of pressure, temperature, and humidity (P_0, T_0, and e_0, respectively). In the mid-latitude regions, N_0 averages $3.06 \cdot 10^{-4}$ in winter and $3.3 \cdot 10^{-4}$ in summer. Parameter b_1, ranging from 0.12 to 0.14 km^{-1}, can be determined from N_0 with allowance for the fact that refractivity at an altitude of 10 km, where $N = 9.2 \cdot 10^{-5}$ is fairly constant. In this case we can derive from equation (1.40) the following formula

$$b_1 = -\frac{1}{10}\ln\left(\frac{9.2 \cdot 10^{-5}}{N_0}\right).$$
(1.41)

As follows from relations (1.39)–(1.41), the dependence $N(h)$ can be found from the near-surface values of pressure, temperature, and humidity. Actually, the vertical profile $N(h)$ can deviate from an exponential shape; the difference, however, is essential only at high zenith angles z_0, when radio waves propagate at small angles to the horizon. The refraction of radio waves propagating in the troposphere is described in more detail in [102, 112, 113].

Let us consider the altitude dependence of the refractive index in the ionosphere. According to [1], the refractivity of high-frequency radio waves in a plasma can be described by the simple formula

$$N = -\gamma N_e f^{-2} , \qquad (1.42)$$

where $\gamma = 40.4$ provided that the electron density N_e and frequency f are expressed in m^{-3} and Hz, respectively. As is evident from this expression, N is negative, and the dependence $N(h)$ coincides with the altitude profile of electron density, $N_e(h)$, dependent on the time of day or night, season, latitude, and solar activity. It should be noted that the refractivity in plasma depends on frequency as f^{-2}. The complex profile $N_e(h)$ can be divided into two regions, one lying above the main ionization maximum ($h > h_m$) and the other lying below it ($h < h_m$). Electron density in the upper ionosphere can satisfactorily be described by the exponential function

$$N_e = N_m \exp[- b_i (h - h_m)], \qquad (1.43)$$

where N_m is the electron density in the main ionization maximum, h_m is the altitude, and b_i is a parameter expressing the rate of the electron density drop with altitude. Below the main ionospheric maximum, the dependence $N_e(h)$ can roughly be approximated by

$$N_e = N_m \left[1 - \left(\frac{h_m - h}{d_i} \right)^2 \right]. \qquad (1.44)$$

This approximation corresponds to a zero electron density at a height $h_m - d_i$, where d_i is the conventional thickness of the lower ionosphere. At $h < h_m$, the dependence $N_e(h)$ shows a complex behavior because of the occurrence of regular ionospheric regions F_1 and E and a sporadically appearing ionized region E_s. It should be noted that this is not taken into account by approximation (1.44). Expression (1.44) is valid for altitudes satisfying the inequality $h_m > h > h_m - d_i$ at lower altitudes, electron density is assumed to be equal to zero. When estimating the parameter d_i, electron densities lower than 10^4 cm^{-3} can be neglected. Expressions (1.43) and (1.44), which describe the dependence $N_e(h)$, contain four parameters (N_m, h_m, b_i, and d_i) dependent on the time of day or night, season, latitude, and solar activity. Parameters N_m and h_m can be measured at ionospheric stations, whereas parameter d_i can be estimated by ground-based high-frequency sounding. Parameter b_i is more difficult for estimation: relevant information can be gained only through the use of incoherent radio systems, which are very scarce.

Inasmuch as the index of refraction depends on altitude in a complex manner, rays are bent mainly in a vertical direction. The direction of refraction is determined by the sign of the altitude derivative dN/dh, since rays are bent towards a medium that has a higher refractive index. Figure 1.1 schematically shows the regions of the tropospheric and ionospheric refraction of radio waves. In the troposphere (the region between lines *1* and *2*), the refracted ray is bent towards the Earth's surface, whereas it is not almost bent in the stratosphere (the region between lines *2* and *3*). In the lower ionosphere (the region between lines *3* and *4*), where electron density rises and refractive index diminishes with altitude, the ray is refracted towards the Earth's surface as well. Above the main ionospheric maximum (the region between lines *4* and *5*), the ray is bent in the opposite direction.

The geometry of the problem of the refraction of radio waves and relevant designations are given in Fig. 1.14. A receiving antenna is placed at point A on the Earth's surface, the center of the Earth is at point O, the arrow tangent to the ray shows the apparent direction towards a spacecraft. In this section, we concentrate on the consideration of the refraction angle ξ between the true direction towards the source of radio waves and the tangent to the ray at the point of signal reception. The refraction angle ξ is equal to the difference $z - z_0$ and shows how greatly the zenith angle of the ray, z_0, differs from the true zenith angle of the spacecraft, z.

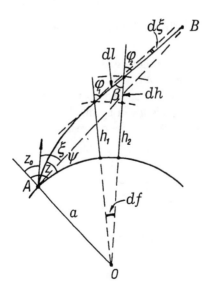

Fig. 1.14. Schematic representation of the refraction angle and the problem of the time delay of radio waves.

Hereafter, we shall often deal with the propagation of radio waves in a spherically symmetric medium; therefore, it would be reasonable to derive here the law of refraction and expressions describing the refractive index and the curvature radius of a refracted ray. To this end, consider a thin layer of the atmosphere between the close altitudes h_1 and h_2. Let $n(h_1)$ be the refractive index inside this layer, and $n(h_2)$ be its value just above the layer. Assuming that this thin layer of the atmosphere is flat, we can use the law of refraction in the form

$$n(h_1)\sin\beta = n(h_2)\sin\varphi_2 . \tag{1.45}$$

From the geometry presented in Fig. 1.14 it follows that

$$(a+h_2)\sin\beta = (a+h_1)\sin\varphi_1 , \tag{1.46}$$

where a is the Earth's radius and φ is the angle made by the ray path with the normal to the atmospheric sphere. The relations (1.45) and (1.46) represent the law of the refraction of radio waves in a spherically symmetric medium:

$$n(h_1)(a+h_1)\sin\varphi_1 = n(h_2)(a+h_2)\sin\varphi_2 . \tag{1.47}$$

This expression is valid for any altitude h, therefore

$$n(h)(a+h)\sin\varphi = n_0 a \sin z_0 , \tag{1.48}$$

where n_0 and z_0 are the refractive index and the zenith angle of the ray near the Earth's surface, that is, at point A in Fig. 1.14.

Let us derive an expression for the curvature radius of the ray in a spherically symmetric medium. The curvature radius is given by the derivative $R_0 = d\xi/dl$, where dl is the length element of the ray. From the geometry presented in Fig. 1.14 it immediately follows

$$d\xi = df + d\varphi , \tag{1.49}$$

$$dh = dl \cos\varphi , \tag{1.50}$$

and

$$df = (a+h)^{-1} dh \tan\varphi . \tag{1.51}$$

Then the following expression for the curvature radius of the ray can be derived:

$$R_0 = \frac{a+h}{\sin\varphi + (a+h)\dfrac{d\varphi}{dh}\cos\varphi} \, . \tag{1.52}$$

Taking into account (1.48), we finally arrive at

$$R_0 = -\frac{n}{\dfrac{dn}{dh}\sin\varphi} \, , \tag{1.53}$$

where n, f, and R_0 depend on the altitude h.

From (1.49) it follows that the refraction angle can be given by the integral

$$\xi = \int_0^H \left(\frac{df}{dh} + \frac{d\varphi}{dh} \right) dh \, , \tag{1.54}$$

where H is the altitude of the source of radio waves. Taking into account (1.51) and deriving $d\varphi/dh$ through the differentiation of expression (1.48), we obtain

$$\xi = -\int_0^H \frac{1}{n}\frac{dn}{dh} \tan\varphi \, dh \, . \tag{1.55}$$

Expressing $\tan\varphi$ through $\sin z_0$ by invoking (1.48) and (1.55), we arrive at the following formula for the refraction angle in a spherically symmetric medium:

$$\xi = -n_0 a \sin z_0 \int_0^H \frac{\dfrac{1}{n}\dfrac{dn}{dh}\,dh}{\sqrt{n^2(a+h)^2 - n_0^2 a^2 \sin^2 z_0}} \, . \tag{1.56}$$

Consider now the refraction of radio waves in the troposphere. Since spacecraft move far above the troposphere, the upper limit in (1.56) can be taken to be infinity, and the refraction angle in the troposphere can be given by

$$\xi_t = -\sin z_0 \int_0^\infty \frac{\dfrac{1}{n}\dfrac{dn}{dh}\,dh}{\sqrt{\left(\dfrac{n}{n_0}\right)^2 \left(1+\dfrac{h}{a}\right)^2 - \sin^2 z_0}} \, . \tag{1.57}$$

The derivative dn/dh differs from zero at altitudes $h \ll a$; therefore, the ratio h/a is much less than unity. According to (1.40), the refractive index, which is close to unity, exponentially decreases with the altitude. Neglecting the terms $(h/a)^2$ and N^2, which are much less than unity, the denominator of the integrand in (1.57) can be transformed according to the expression

$$n\left[\left(\frac{n}{n_0}\right)^2\left(1+\frac{h}{a}\right)^2 - \sin^2 z_0\right]^{\frac{1}{2}} = \left[\cos^2 z_0 + \left(\frac{2h}{a} + 3N - N_0\right)\right]^{\frac{1}{2}}. \qquad (1.58)$$

The right-hand side of this expression can be simplified for $z_0 < 80°$, when the parenthesized terms are much less than $\cos^2 z_0$. In this case the right-hand side of expression (1.58) can be taken to be equal to $\cos z_0$; then, allowing for (1.40), we can derive from (1.57) the following simple formula for the refraction angle in the troposphere:

$$\xi_t = N_0 \tan z_0. \qquad (1.59)$$

Table 1.3. Refraction angles of radio waves in the troposphere

z_0, deg \\ H, km	200	600	> 800
10	11.4	11.8	11.9
20	23.6	24.3	24.6
30	37.5	38.6	39.0
40	54.6	55.9	56.7
50	77.5	79.4	80.5
60	112.2	115.2	116.8
65	138.6	142.3	144.4
70	176.9	181.4	184.5
72	197.6	202.9	206.3
74	223.2	229.2	233.2
76	255.3	262.3	267.1
78	297.3	305.6	311.6
80	354.2	364.9	372.3
81	391.0	402.4	411.6
82	435.7	448.7	459.7
83	491.1	506.3	519.6
84	561.2	579.3	596.0
85	653.2	675.2	696.8
86	777.2	805.1	834.1
87	952.4	989.5	1030
88	1214	1267	1328

The rigorous analysis of the tropospheric refraction of radio waves performed in [114–125] showed that expression (1.59) is valid for zenith angles $z_0 < 80°$. The refraction angle weakly depends on the profile $N(h)$; therefore, the deviation of the real profile from an exponential shape does not influence the refraction angle ξ_t at zenith angles $z_0 < 80°$. In this case the refraction angle depends only on the near-surface value N_0 and zenith angle z_0.

At zenith angles ranging from 80 to 90°, the refraction angle ξ_t depends on the profile $N(h)$ and can be computed by integrating expression (1.57). It should be noted that the calculated values of the refraction angle ξ_t may significantly differ from the true values because of the influence of stratified inhomogeneities.

Detailed information on refraction angles for different meteorological conditions is presented in [118]. Table 1.3, taken from this publication, presents the refraction angles ξ_t (given in angular seconds) for moderate meteorological conditions characterized by $N_0 = 3.29 \cdot 10^{-4}$ and $b_l = 0.126 \ \text{km}^{-1}$. The refraction angles ξ_t are given for only three altitudes of the source of radio waves, H, since ξ_t depends weakly on H.

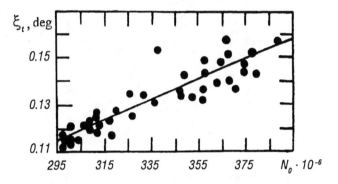

Fig. 1.15. Refraction angle in the troposphere versus the near-surface refractivity of radio waves [121].

In [120–122], the refraction angles were determined by measuring the radio-frequency radiation from various celestial sources. In particular, the author of [121] performed high-accuracy measurements of the tropospheric refraction of centimeter radio waves at high zenith angles under various meteorological conditions. Figure 1.15 shows the dependence of ξ_t on the near-surface values of refractivity, N_0, at a zenith angle $z = 82°$. As is evident from this figure, the tropospheric refraction angle is proportional, in agreement with (1.58), to N_0. The experimental value of ξ_t for $N_0 = 3.29 \cdot 10^{-4}$ is 470 s, whereas ξ_t estimated by formula (1.59) is 480 s, and the tabular value of ξ_t is 460 s. This example shows that formula (1.59) provides an accurate estimation of refraction angles even at the zenith angle $z = 82°$. More information about the tropospheric refraction of radio waves in space communications can be found in [135, 140, 142]. At high zenith angles z_0, the refraction angle ξ_t undergoes slow variations and rapid fluctuations. Slow variations in ξ_t are due to changes in the altitude profile of refractivity induced by varying meteorological conditions. Rapid fluctuations in the arrival angle of radio waves results from stratified and random inhomogeneities (the latter are mainly due to the turbulence of the troposphere).

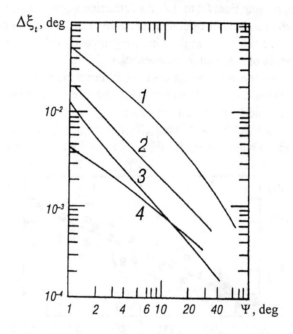

Fig. 1.16. Variations in the refraction angle of radio waves in the troposphere versus the angle ψ.

A detailed analysis of the refraction angle of centimeter radio waves was performed in [65, 106, 120–126]. The dependence $N(h)$ and the refraction angles calculated from the altitude profiles of temperature, pressure, and humidity were used to find the difference $\Delta\xi_t$ between the annual and current values of the refraction angle ξ_t. The difference $\Delta\xi_t$ thus obtained describes slow variations in the refraction angle (Fig. 1.16, curve 1). This curve shows the possible deviations of the refraction angle from average tabular values. As mentioned above, the refraction angle ξ_t is mainly determined by the near-surface value of the refractivity N_0. This allows slow variations in the refraction angle to be corrected. The refraction angle ξ_t can be determined more correctly by taking into account the near-surface values of temperature, pressure, and humidity in accordance with (1.59) and (1.39). Clearly, variations in ξ_t due to stratified and turbulent refractive inhomogeneities are not taken into account in this case. The residual variations in the refraction angle after such correction are small (Fig. 1.16, curve *3*). Investigations into the variations of the arrival angle of the centimeter radio waves transmitted from satellites showed that fluctuations in the refraction angle ξ_t depend on the season, being several times greater in summer than in winter. In Fig. 1.16, curve *2* depicts the average summer dependence of the r.m.s. variation of the refraction angle ξ_t on the ray elevation angle ψ, while curve *4* represents experimental winter fluctuations in the arrival angle of radio waves. The elevation angle ψ, which is taken from the horizon, is equal to $90° - z_0$. It can be assumed that uncontrolled variations in the refraction angle ξ_t are concentrated in the region between curves *2* and *3*.

The refraction of radio waves in the ionosphere was analyzed in [127–131] for the case of a spacecraft-borne transmitter and ground-based receiver. The angle ξ_i of ionospheric refraction was found for various height profiles of electron density, $N_e(h)$. In view of (1.42) and (1.56), the ionospheric refraction angle can be given by the exact formula

$$\xi_i = -\frac{\gamma(a + \Delta h)\sin z_0}{f^2} \int_{\Delta h}^{H} \frac{\frac{1}{n}\frac{dN_e(h)}{dh}dh}{\sqrt{n^2(a+h)^2 - a^2\sin^2 z_0}}, \qquad (1.60)$$

where a is the Earth's radius, $N_e(h)$ is the height profile of electron density, and $\Delta h \approx 50$ km is the height below which $N_e \approx 0$. As follows from (1.60), the refraction angle ξ_i varies with frequency as f^{-2} and strongly depends on the vertical gradient of ionization. Difficulties in gaining information on ionospheric refraction are associated with the necessity to know the variable height profile of electron density, and the gradient dN_e/dh in particular. In [127–131], the authors attempted to find such approximations for the refraction angle ξ_i that would allow the estimation of the ionospheric refraction even if information available on the dependence $N_e(h)$ is scarce.

The integral in expression (1.60) can be divided into two parts, one corresponding to the lower ionosphere, where electron density rises with height, and the other corresponding to the upper ionosphere, where electron density decreases with the increasing altitude h:

$$\xi_i = -\gamma f^{-2}(a + \Delta h)\sin z_0(I_1 + I_2), \qquad (1.61)$$

where

$$I_1 = \int_{\Delta h}^{h_m} \frac{\frac{dN_e}{dh}dh}{\sqrt{(a+h)^2 - a^2\sin^2 z_0}}, \qquad (1.62)$$

and

$$I_2 = \int_{h_m}^{H} \frac{\dfrac{dN_e}{dh}\, dh}{\sqrt{(a+h)^2 - a^2 \sin^2 z_0}}\,. \tag{1.63}$$

Here h_m is the altitude of the main ionospheric maximum. In integrals (1.62) and (1.63), n was taken to be equal to unity; such an approximation is reasonable if $z_0 < 80°$ and $f > 40$ MHz. If the altitude of a spacecraft is 200–400 km, refraction is mainly determined by integral (1.62), and the ray path is bent towards the Earth. But if the altitude of the radio source is higher than 400 km, integral (1.63) should also be taken into account. In the upper ionosphere, the gradient dN_e/dh changes its sign, as a result of which the ray is bent outwards from the Earth. The ionospheric refraction angle ξ_i at $H > 1000$ km is about 30% lower than that at $H \approx 400$ km. In general, the ray path in the ionosphere is bent towards the Earth; therefore, the total refraction angle of radio waves in the Earth's atmosphere is the sum of ξ_t and ξ_i. The refraction angle ξ_i can be calculated from expressions (1.61)–(1.63) and the model dependence $N_e(h)$. At frequencies of about 100 MHz, the refraction angles ξ_t and ξ_i are approximately equal. However, at $f > 1000$ MHz, $\xi_t \gg \xi_i$.

Figures 1.17 and 1.18 show the ionospheric refraction angle ξ_i of 1-m radio waves as a function of the zenith angle z for three satellites, whose altitudes are indicated by numerals along the curves.

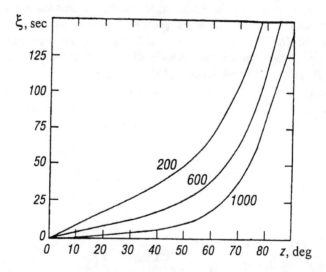

Fig. 1.17. Ionospheric refraction angle vs. zenith angle when the solar activity is high (a summer day).

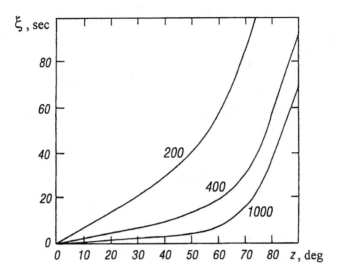

Fig. 1.18. Ionospheric refraction angle vs. the zenith angle when the solar activity is low (a winter night).

The refraction angle of radio waves undergoes slow variations in the ionosphere due to diurnal changes in the electron density $N_e(h)$ and the irregular rapid fluctuations of the tropospheric refraction angle ξ_i caused by the statistical irregularity of the ionospheric electron density [132–134]. Root mean square fluctuations can be estimated by the semi-empirical formula

$$\left\langle \Delta \xi_i^2 \right\rangle^{1/2} = \alpha \lambda^2 \cos^{-1/2} z_m , \qquad (1.64)$$

where z_m is the zenith angle of the ray at altitudes of 200–300 km, and the wavelength λ has the dimension of m. The parameter α averages $0.1'$ and has a maximum value of $0.3'$, which is typical of a disturbed ionosphere. It should be noted that formula (1.64) is valid for $z_m < 80°$.

1.5 Time delay of radio signals in the atmosphere and ionosphere

The distance between a spacecraft and a receiving station, L, can be estimated from the time Δt required for modulated signals to reach the station:

$$L = c \Delta t . \qquad (1.65)$$

Here c is the speed of the electromagnetic waves propagating in a vacuum. The precise measurements of the time interval Δt allow a precise estimation of the distance L. However, the Earth's atmosphere and ionosphere introduce a significant error into ranging, which is because the speed of the wave propagation in the atmosphere and ionosphere differs from c, and rays are bent. As a result, the real distance between transmitting and receiving stations, L_0, is smaller, by a value ΔL, than the distance measured. The necessity to account for the effect of the atmosphere and ionosphere on ranging also arises when one has to determine precisely a satellite's trajectory or the positions of terrestrial objects by the signals from navigation satellites. Even higher accuracy of the L_0 measurements is needed for geodesic purposes. The time service uses satellite signals to compare the readings of widely separated clocks. In all these cases, the time delay of radio waves propagating through the Earth's atmosphere and ionosphere should be taken into account. Analysis of the errors in distance estimates because of the interfering effects of the atmosphere and ionosphere was performed in [118, 129, 136–150].

According to relation (1.38) and Fig. 1.14, the apparent distance L measured along the curved ray AB is given by

$$L = c \int_0^H V_g^{-1} dl = \int_0^H [1 + N(h)] dl , \qquad (1.66)$$

where V_g is the group velocity of radio waves and dl is the length element of the curved ray. The true distance between points A and B is given by

$$L_0 = \int_0^H dl_0 , \qquad (1.67)$$

where dl_0 is the length element of the dashed line AB. Taking into account formulae (1.66) and (1.67), an apparent increase in distance, measured radiometrically, is given by the formula

$$\Delta L = L - L_0 = \int_0^H N(h)\, dl + \left(\int_0^H dl - \int_0^H dl_0 \right) . \qquad (1.68)$$

The parenthesized difference between the two integrals in this expression is negligible when compared with with the first term; therefore,

$$\Delta L = \int_0^H N(h)\, dl . \qquad (1.69)$$

The dependence $N(h)$ for the troposphere and ionosphere was considered in the previous section. Now the differential of the length element dl must be considered. As follows from Fig. 1.14, $dl = dh\cos^{-1}\varphi$. Using formula (1.48) we can find

$$dl = \frac{n(a+h)\,dh}{\sqrt{[n(a+h)]^2 - [n_0 a\sin z_0]^2}}. \tag{1.70}$$

This expression takes into account the refractive bending of a ray. At $z_0 < 80°$, the effect of refraction on ΔL is usually insignificant; therefore, we can simplify expression (1.70) by setting $n = 1$:

$$dl = \frac{(a+h)\,dh}{\sqrt{(a+h)^2 - a^2\sin^2 z_0}}. \tag{1.71}$$

When analyzing the effect of the troposphere on ΔL, expression (1.71) can be simplified as follows. In the region essential for integral (1.69), $a \gg h$. Then, we find from (1.71) that $dh = dl\cos z_0$. The simplification implies that a flat surface is substituted for the spherical Earth's surface. When analyzing the effect of the ionosphere, expression (1.71) can be simplified by assuming that $h = h_m$. Then we have

$$dl = \frac{(a+h_m)\,dh}{\sqrt{(a+h_m)^2 - a^2\sin^2 z_0}} = \frac{dh}{\cos z_m}, \tag{1.72}$$

where h_m is the height of the main ionospheric maximum and z_m is the zenith angle of the ray at this height. Expression (1.72), which takes into account the sphericity of the Earth, is valid for $z_0 < 80°$. This expression directly follows from the geometry given in Fig. 1.14, since

$$\cos z_m = (a+h_m)^{-1}\sqrt{(a+h_m)^2 - a^2\sin^2 z_0}. \tag{1.73}$$

Expression (1.73) relates the zenith angle corresponding to the Earth's surface, z_0, to the zenith angle of the ray near the electron density maximum, z_m.

From expression (1.69) and relations (1.70)–(1.72), one can easily derive the following formulae for the distance correction:

$$\Delta L = \int_0^H \frac{N(h)(a+h)\,dh}{\sqrt{(a+h)^2 - \left(\dfrac{1+N_0}{1+N(h)}\right)^2 a^2\sin^2 z_0}}, \tag{1.74}$$

$$\Delta L = \int_0^H \frac{N(h)(a+h)dh}{\sqrt{(a+h)^2 - a^2 \sin^2 z_0}}, \tag{1.75}$$

$$\Delta L_t = \cos^{-1} z_0 \int_0^\infty N(h)dh, \tag{1.76}$$

$$\Delta L_i = \frac{a+h_m}{\sqrt{(a+h_m)^2 - a^2 \sin^2 z_0}} \int_{\Delta h}^H N(h)dh. \tag{1.77}$$

In formula (1.77), Δh is the altitude of the lower boundary of the ionosphere, which is about 70 km. Formula (1.74) allows the computation of ΔL for all values of the zenith angle z_0, whereas the approximate formula (1.75) is valid only for $z_0 < 85°$. The simple formula (1.76), which gives an obvious dependence of ΔL_t on the zenith angle z_0, is applicable for the analysis of the effect of the troposphere at $z_0 < 80°$. The application of the most rough formula (1.77) is justified because the particular altitude profiles of electron density are usually known only approximately; this does not allow an exact value of the ionospheric correction ΔL_i to be obtained, even if a more exact formula is used.

Let us consider the effect of the troposphere on the apparent range. Assuming that the refractivity varies exponentially with altitude and allowing for (1.40) and (1.76), we can find the following expression

$$\Delta L_t = \frac{N_0}{b_1 \cos z_0}. \tag{1.78}$$

When radio waves propagate vertically, $\Delta L_t = N_0 b_1^{-1}$. From (1.41) it follows that the parameter b_1 can be expressed through N_0; therefore, in this approximation, ΔL_t depends only on the near-surface value of the refractivity, N_0. The correction for the radio wave delay in the troposphere in the mid-latitude regions at $z_0 = 0$ is equal to 2.3–2.4 m in winter and 2.4–2.5 in summer. At higher zenith angles, ΔL_t is greater and reaches $16 \div 21$ m at $z_0 = 80°$. The approximate formula (1.78) is not applicable at $z_0 > 80°$; in this case, however, one can make use of the tabular values of ΔL_t for these zenith angles [118]. The simple formula (1.78), together with formulae (1.39) and (1.41), makes it possible to calculate ΔL_t from the near-surface values of pressure, temperature, and humidity with an error of about ± 20 cm for a vertical ray. Such error is tolerable in the case of positioning ground-based objects by the signals from navigation satellites but is not admissible for precise geodesic measurements or angular measurements with large-baseline radio interferometers. Detailed analysis of the tropospheric correction ΔL_t was performed in [143–150]. The altitude profile of the refractivity, $N(h)$, was analyzed in terms of the dependences known from the physics of the atmosphere: $P(h)$, $T(h)$, and $e(h)$ [146]. If these dependences and their near-surface values P_0, T_0, and e_0 are specified correctly, ΔL_t values at $z_0 = 0$ can be determined with an error of no more than 6 cm. The error is mainly due to uncertainty in the vertical distribution of humidity, $e(h)$.

Let us consider now the effect of humidity on ΔL_t. From (1.76) and (1.39) it follows that

$$\Delta L_t = \Delta L_1 + \Delta L_2 = \alpha_1 \cos^{-1} z_0 \int_0^\infty PT^{-1} dh + \alpha_2 \cos^{-1} z_0 \int_0^\infty eT^{-2} dh , \qquad (1.79)$$

where α_1 and α_2 are constants. The first component of this expression, ΔL_1, corresponds to an absolutely dry atmosphere, while the second component ΔL_2 is due to atmospheric humidity. The component ΔL_1, which gives the major contribution to ΔL_t, makes up 2.25–2.35 m at $z_0 = 0$. This component can be estimated from the near-surface values P_0 and T_0 with an error of about 2 cm. The uncertainty of the determination of correction ΔL_t is mainly due to the variability of atmospheric humidity. At $z_0 = 0$, the component ΔL_2 in the mid-latitude regions is equal to 3–5 cm in winter time and 11–23 cm in summer time. This component can be expressed through the integral humidity of the atmosphere, Q:

$$\Delta L_2 = \frac{\alpha_2}{T_s^2 \cos z_0} \int_0^\infty e \, dh = \frac{\alpha_2 Q}{T_s^2 \cos z_0} . \qquad (1.80)$$

Here T_s is the mean tropospheric temperature at altitudes between 0 and 5 km, which is determined by the near-surface value T_0 and the gradient dT/dh. With a good accuracy, the component ΔL_2 is proportional to the integral atmospheric humidity Q: $\Delta L_2 \approx 6.28 Q$. In the literature, the relationship between ΔL_2 and Q is determined differently, due to the different definitions of the mean temperature T_s. As is shown in [145, 147], the contribution of humidity Q to the component ΔL_2 can correctly be estimated by a radiometric method. The navigation satellites of the global positioning system (GPS) allows a precise estimation of the time delay ΔL_t. By subtracting the component ΔL_1 from the measured value ΔL_t, one can determine ΔL_2 and, consequently, the integral atmospheric humidity Q, which is an important meteorological characteristic. The high efficiency of estimating Q by recording the signals transmitted from the GPS satellites was shown in [151, 152].

Now let us turn to the analysis of the effect of the ionosphere on ΔL. As we are interested in the group delay of radio waves, ΔL_i should be analyzed by formula (1.42) taken with the plus sign. Taking into account (1.75), we get

$$\Delta L_i = \gamma f^{-2} \int_{\Delta h}^{H} \frac{N_e(h)(a+h) dh}{\sqrt{(a+h)^2 - a^2 \sin^2 z_0}} , \qquad (1.81)$$

from which we have for a vertical beam

$$\Delta L_i = \gamma f^{-2} \int_{\Delta h}^{h} N_e(h)dh. \tag{1.82}$$

Expression (1.81) can be simplified by introducing the angle z_m and using formula (1.77):

$$\Delta L_i = \frac{\gamma I_3}{f^2 \cos z_m}. \tag{1.83}$$

Here

$$I_3 = \int_{\Delta h}^{h} N_e(h)dh \tag{1.84}$$

is the electron column density along the vertical ray, and $\cos z_m$ is given by expression (1.73). As follows from (1.83), ΔL_i is proportional to the integral electron density and f^{-2}. Let us find ΔL_i from the experimental data on the electron column density. I_3 can be determined by receiving the satellite signals (relevant methodical details are given in Section 1.8). Measurements performed during the period of low solar activity showed that the integral electron density has a clear-cut diurnal dependence. The summer value of I_3 is $(4-8) \cdot 10^{16}$ m^2 at midnight and $(1-3) \cdot 10^{17}$ m^2 at noon. For vertically propagating 1-m radio waves, these values of the integral electron density correspond to ΔL_i equal to $(18-36)$ m at midnight and $(45-135)$ m at noon. During the periods of a moderate or high solar activity in summer, I_3 is up to $(8-12) \cdot 10^{16}$ m^2 at night or at dawn and $(3-5) \cdot 10^{17}$ m^2 at noon. At $\lambda = 1$ m and $z_0 = 0$, these values of the integral electron density correspond to ΔL_i equal to 36–54 m at night and 130–230 m at noon. In winter, the diurnal dependence of the integral electron density is especially profound, so that, at moderate solar activity, I_3 makes up $(2-5) \cdot 10^{16}$ m^2 at night and before dawn and $(2-4) \cdot 10^{17}$ m^2 at noon, which corresponds to ΔL_i of the vertical beam equal to 9–22 m at night and 90–180 m at noon. The integral electron density depends on the latitude. Relevant investigations carried out in the United States and India showed that the integral electron concentrations at different sites may differ by a factor of two at noon, but insignificantly at night.

Figures 1.19 and 1.20 present the results of the measurements of ΔL_i for $\lambda = 1$ m. Curves *1–4* correspond to satellite altitudes of 1000, 600, 400, and 200 km, respectively. At $H > 1000$ km, ΔL_i depends weakly on altitude and, hence, curves *1* in the figures refer to any altitude higher than 1000 km.

As was shown above, ΔL_i is proportional to λ^2; therefore, the data in Figs. 1.19 and 1.20 can easily be converted to other wavelengths. For 10-cm radio waves propagating vertically, ΔL_i ranges from 0.2 to 2.3 m. The effect of the ionosphere on ΔL_i becomes insignificant only for $\lambda < 3$ cm.

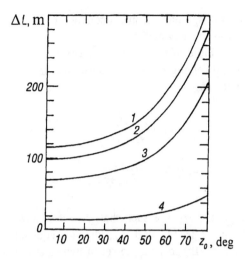

Fig. 1.19. The dependence of the distance difference ΔL on the zenith angle when the solar activity is high (a summer day).

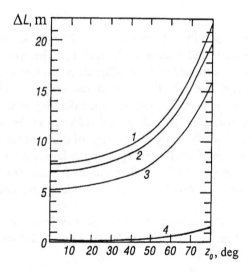

Fig. 1.20. The dependence of the distance difference ΔL on the zenith angle when the solar activity is low (a winter night).

The aforementioned data suggest that the ionospheric variability leads to great variations in ΔL_i and that the theoretical calculations of the ionospheric delay of radio waves are not reliable. In view of this, it has become a common practice in satellite navigation to eliminate ΔL_i by the dual-frequency method.

Let us consider the principle of this method. According to (1.66) and (1.83), the apparent distance between a receiving station and a satellite, $L_{1,2}$, and the true distance between them, L_0, are related as

$$L_1 = L_0 + \frac{\gamma I_3}{f_1^2 \cos z_m},$$

$$L_2 = L_0 + \frac{\gamma I_3}{f_2^2 \cos z_m}, \tag{1.85}$$

where the frequencies f_1 and f_2 are such that $f_2 = mf_1$. Two linear equations (1.85) contain the measurable quantities L_1 and L_2, known frequencies f_1 and f_2, and unknown quantities L_0 and I_3. By solving the set of two equations (1.85), one can obtain the following expression for the true distance between the satellite and the receiving station:

$$L_0 = \frac{m^2 L_2 - L_1}{m^2 - 1}. \tag{1.86}$$

This formula makes it possible to determine L_0 from the known ratio of radio frequencies, m, and the measured distances L_1 and L_2. Therefore, the dual-frequency method allows L_0 to be determined and the influence of the ionosphere to be avoided. It should, however, be noted that actually the dual-frequency method fails to completely avoid the influence of the ionosphere. Indeed, when deriving formula (1.86), it was assumed that the radio frequencies f_1 and f_2 are refracted to the same degree and that formula (1.83) is exact. Actually, however, the rays of different radio frequencies are refracted differently; therefore, formula (1.83) is approximate. A more rigorous analysis showed that the simple formula (1.86) is not applicable for meter radio waves, although it is valid for decimeter radio waves, for which the residual error caused by unaccounted factors is less than 1 cm.

The dual-frequency method allows one of the most important ionospheric characteristics, the integral electron density I_3, to be evaluated. From (1.85) we can easily derive the expression

$$I_3 = \frac{(L_1 - L_2)f_1^2 m^2 \cos z_m}{\gamma(m^2 - 1)}. \tag{1.87}$$

This formula allows the integral electron density I_3 to be obtained from the measurements of L_1 and L_2.

1.6 Phase fluctuations of radio signals in the ionosphere and troposphere

Propagation of radio waves through the atmosphere and ionosphere is accompanied by irregular phase variations. This influences the operation of large antennae and interferometers, impairs the phase stability of communication systems, and can lead to the incorrect positioning of satellites. The geometry of the phase fluctuation problem is shown in Fig. 1.21. The radio waves sent out by the satellite occurring at point B pass through refractive index irregularities in the ionosphere (region *1*) and troposphere (region *2*). Correspondingly, we shall distinguish ionospheric ($\Delta\varphi_\mathrm{i}$) and tropospheric ($\Delta\varphi_\mathrm{t}$) phase fluctuations.

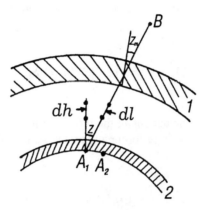

Fig. 1.21. Diagram illustrating the estimation of phase fluctuations.

Let us consider the general relations between phase variations and the refractive index irregularities. In the case of large-scale irregularities, the problem of phase variations can be solved by invoking the concept of rays, since a more stringent wave approach does not essentially increase the accuracy of the solution of this problem [2]. The phase of a wave traveling through the atmosphere can be represented as

$$\varphi = \varphi_0 + \varphi_1 + \delta\varphi = k \int_0^H dl_0 + k \int_0^H N(h,t)dl + k \int_0^H \delta N(h,t)dl . \tag{1.88}$$

Here $k = 2\pi\lambda^{-1}$ is the wavenumber; φ_0 is the phase of radio waves traveling in a vacuum (this phase corresponds to the actual distance to the satellite, L_0); the second term, φ_1, gives the phase shift in the atmosphere and ionosphere without allowing for small-scale irregularities (this phase corresponds to an apparent increase in the range; therefore, $\varphi_1 = 2\pi\lambda^{-1}\Delta L$). Because of slow changes in the altitude profiles of temperature, pressure, and humidity in the troposphere and electron density in the ionosphere, φ_1 undergoes slow but significant changes corresponding to variations ΔL_t and ΔL_i. The third term in formula (1.88) accounts for the effect of small-scale fluctuations in the atmosphere and ionosphere, δN. The distinction between slow variations φ_1 and rapid phase fluctuations $\delta\varphi$ is not strict; this only serves to go over from a stochastic function with stationary increments to a stochastic function with a constant mean value. For instance, if we are interested in the random irregularities of electron density or fluctuations in the tropospheric temperature, while slow changes in the refractive index profile can be excluded from the analysis, then one may use correlation functions to describe statistical irregularities and, hence, phase fluctuations. Such an approximation makes it possible to derive a simple expression for estimating the variance of rapid phase fluctuations.

According to (1.88), the mean square of phase fluctuations is given by the following formula:

$$\langle\delta\varphi^2\rangle = k^2\left\langle\int_0^H \delta N(h_1,t)\,dl_1 \int_0^H \delta N(h_2,t)\,dl_2\right\rangle, \qquad (1.89)$$

where $\delta N(h,t)$ are random fluctuations in the refractive index. Assuming that refractive inhomogeneities are isotropic and performing the integration with respect to altitude, we can derive the expression

$$\langle\delta\varphi^2\rangle = \frac{k^2}{\cos^2 z}\left\langle\int_0^H \delta N(h_1,t)\,dh_1 \int_0^H \delta N(h_2,t)\,dh_2\right\rangle, \qquad (1.90)$$

where z is the zenith angle z_0 of the ray. Introducing the differential coordinate $x = h_1 - h_2$ and the variable $y = \frac{1}{2}(h_1 + h_2)$, we can derive from (1.90)

$$\langle\delta\varphi^2\rangle = \frac{k^2\langle\delta N^2\rangle}{\cos^2 z}\left\langle\int_0^H dy \int_{-\infty}^{+\infty} B(x)\,dx\right\rangle. \qquad (1.91)$$

Here $\langle\delta N^2\rangle$ is the mean square of the refractive index fluctuations and $B(x)$ is the spatial autocorrelation fluctuation function δN normalized to unity. The function $B(x)$ is even and tends to zero at distances much greater than the representative scale of inhomogeneities, $a_{i,t}$, defined by the equation

$$a_{i,t} = \int_0^\infty B(x)\,dx. \qquad (1.92)$$

Substituting (1.92) into (1.91) and integrating over the region of tropospheric and ionospheric inhomogeneities, we arrive at

$$\langle \delta\varphi_t^2 \rangle = \frac{8\pi^2 \langle \delta N^2 \rangle a_t \Delta H_t}{\lambda^2 \cos^2 z_0}, \tag{1.93}$$

$$\langle \delta\varphi_i^2 \rangle = \frac{8\pi^2 \gamma^2 \langle \delta N_e^2 \rangle a_i \Delta H_i \lambda^2}{c^4 \cos^2 z_m}. \tag{1.94}$$

Here $\langle \delta\varphi_t^2 \rangle$ and $\langle \delta\varphi_i^2 \rangle$ are the mean squares of the phase fluctuations caused by the tropospheric and ionospheric refractive index irregularities, respectively. Tropospheric fluctuations depend on the zenith angle near the Earth's surface, z_0, whereas ionospheric fluctuations depend on the zenith angle of radio waves in the region of ionospheric irregularities, z_m (Fig. 1.21). In (1.93), the quantity $\langle \delta N^2 \rangle$ is the mean square of refractive fluctuations in the tropospheric layer that has a thickness $\Delta H_t \approx 6$ km. In (1.94), according to (1.42), $\langle \delta N_e^2 \rangle$ is the mean square of electron density fluctuations, and the region of ionospheric inhomogeneities has a thickness $\Delta H_i \approx 200$ km. When deriving formulae (1.93) and (1.94), it was suggested that the magnitude of fluctuations within the regions $\Delta H_{t,i}$ is constant and that the autocorrelation function $B(x)$ is isotropic. In the case of the real atmosphere, these assumptions are not satisfied; consequently, expressions (1.93) and (1.94), which satisfactorily describe the dependences on wavelength and zenith angle, give only rough estimates of the magnitude of $\delta\varphi_{i,t}$. Generally, phase fluctuations result from both tropospheric and ionospheric irregularities:

$$\langle \delta\varphi^2 \rangle = \langle \Delta\varphi_t^2 \rangle + \langle \Delta\varphi_i^2 \rangle = \frac{A}{\lambda^2 \cos^2 z_0} + \frac{B\lambda^2}{\cos^2 z_m}. \tag{1.95}$$

Here A and B depend on the irregularity parameters δN and $a_{i,t}$ and the thickness of the irregular regions, $\Delta H_{t,i}$ (see expressions (1.93) and (1.94)). As is evident from equation (1.95), the wavelength dependences of the tropospheric and ionospheric irregularities are different.

Let us estimate the tropospheric parameter $\delta\varphi_t$. For definiteness, assume that $\lambda = 4$ cm. According to the comprehensive data on the refractive index fluctuations available in the literature [102–104, 110], the parameter δN averages 10^{-6}, and the representative scale a_t is about 100 m. The thickness of the region of tropospheric irregularities, ΔH_t, is approximately 6 km. If the ray path is vertical, from (1.93) it follows that $\delta\varphi_t = 17°$. Clearly, fluctuations depend on meteorological conditions; therefore, this estimate is approximate. Now let us estimate the ionospheric quantity $\delta\varphi_i$. For definiteness, assume that $\lambda = 1.5$ m, $z_m = 0$, the thickness of the region of ionospheric inhomogeneities, ΔH_i, is about 200 km, and the integral scale of ionospheric inhomogeneities, a_i, is about 5 km.

The relative irregularity of the ionospheric plasma, expressed by the ratio $\delta N_e / N_e$, is about 10^{-2}. This implies that, for altitudes between 200 and 300 km, δN_e is about 10^4 cm^{-3} in the day. With this in mind, we can find from (1.94) that $\delta\varphi_i$ is approximately three radians.

The introduction of the representative scale of irregularities, $a_{i,t}$, would be justified if there were the predominant size of these irregularities, δN, dictated by the mechanism of their formation. Actually, however, the refractive index irregularities in the troposphere and ionosphere widely vary in size from small (the representative scale Λ_{min}) to large (the representative external scale Λ_{max}). In view of this, it would be more correct to introduce the spatial spectrum of refractive index irregularities, $\Phi_N(\mathit{æ})$, which can be given by the power function $\Phi_N \sim \mathit{æ}^{-\alpha}$, where $\mathit{æ} = 2\pi\Lambda^{-1}$ is the spatial wavenumber and α is the power index of the spatial spectrum of the irregularities. Substituting the spatial spectrum $\Phi_N(\mathit{æ})$ for the autocorrelation function $B(x)$ in formula (1.91), one can obtain the expressions similar to (1.93) and (1.94), in which the representative scale $a_{i,t}$ is replaced by the external scale of irregularities, Λ_{max}, and the variance of fluctuations, $\langle\Delta N^2\rangle$, is replaced by the structural coefficient of these fluctuations (see, for instance, [2]). With the spectral approach, we shall arrive at the same formula (1.95) and, therefore, at the same dependences of $\delta\varphi$ on the wavelength and zenith angle. The advantage of the spectral approach becomes apparent when analyzing the temporal spectrum of phase fluctuations, $\Phi_\varphi(F)$ [2]. As follows from the theory of phase fluctuations in a statistically inhomogeneous medium, the power-law spatial spectrum of refractive fluctuations, $\Phi_N \sim \mathit{æ}^{-\alpha}$, corresponds to the power-law spectrum of phase fluctuations, $\Phi_\varphi \sim F^{-m_3}$, where F is the frequency of phase fluctuations. The relationship between the exponents a and m_3 can be given by the simple formula

$$m_3 = \alpha - 1,\tag{1.96}$$

which is derived in Chapter 3. The spectral approach makes it possible to illustrate how the spatial spectrum of fluctuations, Φ_N, can be analyzed in terms of the temporal spectrum of phase fluctuations, Φ_φ. .

Let us consider the experimental estimates of ionospheric phase fluctuations. In [156], the authors studied the phase fluctuations of meter radio waves propagating through the polar ionosphere. Measurements of the two coherent 162- and 324-MHz radio signals transmitted from a navigation satellite were performed at a latitude of 64°. When the ionosphere was quiet, the authors observed irregular phase variations within $\pm(5-10)$ radians and characteristic fluctuation periods lasting from ten seconds to several minutes. In the auroral ovals, the propagation of radio waves was accompanied by irregular burstlike fluctuations (reaching several hundreds of radians and lasting for a few minutes) followed by random phase variations with a standard deviation within several radians. Investigation of the two coherent 150- and 400-MHz radio signals propagating in the ionosphere showed that slow changes in the altitude profile of electron density gave rise to monotonic phase variations occurring at a rate of about 160 radians per min and rapid irregular fluctuations with a standard deviation of 2.5 radians [29]. More detailed information about the ionospheric phase fluctuations can be found in [153–156, 163]. In particular, it has been found that $m_3 \sim 3$.

The comparison of these data with formula (1.95) shows that, at $\lambda < 10$ cm, the ionospheric component of phase fluctuations is small and can be neglected. At $\lambda < 40$ cm, however, phase fluctuations in the ionosphere are considerably higher than in the troposphere.

Consider now tropospheric fluctuations in the phase difference of the signals recorded at two stations, A_1 and A_2, located at a distance ρ one from the other (Fig. 1.21). This case is of special interest for analysis of the interfering effect of the atmosphere on interferometer performance. Assume for simplicity that the signal wavelength is less than 10 cm, so that ionospheric effects can be neglected. If so, slow and large phase variations become more significant than the aforementioned rapid fluctuations. Variations in the phase difference can be expressed through changes in the phase paths, ΔL_1 and ΔL_2, at points A_1 and A_2. If changes in the tropospheric phase paths at these points are different, the phase difference is given by

$$\Delta\varphi = k(\Delta L_1 - \Delta L_2)\cos^{-1} z_0. \tag{1.97}$$

Here ΔL_1 and ΔL_2 are the phase paths for the traces A_1B and A_2B, respectively (Fig. 1.21). Consider the phase difference $\Delta L_1 - \Delta L_2$ as a function of the distance between two points. According to (1.97), the mean square of the phase-difference fluctuations can be given by

$$\langle\Delta\varphi^2\rangle = k^2\langle(\Delta L_1 - \Delta L_2)^2\rangle\cos^{-2} z_0 = k^2 D(\rho)\cos^{-2} z_0, \tag{1.98}$$

where $D(\rho)$ is a structural function of the variations ΔL at two points a distance ρ apart. The theoretical evaluation of the function $D(\rho)$ implies the knowledge of the distribution of the refractive index irregularities in the entire troposphere and, therefore, such evaluation can hardly be performed. On the other hand, reliable information on variations in the phase difference can be obtained experimentally, by means of centimeter-wave interferometers. Thus, in [162], the author used an interferometer with a baseline of 1.6 km for investigating the tropospheric fluctuations of 5-GHz radio waves and found that the standard deviation of fluctuations was 17°, ranging from 9 to 30° from day to day. The results of the interferometer-aided investigations of phase-difference variations in the troposphere are summarized in Table 1.4.

Table 1.4. Values of the function $[D(\rho)]^{1/2}$ for a vertical ray [157–160]

ρ, km	0.1	0.2	0.3	0.37	0.57	0.75	1.1	1.8	2.1	11	35
$D^{1/2}$, cm	0.03	0.09	0.06	0.18	0.25	0.22	0.33	0.12	0.15	0.5	1.5

The scatter of the tabulated $D^{1/2}$ values is explained by the fact that their measurements were taken in different regions without making corrections for the local T_0, P_0, and e_0 values. Table 1.4 and formula (1.98) allow the expected tropospheric fluctuations in the phase difference to be estimated. Thus, for an interferometer baseline of 60÷120 m and 4-cm radio waves, the phase fluctuations that have a characteristic period of 1–2 min average 14÷28°.

Analysis of experimental data allowed the authors of [160–161] to derive the following expression for $D(\rho)$:

$$D = c_1^2 \rho^{5/3}, \tag{1.99}$$

where c_1 is up to $1.7 \cdot 10^{-5} \, \text{cm}^{1/6}$, if ρ is less than 10 km. With ρ increasing, the growth of the fluctuations of an 'electric length' slows down and, at 10 km $< \rho <$ 1000 km, the following formula is valid:

$$D = c_2^2 \rho^{2/3}, \tag{1.100}$$

where the parameter c_2 is equal to $1.3 \cdot 10^{-2} \, \text{cm}^{2/3}$. At still longer baselines (more than 1000 km), electric-length fluctuations may reach 9 cm; this value corresponds to the highest variations of $\Delta L_1 - \Delta L_2$ provided that radio waves propagate vertically through the troposphere. Generally, variations in the phase difference depend on the altitude profiles of temperature, pressure, and humidity. The introduction of the respective corrections makes it possible to reduce variations in the difference $\Delta L_1 - \Delta L_2$ [145–147]. Allowing for the variations ΔL_t caused by varying meteorological conditions was considered in Section 1.5. Accounting for the corrections derived by the radiometric measurements of atmospheric pressure, temperature, and humidity enables one to reduce the uncertainty in the difference $\Delta L_1 - \Delta L_2$ to 1–2 cm. As is evident from Table 1.4, such corrections do not improve the accuracy of interferometric measurements at $\rho < 30$ km, but can considerably improve it at $\rho > 200$ km.

As is known from the theory of interferometers, the shift angle of their directional response diagram, δ, is equal to the ratio of the difference of the electric lengths of the interferometer's arms to the baseline:

$$\delta = D^{1/2} \rho^{-1}. \tag{1.101}$$

This formula defines the attainable accuracy of the determination of angular coordinates with interferometers. From (1.99)–(1.101) it follows that

$$\delta = c_1 \rho^{-1/6} \quad \text{for } \rho < 10 \text{ km}, \\ \delta = c_2 \rho^{-2/3} \quad \text{for } 10 < \rho < 100 \text{ km}. \tag{1.102}$$

Here ρ and δ are expressed in centimeters and radians, respectively. The aforementioned estimates for δ correspond to single measurements. Repeated readings and data averaging may increase the accuracy of angular measurements.

As was shown in [159–161], advanced interferometers ensure the accuracy of angular measurements close to their ultimate values determined by the existence of tropospheric inhomogeneities. Analysis of the effect of the troposphere on the accuracy of angular measurements with interferometers performed in [161] showed that the root-mean-square fluctuations in the electric length of the troposphere rise almost proportionally to the interferometer's baseline up to $\rho \approx 10$ km. Therefore, at $\rho < 10$ km, the angular error δ does not significantly depend on the interferometer baseline, averaging 0.5–1″. At $\rho \approx 2000 - 1000$ km, $D^{1/2}$ is about 4 cm and weakly depends on the baseline. In the latter case, δ tends to diminish as the baseline increases. If ρ is of the order of several thousand kilometers, the attainable accuracy of interferometric angular measurements is as high as $2 \cdot 10^{-3}$ ″.

1.7 Effect of the atmosphere and ionosphere on radio frequency

Changes in the frequency of the radio signals emitted by satellites are described by the Doppler effect. As follows from formula (1.5), which is valid only for radio waves propagating in a vacuum, the Doppler shift in frequency, Δf_0, depends on the wavelength of radio waves and the satellite velocity vector projection onto the satellite–ground station line of sight. The effect of the atmosphere or ionosphere on radio frequency manifests itself in the appearance of an additional component ΔF, so that the total shift in frequency is given by the expression

$$\Delta f = \Delta f_0 + \Delta F . \tag{1.103}$$

Analysis of the effect of media on radio frequency is of great practical significance since the Doppler shift Δf allows the estimation of one of the most important trajectory characteristics, namely, the satellite velocity vector projection onto the satellite–ground station line-of-sight (this projection is referred to as V_2). The effect of media on radio frequency lowers the accuracy of the determination of V_2. The effect of the troposphere on the frequency of radio signals was analyzed in [164–167], and the effect of the ionosphere on the Doppler frequency shift in relation to satellite communications was considered in [168–173].

The effect of a medium on the radio frequency can be analyzed in terms of two different (but giving equivalent results) approaches. The first approach is based on the following generalized expression for the Doppler effect:

$$\Delta f = \lambda_1^{-1} V_3 , \tag{1.104}$$

where V_3 is the velocity vector projection of a satellite onto the refracted ray AB, and λ_1 is the wavelength of radio signals at the location of the satellite.

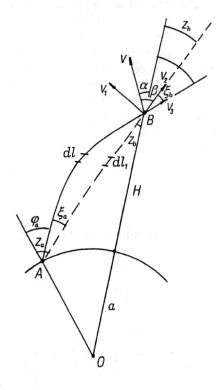

Fig. 1.22. Diagram illustrating the estimation of the frequency shift.

In Fig. 1.22, points A and B indicate the positions of a ground-based station and the satellite, respectively; V_2 and V_1 are the projections of the satellite velocity vector, $\overline{V}$, onto the straight line AB (dashed line) and its normal, respectively; and V_3 is the projection of vector $\overline{V}$ onto the refracted ray shown in this figure by the solid line. Let us introduce the angle β between the velocity vector $\overline{V}$ and direction AB and the angle ξ_b between the direction AB and the ray at point B. The projections V_1, V_2, and V_3 can be expressed through the vector $\overline{V}$ and the angles β and ξ_b:

$$V_2 = V\cos\beta,$$

$$V_1 = V\sin\beta, \qquad\qquad\qquad (1.105)$$

$$V_3 = V\cos(\beta+\xi_b) \approx V_2 - \xi_b V_1.$$

The last formula takes into account that $\xi_b \ll 1$. Let the refractive index at point B be equal to $1 + N_b$, then

$$\lambda = \lambda_1 (1 + N_b). \tag{1.106}$$

Substituting (1.105) and (1.106) into (1.104) yields

$$\Delta f = \lambda^{-1} (V_2 + V_2 N_b - \xi_b V_1 - \xi_b V_1 N_b). \tag{1.107}$$

A comparison of (1.103) and (1.107) leads to the following expression for the sum of the tropospheric and ionospheric frequency shifts:

$$\Delta F = \lambda^{-1} [V_2 N_b - \xi_b V_1 (1 + N_b)], \tag{1.108}$$

where ξ_b and N_b are, respectively, the refraction angle and refractivity at point B. At $\lambda < 10$ cm, the effect of the troposphere prevails over the effect of the ionosphere. Assuming that $N_b = 0$ in formula (1.108), we have

$$\Delta F_t = -\lambda^{-1} \xi_b V_1. \tag{1.109}$$

Conversely, at $\lambda > 1$ m, the effect of the ionosphere prevails over the effect of the troposphere. In this case we can derive from (1.108) and (1.42) the following expression for the ionospheric frequency shift:

$$\Delta F_i = -\gamma N_e \lambda c^{-2} V_2 - \xi_b \lambda^{-1} V_1. \tag{1.110}$$

In formula (1.109), ξ_b is the tropospheric refraction angle. In formula (1.110), ξ_b is the ionospheric refraction angle at point B and N_e is the electron density at this point. The refraction angle at point B, ξ_b, can be found through the refraction angle at point A, ξ_a. Then we can derive from (1.48) the following equation

$$(1 + N_0) a \sin \varphi_a = (1 + N_b)(a + H) \sin \varphi_b. \tag{1.111}$$

It follows from the geometry in Fig. 1.22 that

$$\varphi_a = z_a - \xi_a,$$

$$\varphi_b = z_b + \xi_b, \tag{1.112}$$

$$a \sin z_0 = (a + H) \sin z_0.$$

If $N_0 \ll 1$ and $N_b \ll 1$, we can obtain from (1.111) and (1.112) the following approximate formula

$$\xi_b = \frac{-\xi_a a \cos z_a}{\sqrt{(a+H)^2 - a^2 \sin^2 z_a}}. \tag{1.113}$$

Formulae (1.109) and (1.110) allow the ionospheric and tropospheric components of the frequency shift ΔF to be found from the values of the ionospheric and tropospheric refraction at the location of a satellite. Similarly, formulae (1.112) and (1.113) allow these components to be found from the refraction angle at the location of a ground-based station.

In an alternative approach, ΔF is analyzed in terms of the relationship between frequency and phase:

$$\Delta f = \frac{1}{2\pi} \frac{d\varphi}{dt} = \lambda^{-1} \frac{d}{dt} \int_0^H [1 + N(h,t)] \, dl. \tag{1.114}$$

Here $N(h,t)$ is the height profile of refractivity in the atmosphere or ionosphere and dl is the length element of a ray. In a vacuum, the Doppler shift in frequency is equal to

$$\Delta f_0 = \lambda^{-1} \frac{d}{dt} \int_0^H dl_1, \tag{1.115}$$

where dl_1 is the length element of the straight line AB (Fig. 1.22). By subtracting (1.115) from (1.114), we can derive the expression for ΔF:

$$\Delta F = \lambda^{-1} \frac{d}{dt} \left[\int_0^H [1 + N] \, dl - \int_0^H dl_1 \right]. \tag{1.116}$$

At frequencies higher than 50 MHz and zenith angles lower than 85°, the difference between dl and dl_1 can be neglected. Then we can derive from (1.116) the following expression:

$$\Delta F = \lambda^{-1} N_b \frac{dH}{dt} + \lambda^{-1} \int_0^H \frac{dN}{dt} \, dl + \lambda^{-1} \int_0^H N \frac{d}{dt} (dl). \tag{1.117}$$

When deriving formula (1.117), it was assumed that refractivity depends only on height and time, whereas its horizontal gradient was neglected.

Let us consider first a simple case of the troposphere influence on radio frequency provided that changes in the tropospheric refractivity are slow, small, and tend to zero at point B. In this case we can obtain from (1.117):

$$\Delta F_{t} = \lambda^{-1} \int_{0}^{\infty} N(h) \frac{d}{dt}(dl). \qquad (1.118)$$

Here the upper limit of the integral is taken to be infinity, since the function $N(h)$ vanishes at $h > 50$ km. Using (1.118) and (1.40), we arrive at the following formula:

$$\Delta F_{t} = \frac{N_0 \sin z_{a}}{\lambda b_1 \cos^2 z_{a}} \left(\frac{dz_{a}}{dt} \right). \qquad (1.119)$$

According to (1.41), b_1 can be expressed through N_0; therefore, the troposphere frequency shift ΔF_{t} depends only on N_0, z_{a}, and dz_{a}/dt. It should be noted that the simple formula (1.119) is valid only for $z_{a} < 85°$. A more precise, albeit cumbersome, formula for ΔF_{t}, which is valid at $z_{a} > 85°$, can be found in [164]. The calculations based on formula (1.119) and the experimental data presented in [164, 166] indicate that the tropospheric shift ΔF_{t} is noticeable only at large zenith angles. For instance, at a frequency of 108 MHz and $z_{a} = 85°$, $\Delta F_{t} \approx 0.1$ Hz [164]. Experimental data on the effect of the troposphere on the accuracy of the determination of the velocity vector projection V_2 show that, at $f = 400$ MHz, the tropospheric frequency shifts are equal to 0.3 and 0.8 Hz at zenith angles z_{a} equal to 83 and 86°, respectively [166]. In [166, 167], the authors treated the problem of the retrieval of the height profile $N(h)$ from the experimental dependence $\Delta F_{t}(z_{a})$. As follows from (1.119) and (1.109), ΔF_{t} is proportional to dz_{a}/dt or V_1; therefore, it depends on the satellite altitude. For geostationary satellites, $dz_{a}/dt = 0$ and $V_1 = 0$ and, hence, $\Delta F_{t} = 0$. This inference is valid only when slow variations in phase and frequency are caused by slow changes in the atmospheric parameters P, T, and e. For a vertically ascending rocket, $dz_{a}/dt = 0$ and $V_1 = 0$; therefore, again, the frequency shift in the troposphere is equal to zero.

Refer now to the effect of the ionosphere on radio frequency. There are two aspects of this problem. For technical reasons, the effect of the ionosphere on radio frequency should be minimized. On the other hand, the radiophysical investigations of the ionosphere are based on the recording and analysis of ΔF_i. The effect of the ionosphere on radio frequency was analyzed in [168–173].

Let us consider the effect of the ionosphere on the accuracy of frequency measurements. For a stationary ionosphere, expressions (1.117) and (1.42) yield the following equation

$$\Delta F_{i} = -\gamma c^{-2} \lambda N_e(H) \frac{dH}{dt} - \gamma c^{-2} \lambda \int_{0}^{H} \frac{dN_e(h,t)}{dt} dl - \gamma c^{-2} \lambda \int_{0}^{H} N_e(h,t) \frac{d}{dt}(dl), \qquad (1.120)$$

where dl is determined by formulae (1.70) and (1.71), and $N_e(H)$ is the electron density at the position of a satellite. For geostationary satellites, we can derive the following equation from (1.120) and (1.110):

$$\Delta F_i = -\gamma c^{-2} \lambda \int_0^H \frac{dN_e(h,t)}{dt} dl .$$
(1.121)

As can be seen from this equation, changes in frequency are due only to a nonstationarity of the vertical profile of electron density in the ionosphere. For a vertically ascending rocket, $V_1 = 0$ and $\xi_b = 0$; therefore, in this case the ionosphere can be considered stationary. Then expressions (1.110) and (1.117) yield the following formula

$$\Delta F_i = -\gamma c^{-2} \lambda N_e(H) \frac{dH}{dt} .$$
(1.122)

As is evident from this formula, ΔF_i depends on the electron density in the vicinity of a spacecraft and its speed. If the frequency is high ($f > 500$ MHz) and the ionosphere can be considered stationary, the first and second terms in formula (1.120) can be neglected to give the following expression:

$$\Delta F_i = -\gamma c^{-2} \lambda \int_0^H \frac{d}{dt} [N_e(H) dl] .$$
(1.123)

In this case, ΔF_i is determined by the gradient of the electron column density along the satellite–ground station path.

In earlier publications [168–197], the frequency shift ΔF_i was numerically treated in terms of the model distribution of the electron density, $N_e(h)$. This analysis showed that the contribution of the ionosphere, ΔF_i, to the Doppler shift in frequency, Δf_0, is sizeable for meter and decimeter radio waves but is quite low for centimeter radio waves. Based on these results, the dual-frequency technique was developed to reduce the effect of the ionosphere on the accuracy of the Δf_0 determination [168, 170–173]. The technique uses a satellite that sends out two coherent radio frequencies, f_1 and f_2, which are received at a ground-based station as two Doppler-shifted frequencies given by the expressions

$$\Delta f_1 = \alpha_0 f_1 + \alpha_1 f_1^{-1},$$
$$\Delta f_2 = \alpha_0 f_2 + \alpha_1 f_2^{-1}.$$
(1.124)

Here the first terms correspond to the actual Doppler shift in frequency, Δf_0; therefore, $\alpha_0 = V_2 c^{-1}$. The second terms correspond to ΔF_i; therefore, $\alpha_1 = -\gamma c^{-1} dI/dt$, where I

is the integral electron density. Equations (1.124) make it possible to eliminate the unknown quantity α_1 and determine α_0 and, consequently, V_2:

$$V_2 = \frac{c(m\Delta f_2 - \Delta f_1)}{f_1(m^2 - 1)}, \qquad (1.125)$$

where $m = f_2 f_1^{-1}$. Measuring Δf_1 and Δf_2 makes it possible to exclude the effect of the ionosphere and correctly determine the velocity vector projection V_2. The opinion that expression (1.125) allows the complete avoidance of the effect of the ionosphere on the accuracy of the V_2 determination is not true. Actually, the dual-frequency technique fails to completely eliminate the influence of the ionosphere because of minor effects produced by the ionospheric plasma. The view that the effect of the ionosphere is absent is due to assumptions that the refraction of the radio frequencies f_1 and f_2 is the same, i.e., $dl = dl_1$ (Fig. 1.22), and the relationship between the refractivity N and the electron density N_e is given by the approximate formula (1.42). More detailed analysis performed in [168–173] showed that the dual-frequency technique does make it possible to avoid the effect of the ionosphere and thereby ensures a satisfactory accuracy of measurements at radio frequencies higher than 700 MHz.

Thus, in Sections 1.5 and 1.6, we treated some basic equations describing the effects of the atmosphere and ionosphere on the measurements of satellite velocity and range. Most of the pertinent information was acquired through the use of the recently developed precise GPS and GLONASS (Global Orbiting and Navigation Satellite System) [174]. The potentialities of the GPS satellites in the study of polar motion, tectonic framework, and the atmosphere and ionosphere monitoring were described in the comprehensive review [175].

1.8 Techniques for monitoring the Earth's ionosphere with satellite signals

The fact that the ionosphere alters all the characteristics of radio signals, including their amplitude, frequency, phase, group delay, and polarization, was used to investigate ionospheric plasma. Ionospheric effects depend on the wavelength of radio signals and satellite trajectories. This section is devoted to methods for studying the ionosphere using the signals emitted by satellites and probe rockets. The respective experimental data are not considered here, since there is voluminous literature on this problem, part of which is included in the list of references to this chapter.

Let us consider these methods. As follows from (1.103), the ionospheric frequency shift ΔF_i is a minor component of the total Doppler shift in frequency, Δf_0. The dual-frequency method allows the estimation of ΔF_i by eliminating Δf_0. Indeed, if a spacecraft transmits two coherent radio frequencies, f_1 and $f_2 = m f_1$, then a ground-based receiving station can generate and record the differential frequency $F_d = m\Delta f_1 - \Delta f_2$.

From (1.124) follows the expression for F_d, which does not contain the Doppler shift in frequency, Δf_0:

$$F_d = \alpha_1 f_2^{-1}(m^2 - 1).$$ (1.126)

For a vertically ascending probe rocket, from the comparison of (1.122) and (1.124) it follows that

$$\alpha_1 = -\frac{\gamma}{c}\frac{dH}{dt}N_e(H)$$ (1.127)

and, hence,

$$N_e = \frac{cf_2 F_d}{\gamma \dfrac{dH}{dt}(m^2 - 1)}.$$ (1.128)

It can be seen that the differential frequency F_d is proportional to the electron density at the rocket position; therefore, the frequency method can be employed for elucidating the height profile of the electron density. It should be noted that this parameter greatly varies with height; for instance, N_e ranges from 10^2 to 10^6 as the height increases from 60 to 400 km. In view of this, it is desirable that the frequency method uses from three to five coherent radio frequencies, rather than two. The highest frequency used in the method, which must be such that the effect of the ionosphere could be neglected, is used to measure the spacecraft speed, dH/dt, whereas the minimum frequency of the method allows even small values of the electron density N_e in the lower ionosphere to be estimated. The details of this method and its use for the investigation of the ionosphere from vertically ascending rockets are described in [176–182, 185, 229].

When two coherent radio waves are emitted from a geostationary satellite, the differential frequency is proportional to the rate of changes in the integral electron density. In this case it follows from (1.121) and (1.124) that

$$\alpha_1 = -\frac{\gamma}{c}\frac{dI}{dt},$$ (1.129)

from whence

$$\frac{dI}{dt} = \frac{cf_2 F_d}{\gamma(m^2 - 1)}.$$ (1.130)

Here I is the integral electron density along the satellite–ground station path, whose relation to the vertical integral electron density I_3 is given by $I = I_3 \cos^{-1} z_m$. It can be seen from (1.130) that variations in F_d can efficiently be used to study the rate of change of electron density dI/dt. By integrating (1.130) with respect to time, one can determine the electron column density provided that some relevant data concerning I_3 are available (this is necessary in order to determine the integration constant). Given the electron concentration at a certain moment, for instance, at noon, one can determine diurnal variations in I_3 by invoking (1.130).

Let us now consider the case when the zenith angle of a ray, z_0, changes so that it reaches its minimum value $z_{\min}$ at the moment t_m. In this case the ionosphere can be investigated by recording the differential phase of two coherent signals, $\Phi = m\varphi_1 - \varphi_2$, whereas rapid phase changes due to the Doppler effect are neglected. Integrating (1.130) with respect to time yields the following expression for the differential phase of the two coherent signals:

$$\Phi + \Phi_0 = S_1 \int_0^H \frac{N_e(h)dh}{\cos z(h)}, \qquad (1.131)$$

where

$$S_1 = \frac{2\pi\gamma(m^2 - 1)}{cf_2}.$$

Here Φ_0 is the unknown initial phase difference of the two coherent signals, and $z(h)$ is the altitude dependence of the zenith angle of the ray path. Expression (1.131) is basic for determining the integral electron density through the differential phase of two coherent signals (this method is described in detail in [186–194, 210, 211, 229]). Some modifications of this method involve different procedures for evaluating the integral electron density from the experimental function $\Phi(t)$ when the initial phase difference Φ_0 is unknown. Let us introduce the mean quantity $\langle \cos^{-1} z \rangle$. Then expression (1.130) transforms to the form

$$\Phi + \Phi_0 = S_1 \langle \cos^{-1} z \rangle I_3, \qquad (1.132)$$

where

$$\langle \cos^{-1} z \rangle = \frac{1}{I_3} \int_0^H \frac{N_e(h)dh}{\cos z(h)}. \qquad (1.133)$$

Analysis of expression (1.133) for the model dependence $N_e(h)$ showed that $\langle \cos z \rangle = \cos z_m$ to a satisfactory accuracy. Here z_m is the zenith angle of the ray paths near the main maximum of ionization, whose relationship with the zenith angle at the location of a ground-based receiving station is given by expression (1.73). Using the designation $S = S_1 \cos^{-1} z_m$, we can derive from formula (1.132) the following equation which relates the differential phase Φ and the integral electron density in a vertical direction, I_3:

$$\Phi + \Phi_0 = SI_3. \qquad (1.134)$$

As is evident from (1.134), in order to obtain I_3, one has to find preliminarily the unknown initial phase Φ_0.

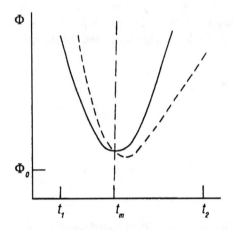

Fig. 1.23. Diagram illustrating the estimation of Φ_0.

Figure 1.23 schematically shows two smoothed dependences $\Phi(t)$, which were measured between the times t_1 and t_2. The time t_m corresponds to the minimum zenith angle z_{min}, when the elevation of the satellite above the horizon was maximum. If the horizontal gradients of the electron density are flat, and the instability of the ionosphere is low and hence can be neglected, the function $\Phi(t)$ is symmetric relative to the dashed straight line $t = t_m$ (this simple case is represented by the solid line in Fig. 1.23). Such a situation is typical of the noon hours in low-latitude regions. In the morning and at night, variations in I_3 are considerable. Ionospheric inhomogeneities and horizontal gradients distort the dependence $\Phi(t)$, so that it becomes asymmetric relative to the dashed line $t = t_m$ in Fig. 1.23. Actually, the dependences $\Phi(t)$ are not smooth because they are impressed by the phase fluctuations $\delta\Phi$ caused by plasma inhomogeneities and receiver noise. This complicates the determination of Φ_0 and I_3 with the use of the simple equation (1.134). However, there are several methods for estimating I_3 from the differential phase [189–192, 229]. Let us describe, after the authors of [229], one of the methods that are used to find Φ_0. The differentiation of formula (1.134) with respect to time yields the following expression

$$\dot{\Phi} = S\dot{I} + I\dot{S} .$$ (1.135)

If we take derivatives for a small time interval around $t \approx t_m$, then

$$\dot{\Phi}_m = S_m \dot{I}_m ,$$ (1.136)

since, at $t \approx t_m$, the zenith angle $z_0 = z_{min}$ and, hence, the derivative $\dot{S} = 0$. Relation (1.136) allows one to determine the nonstationarity of the integral electron density $\dot{I}_3$ from the experimental values for $\dot{S}_m$ and $\dot{\Phi}_m$ measured within a small time interval around t_m. The nonstationarity is assumed to be constant during the period of measurements, $t_1 t_2$:

$$\dot{I}_3 = S_m^{-1} \dot{\Phi}_m .$$ (1.137)

From (1.137) and (1.135) follows

$$\dot{\Phi} = SS_m^{-1} \dot{\Phi}_m + I_3 \dot{S} .$$ (1.138)

After deriving I_3 from equation (1.134) and substituting the resultant expression into (1.138), we arrive at

$$\dot{\Phi} = SS_m^{-1} \dot{\Phi}_m + \dot{S} S^{-1} (\Phi + \Phi_0) .$$ (1.139)

In this equation, S, S_m, and $\dot{S}$ are known from the trajectory parameters of a satellite, while $\dot{\Phi}$, $\dot{\Phi}_m$, and Φ are experimental data from the interval $t_1 t_2$. The unknown initial phase can be determined by using (1.139) and varying Φ_0 under the condition of the minimum of the sum $\sum (\dot{\Phi}_i - \dot{\Phi})$ where $\dot{\Phi}_i$ are experimental values and $\dot{\Phi}$ are derivatives calculated by formula (1.139). Given Φ_0, the integral electron density I_3 can be found from equation (1.134).

When a satellite transmits linearly polarized radio waves, electron density can be found by invoking the Faraday effect. In this polarization method, the propagation of linearly polarized radio waves through a plasma occurring in a magnetic field is accompanied by a rotation of the polarization plane by angle Ω, which can be given by the formula

$$\Omega = Kf^{-2} \int_0^H N_e(h,t) H_0(h) \cos \theta_1 dl ,$$ (1.140)

where H_0 is the geomagnetic field strength, and θ_1 is the angle made by vectors H_0 and dl [1]. Formula (1.140), which gives the Faraday effect in plasmas, is valid for frequencies higher than 40 MHz and angles θ_1 lower than 80°. Parameter K is equal to 1.35×10^{-6}, if Ω, frequency, H_0, N_e, and dl have the dimensions of degree, MHz, gauss, cubic cm, and cm, respectively. Since N_e, θ, and dl depend on time, angle Ω depends on time as well. For this reason, if a receiving antenna is linearly polarized, one should expect polarization fades.

Expression (1.140) shows that the parameters of the ionosphere can be determined from the dependence $\Omega(t)$. The results of such studies are presented in [183–185, 193–199, 204, 206]. In contrast to the angle Ω, the frequency of polarization fades, F_Ω, can easily be found. In the case of a linearly polarized receiving antenna, the rotation of the polarization plane of a radio signal is perceived as its amplitude modulation with frequency F_Ω. The differentiation of (1.140) with respect to time yields the following expression for the frequency of polarization fades:

$$F_\Omega = \frac{K}{2\pi f^2}\left[H_0 \cos\theta_1 \frac{dH}{dt} N_e(H) + \frac{d}{dt}\int_0^H H_0 \cos\theta_1 N_e(h,t)dl\right]. \tag{1.141}$$

Here the upper integration limit H is taken to be time-independent. Let us analyze the relationship between F_Ω, the satellite trajectory, and ionospheric parameters.

If a probe rocket is launched vertically, $\cos\theta_1\, dl$ does not depend on time, and the ionosphere can be considered stationary throughout the period of observation. Hence, only the first term of expression (1.141) is significant, so that

$$F_\Omega = \frac{KH_0 \cos\theta_1}{2\pi f^2}\frac{dH}{dt}N_e(H). \tag{1.142}$$

This formula allows the vertical profile of electron density, $N_e(H)$, to be obtained from the measured frequency of polarization fades, F_Ω, the known magnetic field strength H_0, and the speed of the vertically ascending rocket. The speed dH/dt can be determined from the Doppler effect measured at a high frequency, for which the effect of plasma can be neglected. Some specific features of the Faraday effect and the results of its application for determining the dependence $N_e(H)$ in the case of vertically ascending probe rockets are described in [183–185].

If radio signals are emitted from a geostationary satellite, $N_e(H)$ is negligible, $\cos\theta_1\, dl$ is almost independent of time, and only the second term of formula (1.141) is significant. In this case F_Ω is determined by the instability of the ionosphere:

$$F_\Omega = \frac{K}{2\pi f^2}\frac{d}{dt}\int_0^H H_0 \cos\theta_1 N_e(h,t)dl. \tag{1.143}$$

Formula (1.143) can be transformed by introducing the following effective parameter

$$\left\langle \frac{H_0 \cos\theta_1}{\cos z} \right\rangle = I_3^{-1} \int_0^H \frac{H_0 \cos\theta_1 N_e(h,t)dh}{\cos z}, \qquad (1.144)$$

where $N_e(h,t)$ is the height profile of the electron density in the ionosphere. The term $\langle H_0 \cos\theta_1 \cos^{-1} z \rangle$ corresponds to H_0, θ_1, and z values at the altitude of the main ionospheric maximum, i.e., for $h = 300$ km. From (1.143) and (1.144) follows

$$F_\Omega = \frac{K}{2\pi f^2} \left\langle \frac{H_0 \cos\theta_1}{\cos z} \right\rangle \frac{dI_3}{dt}. \qquad (1.145)$$

This expression shows that the frequency of polarization fades, F_Ω, can provide information on the rate of change of integral electron density. Integrating (1.145) with respect to time, we can derive an expression for integral electron density that contains an unknown integration constant. In the case of a geostationary satellite, the integral electron density I_3 can be found provided that the integration constant is known from independent data. If, for instance, I_3 is known for 12:00 local time, then diurnal variations in I_3 can be found from the Faraday effect. The results of ionospheric studies with the use of geostationary satellites and the Faraday effect are described in [195, 204, 206, 207].

If signals are transmitted from a satellite moving in a moderately elliptical orbit, changes in H can be neglected, and expressions (1.141) and (1.144) again yield (1.145). Designating

$$C = \frac{K}{2\pi f^2} \left\langle \frac{H_0 \cos\theta_1}{\cos z} \right\rangle \qquad (1.146)$$

and integrating expression (1.145) over time, we obtain

$$\Omega + \Omega_0 = CI_3. \qquad (1.147)$$

Here Ω is the angle of rotation of the polarization plane, Ω_0 is the initial angle (or integration constant), and C is the known function of time. Equation (1.147) allows I_3 to be determined provided that Ω_0 is known. This equation is similar to (1.134); therefore, Ω_0 can be found just as F_Ω has been determined. The authors of [193, 196–199] described the results of ionospheric investigations based on the analysis of the Faraday effect for satellite signals. Techniques for investigating the ionosphere with the use of two coherent frequencies and the Faraday effect have much in common. Moreover, the Doppler and Faraday effects are frequently used together to increase the reliability of experimental data [186, 193, 194, 229]. A substantial body of experimental data for the integral electron density is presented in [200–209].

The group-delay effect also enables the measurement of integral electron density. If a satellite transmits two radio frequencies f_1 and f_2, one can determine the difference between two apparent distances, $\Delta L = L_1 - L_2$, to an accuracy dependent on the bandwidth of the modulating signal. According to (1.85), the difference $\Delta L = L_1 - L_2$ is proportional to the integral electron density I_3. From (1.85), one can easily derive the following expression

$$I_3 = \frac{\Delta L f_1^2 f_2^2 \cos z_m}{\gamma (f_2^2 - f_1^2)}. \tag{1.148}$$

The application of the effect of the radio wave delay for ionospheric monitoring became possible with the development of the precise satellite-borne navigation systems, GPS and GLONASS. Each of these systems includes 21 satellites orbiting in three planes, so that the simultaneous reception of signals from at least four satellites is continually possible at any ground-based station. The satellites transmit two radio frequencies ($f_2 = 1575$ MHz and $f_1 = 1227$ MHz) that undergo a pseudo-random phase modulation, which allows the folding of these signals at a receiving station and the retrieval of the highly stable monochromatic signal. In addition to the determination of L_1 and L_2, this technique permits precise phase measurements, which increases the accuracy of th I_3 evaluation. The monitoring of integral electron density by measuring the time delay of radio waves was proved to be advantageous in practice since, first, one does not need to know the integration constant and, second, I_3 can be found at any point on the Earth at any time of day or night. The high efficiency of the monitoring of integral electron density through the use of GPS was demonstrated in [214–217].

The aforementioned research methods are in use for determining either the time dependence of integral electron density, $I_3(t)$, or its height dependence $N_e(H)$. The ionosphere is a highly irregular medium; therefore, $I_3(t)$ and $N_e(H)$ characterize its average state. In other words, these characteristics can be taken as an average background, against which plasma irregularities are impressed. Electron density irregularities, whose sizes may range from hundreds of meters to hundreds of kilometers, lead to rapid fluctuations in the amplitude and phase of radio waves (the theoretical consideration of amplitude and phase fluctuations was performed in Sections 1.2 and 1.6, respectively). To elucidate the latitude distribution of plasma inhomogeneities, fluctuations in the amplitude of radio waves were studied in various regions of the Earth [15–45, 212, 213]. Experimental data on phase fluctuations, which are not so voluminous, show the presence of two zones of increased plasma inhomogeneity, the first lying near the geomagnetic equator (within the geomagnetic latitudes ±20°) and the second lying near the northern auroral oval [29, 153–156, 213].

The height distribution of electron density irregularities can be studied by the method of the spaced reception of radio signals, which is able to establish the relationship between the speed of a diffraction pattern along the Earth's surface, the altitude of ionospheric irregularities, and the satellite velocity. Assume, for simplicity, that fluctuations in the amplitude and phase of radio waves are recorded at a zenith angle $z_0 \approx 0$ at two receiving stations A_1 and A located a distance g_1 apart (Fig. 1.24).

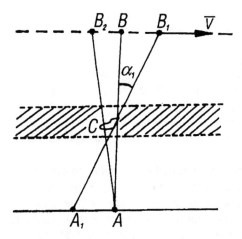

Fig. 1.24. Diagram illustrating the spaced signal reception technique.

Points B and B_1 in this figure correspond to the satellite positions at the time instants t and $t + \Delta T_1$ (the satellite altitude is H). Assume also that the line A_1A is parallel to the satellite velocity vector V. Irregular plasma (shown by the hatched region in Fig. 1.24) is concentrated at a mean height h_n. The letter C marks an intense plasma irregularity. If the satellite covers a distance $BB_1 = V\Delta T_1$ in a definite time T_1, the diffraction pattern produced by the intense plasma irregularity C shifts to point A_1. Since the speed of ionospheric winds is much lower than that of the satellite, the irregularity C does not change its position during the time period ΔT_1; therefore, $A_1A = V_d\Delta T_1$, where V_d is the velocity of the diffraction pattern along the Earth's surface. From Fig. 1.24 it follows that

$$BB_1 = V\Delta T_1 = (H - h_n)\tan\alpha_1$$

and

$$AA_1 = V_d\Delta T_1 = h_n \tan\alpha_1,$$

from whence

$$V_d = \frac{Vh_n}{H - h_n}. \tag{1.149}$$

The time delay of fades at point A_1 relative to the time of their observation at point A is equal to $\tau = g_1 V_d^{-1}$, where $g_1 = AA_1$. Then we can derive from (1.149)

$$h_n = \frac{Hg_1}{V\tau + g_1}.$$ (1.150)

It can be seen that τ being given, the height of plasma irregularities, h_n , can easily be estimated. The time τ can be found by analyzing the correlation function of amplitude or phase fluctuations recorded at points A and A_1. As follows from (1.49), formula (1.150) is valid for $H \approx h_n$. Experiments showed that ionospheric irregularities are concentrated at heights of $200 - 300$ km, that is, in the main ionospheric layer with the peak electron density.

The spaced reception of signals makes it possible to determine the characteristic scales of large plasma irregularities, Λ. The effect of the irregularity C on the phase or amplitude of radio waves at point A begins as soon as the satellite appears at point B_2 and ends as soon as the satellite is at point B_1. If a plasma irregularity of the scale Λ traverses the ray AB in a time of τ, then, as follows from Fig. 1.24, $V\Delta t H^{-1} = \Lambda h_n^{-1}$. Allowing for (1.150), we have

$$\Lambda = \frac{Vg_1\Delta t}{V\tau + g_1}.$$ (1.151)

This expression shows how the technique of spaced signal reception allows the size-scale Λ of plasma irregularities to be found from the observation time Δt and the time delay τ of characteristic fluctuations in the amplitude or phase of radio waves.

Ionospheric fluctuations have different sizes; this poses the problem of obtaining the spatial spectra of plasma irregularities. As has already been noted in Sections 1.2 and 1.6, the temporal spectra of amplitude or phase fluctuations allow one to obtain the spatial spectrum index α of fluctuations in electron density. Experiments show that $\alpha \approx 4$. It should be emphasized that the estimation of α through the temporal spectra of amplitude or phase fluctuations implies that the spatial spectrum of fluctuations in electron density is given by the power function (1.15), so that relations (1.17) and (1.96) are valid. Spaced observations showed that the average power-law spectrum Φ_N is impressed by the components that indicate the occurrence of ionospheric inhomogeneities of the characteristic scale Λ. Usually, the spatial spectra of plasma inhomogeneities contain the components of four characteristic scales, $\Lambda = 2 - 4$ km, $\Lambda = 14 - 16$ km, $\Lambda = 28 - 32$ km, and $\Lambda = 100 - 120$ km. Plasma inhomogeneities are anisotropic: they are oblong and oriented preferentially along the geomagnetic field lines.

The spaced reception of radio signals considerably improves the accuracy of estimation of the integral electron density I_3 through the measurements of differential phase. According to (1.134), to obtain I_3 by this method, one should know the initial phase Φ_0. If the ionospheric region to be studied has no steep gradients of the electron density, Φ_0

can be found through recording the satellite signals at only one ground-based station. However, the high-latitude ionosphere is characterized by considerable gradients of the electron density along latitude lines; in this case, to obtain I_3, two receiving stations should be spaced along a latitude line. The spaced reception of radio signals at these stations yields two equations of type (1.134) containing two unknowns, namely, the initial phases Φ_{01} and Φ_{02} (it is obvious that they can easily be found by solving the two equations). The authors of [208, 210, 211] demonstrated the efficiency of the two latitude-spaced stations technique for investigating the latitude distribution of the integral electron density.

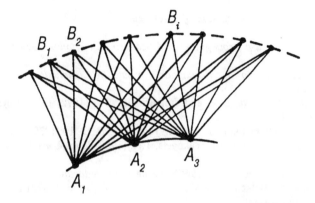

Fig. 1.25. Diagram illustrating ionospheric tomography.

The development of the spaced reception technique led to so-called ionospheric tomography [218–225, 230]. The principle of such tomography is illustrated in Fig. 1.25. A satellite moving in a polar orbit (the satellite positions at various time instants are shown by points B_i) emits two coherent radio waves. A few ground-based receiving stations A_j, spaced latitudinally, record the differential phase Φ or the distance difference ΔL. According to (1.134) or (1.148), these differences are proportional to the electron column density along the straight rays B_iA_j (Fig. 1.25) provided that the radio frequencies are sufficiently high for the refractive bending of rays to be neglected. The problem of ionospheric tomography lies in obtaining the altitude profiles of the electron density, $N_e(H)$, at various latitudes χ from the linear integrals of the electron density along the rays B_iA_j. The dependence $N_e(h, \chi)$ can be retrieved by constructing a space network with mesh sizes Δh and $\Delta\chi$ and introducing a set of linear equations instead of linear integrals along the rays. Actually, the retrieval of function $N_e(h, \chi)$ from the Φ and ΔL values measured at several latitude-spaced receiving stations is a very complex technical and calculation problem. Nevertheless, the efficiency of ionospheric tomography has been demonstrated by three research groups, which analyzed the experimental dependences $N_e(h, \chi)$ and found that ray tomography is a reliable technique to retrieve large-scale inhomogeneities and some features of the function $N_e(h, \chi)$ [225]. Thus, it is beyond doubt that ray tomography is an efficient approach for investigating global structures and inhomogeneities in the ionosphere.

This section gives only a brief account of some methods for studying the ionosphere through the use of satellite signals. Relevant experimental data are reviewed in detail in [194, 226–229].

References

1. Ginsburg, W.L. (1970) *Propagation of Electromagnetic Waves in Plasma*, N.Y.: Pergamon Press.
2. Ishimaru, A. (1978) *Wave Propagation and Scattering in Random Media*, N.Y.: Academic Press.
3. Ippolito, L.J. (1986) *Radiowave Propagation in Satellite Communications*, N.Y.: Van Nostrand Reinhold Co.
4. Komrakov, G.P., Skrebkova, L.A., and Tolmacheva, A.V. (1972) *Ionosfernye Issledovaniya*, 20: 55 (in Russian).
5. Belkovich, V.V. and Benediktov, E.A. (1969) *Radiofizika*, **12**, 10: 1439 (in Russian).
6. Belkina, L.M., Bocharov, V.I., and Yakovets, T.K. (1968) *Ionosfernye Issledovaniya*, 16: 115 (in Russian).
7. Kochenova, N.A. and Fligel, M.D. (1969) *Ionosfernye Issledovaniya*, 17: 150 (in Russian).
8. Beloborodova, A.A., Belkovich, V.V., Benediktov, E.A., *et al.* (1972) *Ionosfernye Issledovaniya*, 20: 75 (in Russian).
9. Driatskii, V.M. (1972) *Ionosfernye Issledovaniya*, 20: 65 (in Russian).
10. Sellers, B. and Hansen, F.A. (1977) *Radio Sci.*, **12**, 5: 779.
11. Brown, R.R. (1978) *J. Geophys. Res.*, **83**, A3: 1169.
12. Chytil, B. (1967) *J. Atmosph. Terr. Phys.*, **29**, 9: 1175.
13. Bischoff, K. and Chytil, B. (1969) *Planet. Space Sci.*, **17**, 5: 1059.
14. Whitney, H.E., Aarons, J., and Malik, C. (1969) *Planet. Space Sci.*, **17**, 5: 1069.
15. Crane, R.K. (1977) *Proc. IEEE*, **65**, 2: 180.
16. Aarons, I., Whitney, H., and Allen, R. (1971) *Proc. IEEE*, **59**, 2: 54.
17. Basu, S., Basu, D., Mullen, I.P., and Bushby, A. (1980) *Geophys. Res. Lett.*, **7**: 259.
18. Aarons, I., Whitney, H., Mackenzie, E., and Basu, S. (1981) *Radio Sci.*, **16**, 5: 939.
19. Frihagen, I. (1969) *J. Atmosph. Terr. Phys.*, **31**, 1: 81.
20. Sinclair, I. and Kelliher, R.F. (1969) *J. Atmosph. Terr. Phys.*, **31**, 1: 201.
21. Aarons, J. and Mullen, I.P. (1969) *J. Geophys. Res.*, **74**, 3: 884.
22. Koster, I.R. (1972) *Planet. Space Sci.*, **20**, 6: 1999.
23. Bandyopadhyay, P. and Aarons, I. (1970) *Radio Sci.*, **5**, 6: 931.
24. Kelliher, R.F. and Sinclair, I. (1970) *J. Atmosph. Terr. Phys.*, **32**, 7: 1259.
25. Titheridge, I.E. and Stuart, G.F. (1968) *J. Atmosph. Terr. Phys.*, **30**, 1: 85.
26. Allen, R.S. (1969) *J. Atmosph. Terr. Phys.*, **31**, 2: 289.
27. Aarons, I., Allen, R.S., and Elkins, I. (1967) *J. Geophys. Res.*, **72**, 11: 2891.
28. Aarons, I. (1967) *J. Atmosph. Terr. Phys.*, **29**, 6: 1619.
29. Crane, R.K. (1976) *J. Geophys. Res.*, **81**, 13: 2041.
30. Elkins, T.I. (1969) *J. Geophys. Res.*, **74**, 16: 4105.
31. Rufenach, C.L. (1972) *J. Geophys. Res.*, **77**, 8: 4761.

32. Singleton, D.G. (1974) *J. Atmosph. Terr. Phys.*, **36**, 1: 113.
33. Rufenach, C.L. (1974) *J. Geophys. Res.*, **79**, 10: 1562.
34. Rufenach, C.L. (1975) *Radio Sci.*, **10**, 2: 155.
35. Rino, C.L., Livingston, R.C., and Whitney, H.E. (1976) *J. Geophys. Res.*, **81**, 13: 2051.
36. Whitney, H.E. and Basus, B. (1977) *Radio Sci.*, **12**, 1: 123.
37. Kung Chie, Y. and Chaohan, L. (1982) *Proc. IEEE*, **70**, 4: 324.
38. Aarons, I. (1982) *Proc. IEEE*, **70**, 4: 360.
39. Mullen, I.P., Mackenzie, E.M., and Basu, S. (1985) *Radio Sci.*, **20**, 3: 357.
40. Rastogi, R.G. and Koparkar, P.V. (1990) *J. Atmosph. Terr. Phys.*, **52**, 1: 69.
41. Franke, S.I., Liu, C.H., and Fang, D.I. (1984) *Radio Sci.*, **19**, 3: 695.
42. Franke, S.I. and Liu, C.H. (1985) *Radio Sci.*, **20**, 3: 403.
43. Karasawa, Y., Yasukawa, K., and Yamada, M. (1985) *Radio Sci.*, **20**, 3: 643.
44. Aarons, I. (1985) *Radio Sci.*, **20**, 3: 397.
45. Aarons, I. (1987) *Radio Sci.*, **22**, 1: 100.
46. Van Vleck, I.H. (1947) *Phys. Rev.* **71**, 7: 425.
47. Barrett, A.H. and Chung, V.K. (1962) *J. Geophys. Res.* **67**, 9: 4259.
48. Zhevakin, S.A. and Naumov, A.P. (1963) *Radiofizika*, **6**, 4: 674 (in Russian).
49. Zhevakin, S.A. and Naumov, A.P. (1964) *Radiotekhnika i Elektronika*, **9**, 8: 1327 (in Russian).
50. Aganbekyan, K.A., Zrazhevskii, A.Y., Kolosov, M.A., and Sokolov A.V. (1971) *Radiotekhnika i Elektronika*, **16**, 9: 1564 (in Russian).
51. Zrazhevskii, A.Y. (1976) *Radiotekhnika i Elektronika* (Radio Engineering and Electronic Physics), **21**, 5: 951 (in Russian)*.
52. Zrazhevskii, A.Y. and Iskhakov, I.A. (1978) *Radiotekhnika i Elektronika* (Radio Engineering and Electronic Physics), **23**, 7: 1338 (in Russian)*.
53. Van Vleck, I.H. and Weisskopf, V.F. (1945) *Rev. Mod. Phys.*, **17**, 2/3: 227.
54. Van Vleck, I.H. (1947) *Phys. Rev.*, **71**, 7: 413.
55. Meeks, M.L. and Lilley, A.E. (1963) *J. Geophys. Res.*, **68**, 6: 1683.
56. Naumov, A.P. (1965) *Radiofizika*, **8**, 4: 668 (in Russian).
57. Zhevakin, S.A. and Naumov, A.P. (1965) *Radiotekhnika i Elektronika*, **10**, 6: 987 (in Russian).
58. Wulfsberg, K.N. (1967) *Radio Sci.*, **2**, 3: 319.
59. Rober, E.E., Mitchell, R.L., and Carter, C.I. (1970) *IEEE Trans. on Ant. and Prop.*, **AP-18**, 4: 472.
60. Vlasov, A.A., Kadygrov, E.N., and Shaposhnikov, A.N. (1990) *Issledovaniya Zemli iz Kosmosa*, 1: 36 (in Russian).
61. Zhevakin, S.A. and Troitskii, V.S. (1959) *Radiotekhnika i Elektronika*, **4**, 1: 21 (in Russian).
62. Zhevakin, S.A. and Naumov, A.P. (1966) *Radiofizika*, **9**, 3: 433 (in Russian).
63. Kislyakov, A.G. and Stankevich, K.S. (1967) *Radiofizika*, **10**, 9/10: 1224 (in Russian).
64. Zhevakin, S.A. and Naumov, A.P. (1967) *Radiofizika*, **10**, 9/10: 1214 (in Russian).
65. Crane, R.K. (1971) *Proc. IEEE*, **59**, 2: 173.
66. Dmitrenko, D.A. and Dmitrenko, L.V. (1973) *Radiofizika*, **16**, 12: 1817 (in Russian).
67. Jacobs, E. and Stacey, I.M. (1974) *IEEE Trans. Aerospace and Electronic Systems*, **10**, 1: 144.
68. Ippolito, L.I. (1971) *Proc. IEEE*, **59**, 2: 189.
69. Sokolov, A.V. (1974) Propagation of Millimeter Radiowaves in the Earth's Atmosphere, in: *Itogi Nauki i Tekhniki, Radiotekhnika*, Moscow: VINITI, vol. **5** (in Russian).
70. Ippolito, L.I. (1975) *IEEE Trans. on Aerospace and Electronic Systems*. **AES-II**, 6: 1067.
71. Naumov, A.P. and Zinicheva, M.B. (1980) *Radiotekhnika i Elektronika* (Radio Engineering and Electronic Physics), **25**, 5: 919 (in Russian)*.

72. Liebe, H.I. (1981) *Radio Sci.*, **16**, 6: 1183.
73. Swith, E.K. (1982) *Radio Sci.*, **17**, 6: 1455.
74. Liebe, H.I. (1985) *Radio Sci.*, **20**, 5: 1069.
75. Iskhakov, I.A., Zrazhevskii, A.Y., and Aganbekyan, K.A., (1986) in: *Electromagnetic Waves in the Atmosphere and Space*, Sokolov, A.V. and Semenov, A.A. (Eds.), Moscow: Nauka (in Russian).
76. Hogg, D.C. and Shing, C.T. (1975) *Proc. IEEE*, **63**, 9: 1308.
77. Crane, R.T. (1977) *Proc. IEEE*, **65**, 3: 459.
78. Bartolome, P.I. (1977) *Proc. IEEE*, **65**, 3: 475.
79. Ienkinson, G.F. (1977) *Proc. IEEE*, **65**, 3: 480.
80. Pratt, T. and Browning, D.I. (1977) *URSI Commission F, Comptes Rendus Proc.*, p. 357.
81. Allnutt, I.E. and Shutie, P.F. (1977) *URSI Commission F, Comptes Rendus Proc.*, p. 371.
82. Howell, R.G. and Thiriwell, I. (1977) *URSI Commission F, Comptes Rendus Proc.*, p. 339.
83. Pratt, T. and Browning, D.I. (1977) *URSI Commission F, Comptes Rendus Proc.*, p. 361.
84. Dijk, I. and Maanders, E.I. (1977) *URSI Commission F, Comptes Rendus Proc.*, p. 375.
85. Setzer D.E. (1970) *Bell Syst. Techn. J.*, **49**, 8: 1873.
86. Bodtman, W.F. and Ruthroff, G.L. (1974) *Bell Syst. Techn. J.*, **53**, 7: 1329.
87. Malinkin, V.G. (1981) *Radiotekhnika i Elektronika* (Radio Engineering and Electronic Physics), **26**, 1: 59 (in Russian)*.
88. Andrews, I.H. *et al.* (1982) *Radio Sci.*, **17**, 6: 1349.
89. Tsolaskis, A. and Stutzman, W.L. (1983) *Radio Sci.*, **18**, 6: 1287.
90. Stutzman, W.L., Bostian, C.W., Tsolakis, A., and Pratt, T. (1983) *Radio Sci.*, **18**, 5: 720.
91. Slobin, S.D. (1982) *Radio Sci.*, **17**, 6: 1443.
92. Cox, D.C. and Arnold, H.W. (1982) *Proc. IEEE*, **70**, 5: 458.
93. Dutton, E.I., Kobayashi, H.K., and Dougherty, H.T. (1982) *Radio Sci.*, **17**, 6: 1360.
94. Crane, R.K. (1982) *Radio Sci.*, **17**, 6: 1371.
95. Macchiarella, G. (1985) *Radio Sci.*, **20**, 1: 35.
96. Kanellopoulos, I.D. and Koukolas, S.G. (1987) *Radio Sci.*, **22**, 4: 549.
97. Sokolov, A.V., Sukhonin, E.V., Babkin, Y.S., and Iskhakov, I.A. (1986) in: *Electromagnetic Waves in the Atmosphere and Space*, Sokolov, A.V. and Semenov, A.A. (Eds.), Moscow: Nauka, p. 96 (in Russian).
98. Andreyev, G.A., Zrazhevskii, A.J., Kutuza, B.G., *et al.* (1985) Propagation of Millimeter and Submillimeter Waves in the Troposphere, in: *Problems of Modern Radio Engineering and Electronics*, Kotelnikov, V.A. (Ed.), Moscow: Nauka, p. 151 (in Russian).
99. Sukhonin, E.V. (1990) Prognostication of Millimetre Radio Waves Attenuation in the Atmosphere, in: *Itogi Nauki i Tekhniki, Radiotekhnika*, Moscow: VINITI, **41**: 3 (in Russian).
100. Pozhidaev, V.N. (1992) *Radiotekhnika i Elektronika* (J. Communications Technology and Electronics), **37**, 10: 1764 (in Russian)*.
101. Van de Hulst, H.C. (1957) *Light Scattering by Small Particles*, N.Y.: John Wiley & Sons.
102. Kazakov, L.Y. and Lomakin, A.N. (1976) *The Irregularities of the Refraction Index in the Earth Troposphere*, Moscow: Nauka (in Russian).
103. Gossard, E.E. (1977) *Radio Sci.*, **12**, 1: 89.
104. Lawrence, R.S., Clifford, S.F., and Ochs, G.R. (1977) *URSI Commission F, Comptes Rendus Proc.*, p. 400.
105. Stankevich, K.S. (1974) *Radiofizika* (Radiophysics and Quantum Electronics), **17**, 5: 666 (in Russian)*.
106. Crane, R.K. (1977) *URSI Commission F, Comptes Rendus Proc.*, p. 415.
107. Hodge, D.B., Theobold, D.M., and Devasirvatham, D.M. (1977) *URSI Commission F, Comptes Rendus Proc.*, p. 421.

108. Vogel, W.I., Straiton, A.W., and Fanin, B.M. (1977) *Radio Sci.*, **12**, 5: 757.
109. Banjo, O.P. and Vilar, E. (1986) *IEEE Trans. Commun.*, **34**, 8: 774.
110. Andreas, E.L. (1989) *Radio Sci.*, **24**, 5: 667.
111. Cox, D.C., Arnold, H.W., and Hoffman, H.H. (1981) *Radio Sci.*, **16**, 5: 885.
112. Bean, B.R. and Dutton, E.G. (1966) *Radio Meteorology*, Washington: United States Department of Commerce.
113. Hill, R.I., Lawrence, R.S., and Priestley, I.T. (1982) *Radio Sci.*, **17**, 5: 1251.
114. Skrypnik, G. (1965) *Radiofizika*, **8**, 3: 485 (in Russian).
115. Armand, N.A. and Kolosov, M.A. (1965) *Radiotekhnika i Elektronika*, **10**, 8: 1401 (in Russian).
116. Shabelnikov, A.V. (1968) *Radiotekhnika i Elektronika*, **13**, 12: 2115 (in Russian).
117. Andrianov, V.A., Armand, N.A., and Vetrov, V.N. (1973) *Radiotekhnika i Elektronika*, **18**, 4: 673 (in Russian).
118. Kolosov, M.A. and Shabelnikov, A.V. (1976) *The Radio Waves Refraction in the Atmospheres of the Earth, Mars and Venus*, Moscow: Sovetskoye Radio (in Russian).
119. Bean, B.R. (1964) *Troposphere Refraction, Advances in Radio Res.*, Saxton. I. (Ed.), N.Y.: Acad. Press.
120. Anderson, W.L., Beyers, N.I., and Rainey, R.I. (1960) *IRE Trans.*, **AP-8**, 5: 724.
121. Anway, A. (1963) *J. Res. NBS*, **67D**, 2: 153.
122. Efanov, V.A., Kolosov, M.A., Moiseev, I.G., et al. (1978) *Radiotekhnika i Elektronika* (Radio Engineering and Electronic Physics), **23**, 9: 1969 (in Russian)*.
123. Crane, R.K. (1976) *Meth. Exp. Phys., Astrophysics*, **12**: 186.
124. Sergienko, V.I., Modestov, G.I., Myasnikov, Y.S., et al. (1989) *Radiotekhnika i Elektronika* (Soviet J. Communications Technology and Electronics), **34**, 9: 1817 (in Russian)*.
125. Shabelnikov, A.V. (1986) in: *Electromagnetic Waves in the Atmosphere and Space*, Sokolov, A.V. and Semenov, A.A. (Eds.), Moscow: Nauka, p. 25 (in Russian).
126. Vilar, E. and Howard, S. (1986) *IEEE Trans. Antennas and Propag.*, **34**, 1: 2.
127. Weisbrod, S. and Colin, L. (1960) *IRE Trans.*, **AP-8**, 1: 107.
128. Lowen, R.W. (1962) *J. Geophys. Res.*, **67**, 6: 2339.
129. Gdalevich, G.L., Gringauz, K.T., Rudakov, V.A., and Rytov, S.M. (1963) *Radiotekhnika i Elektronika*, **8**, 6: 942 (in Russian) .
130. Titheridge, I.E. (1964) *J. Atmosph. Terr. Phys.*, **26**, 2: 159.
131. Voronin, A.L. (1964) *Geomagnetizm i Aeronomiya*, **4**, 3: 531 (in Russian).
132. Vitkevich, V.V. and Kokurin, Y.L. (1957) *Radiotekhnika i Elektronika*, **2**, 7: 826 (in Russian).
133. Lawrence, R.S., Iespersen, I., and Lamb, R.C. (1961) *J. Res. NBS*, **D65**, 2: 333.
134. Bramley, E.N. (1974) *J. Atmosph. Terr. Phys.*, **36**, 9: 1503.
135. Millman, G.H. (1958) *Proc. IRE*, **46**, 8: 1492.
136. Stephen, M. and Harris, A. (1961) *IRE Trans.*, **AP-9**, 2: 207.
137. Freeman, I.I. (1962) *J. Res. NBS*, **66D**, 6: 695.
138. Sweezy, W.B. and Bean, B.R. (1963) *J. Res. NBS*, **67D**, 2: 354.
139. Rosa, A.V. (1969) *IEEE Trans.*, **AP-17**, 5: 628.
140. Takahashi, K. (1970) *IEEE Trans. Aerospace and Electronic Systems*, **AES-6**, 6: 770.
141. Hopfield, H.S. (1971) *Radio Sci.*, **6**, 3: 357.
142. Tyagi, T.R., Ghosh, A.B., Mitra, A.P., and Somayajuli, I.V. (1972) *Space Res.*, **12**, 12: 1195.
143. Le Vine, D.M. (1972) *Radio Sci.*, **7**, 6: 625.
144. Mathur, N.C., Gossi, M.D., and Pearlman, M.R. (1970) *Radio Sci.*, **5**, 10: 1253.
145. Shaper, L.W., Staelin D.H., and Waters, I.W. (1970) *Proc. IEEE*, **58**, 2: 272.
146. Askne, I. and Nordius, H. (1987) *Radio Sci.*, **22**, 3: 379.

147. Zinicheva, M.B. and Naumov, A.P. (1987) *Radiofizika* (Radiophysics and Quantum Electronics), **30**, 9: 1163 (in Russian)*.
148. Klobuchar, I. (1986) *IEEE Posit. Locat. and Navig. Symp. Res.*, p. 280.
149. Coster, A.I., Buonsanto M., Gaposchkin E.M., *et al.* (1990) *Adv. Space Res.*, **10**, 8: 105.
150. Coco, D. Cocker, C., Dahlke, S., and Clynch, I. (1991) *IEEE Trans. on AES*, **27**, 6: 931.
151. Bevis, M., Businger, S., Herring, T., *et al.* (1992) *J. Geophys. Res.*, **97**: 15787.
152. Businger, S., Chiswell, S., Bevis, M., *et al.* (1996) *Bulletin of the American Meteorological Society*, **77**, 1: 5.
153. Spoelstra, T.A. and Kelder, H. (1984) *Radio Sci.*, **19**, 3: 779.
154. Johnson, A. (1985) *Radio Sci.*, **20**, 3: 339.
155. Cannon, W.H. (1987) *Radio Sci.*, **22**, 1: 141.
156. Porcello, L.I. and Hughes L.R. (1968) *J. Geoph. Res.*, **73**, 19: 6337.
157. Stokii, A.A. (1973) *Radiofizika*, **16**, 5: 806 (in Russian).
158. Stokii, A.A. (1976) *Radiofizika* (Radiophysics and Quantum Electronics), **19**, 11: 1678 (in Russian)*.
159. Armand, N.A. and Lomakin, A.N. (1976) *Radiotekhnika i Elektronika* (Radio Engineering and Electronic Physics), **21**, 1: 11 (in Russian)*.
160. Dravskich, A.F. and Finkel'stein, A.M. (1979) *Astronomicheskii Zhurnal* (Soviet Astronomy J.), **23**, 5: 620 (in Russian)*.
161. Dravskich, A.F., Stokii, A.A., Finkel'stein, A.M., and Fridman, P.A. (1977) *Radiotekhnika i Elektronika* (Radio Engineering and Electronic Physics), **22**, 11: 51 (in Russian)*.
162. Hinder, R.A. (1970) *Nature*, **225**: 614.
163. Andrianov, V.A., Arkhangel'skii, V.A., Bobrov, V.V., *et al.* (1990) *Radiotekhnika i Elektronika* (Soviet J. Communications Technology and Electronics), **35**, 5: 1081 (in Russian)*.
164. Hopfield, H.S. (1963) *J. Geoph. Res.*, **68**, 18: 5157.
165. Armand, N.A., Andrianov, V.A., Kopilevich, D.I., *et al.* (1986) *Radiotekhnika i Elektronika* (Soviet J. Communications Technology and Electronics), **31**, 12: 2305 (in Russian)*.
166. Armand, N.A., Andrianov, V.A., and Smirnov, V.M. (1987) *Radiotekhnika i Elektronika* (Soviet J. Communications Technology and Electronics), **32**, 4: 673 (in Russian)*.
167. Pavelyev, A.G. (1984) *Radiotekhnika i Elektronika* (Radio Engineering and Electronic Physics), **29**, 9: 1658 (in Russian)*.
168. Guier, W.H. (1961) *Proc. IRE*, **49**, 11: 1680.
169. Bennett, I.A. (1967) *J. Atmosph. Terr. Phys.*, **29**, 7: 887.
170. Willman, I.F. (1965) *IEEE Trans. Aerospace Electronic Syst.*, **1**, 3: 283.
171. Tucker, A.I. and Fannin, B.M. (1968) *J. Geoph. Res.*, **73**, 13: 4325.
172. Willman, I.F. and Tucker, A.I. (1968) *J. Geoph. Res.*, **73**, 1: 385.
173. Soicher, H. (1977) *IEEE Trans. Antennas and Propagation*, **AP-25**, 5: 705.
174. Procedings of ION GPS. (1992–1997) *Annual International Meeting of the Satellite Division of the Institute of Navigation.*
175. Beutler, G., Rothacher, M., Springer, T., *et al.* (1999) *International GPS Service: Information and Resources.*
176. Gringauz, K.I. (1958) *Doklady Akademii Nauk SSSR*, **120**, 6: 1234 (in Russian).
177. Gringauz, K.I. and Rudakov, V.A. (1961) *Iskustvennye Sputniki Zemli*, 6: 48 (in Russian).
178. Rudakov, V.A. (1964) *Kosmicheskie Issledovaniya*, **2**, 6: 946 (in Russian).
179. Berning, W.W. (1960) *J. Geophys. Res.*, **65**, 9: 2589.
180. Jackson, I.E. and Bauer, S.I. (1961) *J. Geophys. Res.*, **66**, 9: 3055.
181. Bauer, S.I., Bluml, L.I., and Donley, I.L., *et al.* (1964) *J. Geophys. Res.*, **69**, 1: 186.
182. Jackson, I.E. and Seddon, I.C. (1958) *J. Geophys. Res.*, **63**: 197.
183. France, L.A. and Williams, E.R. (1976) *J. Atmosph. Terr. Phys.*, **38**: 957.

184. Smith, L.J. and Gilchrist, B.E. (1984) *Radio Sci.*, **19**, 3: 913.
185. Danilov, A.D. and Smirnov, N.V. (1994) *Geomagnetizm i Aeronomiya*, **34**, 6: 74 (in Russian).
186. Mityakova, E.E., Mityakov, N.A., and Rapoport, V.O. (1960) *Radiofizika*, **3**, 6: 949 (in Russian).
187. Mityakov, N.A., Mityakova, E.E., and Cherepovetskii, V.A. (1963) *Geomagnetizm i Aeronomiya*, **3**, 5: 816 (in Russian).
188. Mityakov, N.A. and Mityakova, E.E. (1963) *Geomagnetizm i Aeronomiya*, **3**, 5: 858 (in Russian).
189. Mityakov, N.A., Mityakova, E.E., and Cherepovetskii, V.A. (1968) *Radiofizika*, **11**, 9: 1318 (in Russian).
190. Carriott, O.K. (1960) *J. Geoph. Res.*, **65**, 4: 1139.
191. Mendonca, F. (1962) *J. Geophys. Res.*, **67**, 6: 2315.
192. Tuhi, R.T. (1974) *J. Atmosph. Terr. Phys.*, **36**, 7: 1157.
193. Golton, E. (1962) *J. Atmosph. Terr. Phys.*, **24**, 3: 554.
194. Getmantsev, G.G., Gringauz, K.I., Erukhimov, L.M., *et al.* (1968) *Radiofizika*, **11**, 5: 649 (in Russian).
195. Titheridge, I.E. (1966) *J. Atmosph. Terr. Phys.*, **28**, 12: 1135.
196. Liszka, L. (1967) *J. Atmosph. Terr. Phys.*, **29**, 10: 1243.
197. Taylor, G.N. and Earnshaw, R.D. (1970) *J. Atmosph. Terr. Phys.*, **32**, 10: 1675.
198. Lawrence, R.S., Posakony, D.I., Garriott, O.K., and Hall, S.C. (1963) *J. Geophys. Res.*, **68**, 7: 1889.
199. Carriott, O.K., Rosa, A.V., and Ross, W.I. (1970) *J. Atmosph. Terr. Phys.*, **32**, 4: 705.
200. Rastogi, R.G. and Sharma, R.P. (1971) *Planet. Space. Sci.*, **19**, 11: 1505.
201. Walker, G.O. (1971) *J. Atmosph. Terr. Phys.*, **33**, 7: 1041.
202. Titheridg, I.E. (1973) *J. Atmosph. Terr. Phys.*, **35**, 5: 981.
203. Merrill, R.G. and Zawrence, R.S. (1969) *J. Geophys. Res.*, **74**, 19: 4661.
204. Davies, K., Fritz, R.B., and Gray, T.B. (1976) *J. Geophys. Res.*, **81**, 16: 2825.
205. Essex, E.A. (1978) *J. Atmosph. Terr. Phys.*, **40**, 9: 1019.
206. Davies, K., Donnelly, R., Grubb, R., *et al.* (1979) *Radio Sci.*, **14**, 1: 85.
207. Soicher, H. and Gorman, F. (1985) *Radio Sci.*, **20**, 3: 383.
208. Leitinger, R., Hartman, G., Lohmar, F., and Putz, E. (1984) *Radio Sci.*, **19**, 3: 789.
209. Earnshaw, R.D. and Taylor, G.N. (1968) *J. Atmosph. Terr. Phys.*, **30**, 7: 1369.
210. Bryunelli, B.E., Chernyshov, M.Y., and Chernyakov, S.M. (1992) *Geomagnetizm i Aeronomiya*, **32**, 5: 82 (in Russian).
211. Chernyakov, S.M., Tereshchenko, E.D., Kunitsyn, V.E., and Bryunelli, B.E. (1992) *Geomagnetizm i Aeronomiya*, **32**, 4: 94 (in Russian).
212. Rino, C.L. and Owen, I. (1984) *Radio Sci.*, **19**, 3: 891.
213. Basu, S., Basu, S., and MacKenzie, E. (1985) *Radio Sci.*, **20**, 3: 347.
214. Royden, H.N., Miller, R.B., and Buennagel, L.A. (1984) *Radio Sci.*, **19**, 3: 798.
215. Lanyi, G.E. and Roth, T. (1988) *Radio Sci.*, **83**, 4: 483.
216. Sardon, E., Rius, A., and Zarraoa, N. (1994) *Radio Sci.*, **29**, 3: 577.
217. Zarraoa, N. and Sardon, E. (1996) *Ann.Geophysicae*, **14**: 11.
218. Yeh, K.C. and Raymund, A. (1991) *Radio Sci.*, **26**, 6: 1361.
219. Saenko, Yu.S., Shagimuratov, I.I., *et al.* (1991) *Geomagnetizm i Aeronomiya*, **31**, 3: 558 (in Russian).
220. Oraevskii, V.N., Kunitsyn, V.E., and Shagimuratov, I.I. (1995) *Geomagnetizm i Aeronomiya*, **35**, 1: 117 (in Russian).
221. Austen, R.I., Franke, S.I., and Liu, C.H. (1988) *Radio Sci.*, **23**, 3: 299.
222. Raymond, T.D., Austen, I.R., Franke, S.I., *et al.* (1990) *Radio Sci.*, **25**, 5: 771.

223. Kunitsyn, V.E. and Tereshchenko, E.D. (1992) *IEEE Antennas and Propag. Magazine*, **34**, 5: 22.
224. Kunitsyn, V.E., Tereshchenko, E.D., Andreeva, E.E., and Rasinkov, O.C. (1994) *Int. J. Imaging Systems and Technology*, **5**: 128.
225. Foster, I.C., Buonsanto, M.I., Holt, I.M., *et al.* (1994) *Int. J. Imaging Systems and Technology*, **5**: 148.
226. Alpert, I.L. (1976) *Space Sci. Rev.*, **18**, 5/6: 551.
227. Evans, I.V. (1977) *Rev. Geophys. and Space Phys.*, **15**, 3: 325.
228. Davies, K. (1980) *Space Sci. Rev.*, **25**: 357.
229. Solodovnikov, G.K., Sinel'nikov, V.M., and Krokhmalnikov, E.V. (1988) *Remote Sensing of the Earth's Ionosphere by the Radio Signals of Space Probes*, Moscow: Nauka (in Russian).
230. Kunitsyn, V.E. and Tereshchenko, E.D. (1991) *Tomography of the Ionosphere*, Moscow: Nauka (in Russian).

Chapter 2

Investigations of planetary atmospheres and ionospheres by the radio occultation technique

2.1 Direct problem of the radio occultation investigations of planetary atmospheres and ionospheres

Launching spacecraft to planets offers an opportunity to study their atmospheres and ionospheres by the radio occultation technique. In short, this technique is as follows. When a spacecraft disappears behind a planet and then reappears, radio waves pass through the planetary atmosphere and undergo changes in their amplitude and frequency. These changes, recorded at Earth-based stations, provide information about the atmosphere and ionosphere in particular regions of the planet. There are direct and reverse radio occultation problems. Direct problem implies that the atmospheric model is known, and one has to find changes in the amplitude, frequency, or phase of radio waves for a given trajectory of a spacecraft. In other words, a direct problem serves to study unknown planetary atmospheres. The authors of [1, 2] formulated the direct radio occultation problem, derived approximate formulae relating changes in the field strength and frequency of radio waves induced by an atmosphere, and estimated corresponding effects in the thin atmosphere and ionosphere of Mars. This problem was also treated as applied to the dense Venusian [3] and Jovian [4, 5] atmospheres.

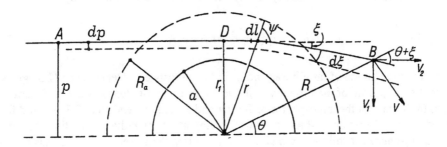

Fig. 2.1. The geometry of the radio occultation investigations of planetary atmospheres.

The direct radio occultation problem is formulated here according to [1–5]. Let a spacecraft with a transmitter and an antenna transmitting radio signals towards the Earth (this direction is shown in Fig. 2.1 by the respective arrow) goes behind a planet at point B; the atmosphere is assumed to be spherically symmetric, that is, its refractive index n depends on the radius r; the spacecraft trajectory is defined by the coordinates R and θ or

73

by its velocity vector V; the absorption of radio waves is assumed to be low; the attenuation of radio waves is mainly determined by the refraction, or bending, of the ray path. One has to find changes in the energy flux and frequency of radio signals received on Earth.

Consider first the refraction of radio waves during the remote sensing of an atmosphere. The ray AB, representing a straight line beyond the atmosphere, begins to bend in the vicinity of point D in the atmosphere to be finally refracted by a refraction angle ξ (Fig. 2.1). To determine this angle, let us examine two ray segments, AD and DB, by invoking the results of Section 1.4. Point D in Fig. 2.1 corresponds to the closest approach of the ray to the planet's surface. From (1.48) and Fig. 2.1 it follows that

$$p = n(r)r \sin \psi = n(h_1)(a + h_1) = R\sin(\Theta + \xi).$$ (2.1)

Here a is the planet's radius, $h_1 = r_1 - a$ is the minimal ray altitude above the planet surface, p is the impact parameter, and ψ is the angle between the ray element and radius-vector r. When deriving expression (2.1) it was taken into account that the angle ψ at point D is equal to 90°. Taking into account (1.56) and (2.1), we arrive at the following formula for the refraction angle:

$$\xi = -2p \int_{r_1}^{\infty} \frac{1}{n} \frac{dn}{dr} \frac{dr}{\sqrt{r^2 n^2 - p^2}}.$$ (2.2)

Factor 2 in this expression is to account for the two path segments, AD and DB. The height dependence of the refractivity, $N(h) = n - 1$, can be approximated by

$$N(h) = \exp(-a_1 h^2 - b_1 h - c_1),$$ 2.3)

where h is the altitude of an arbitrary point in the atmosphere. Dependence (2.3) gives a correct height profile of the refractivity in the atmospheres of Venus, Mars, and Earth for any h. In the case of the atmospheres of Jupiter and Saturn, dependence (2.3) is valid for a limited range of altitudes, which is, however, important for radio occultation studies of those atmospheres. It follows from (2.2) and (2.1) that the integrand in (2.2) has an infinite value at the lower integration limit. This difficulty can be avoided by calculating the integral by parts:

$$\xi = -2p \int_{r_1}^{\infty} \frac{\left[n^2 \dfrac{dn}{dr} + 5nr \left(\dfrac{dn}{dr} \right)^2 + 2r^2 \left(\dfrac{dn}{dr} \right)^3 - n^2 r \dfrac{d^2 n}{dr^2} \right] \sqrt{n^2 r^2 - p^2}\, dr}{n^3 r^2 \left(n + r \dfrac{dn}{dr} \right)^2}.$$ (2.4)

It should be emphasized that a_1 in formula (2.3) is a parameter but not the planet's radius, symbolized by a.

Formula (2.4), which allows the computer-aided calculation of the refraction angle, has two related parameters characterizing the ray path, namely, the impact parameter p and the minimal ray altitude h_1. From equation (2.1), which is valid for a spherically symmetric medium, follows

$$p = n(r_1)r_1 \tag{2.5}$$

and then

$$p = (a + h_1)\left[1 + \exp(-a_1 h_1^2 - b_1 h_1 - c_1)\right]. \tag{2.6}$$

Formula (2.6), which describes the relation between the parameters p and h_1, and formula (2.4) allow one to find the refraction angle as a function of the minimal ray altitude h_1 or the impact parameter p.

Thin atmospheres can be analyzed in an isothermal approximation, which suggests that the altitude dependence of the refractivity is expressed by (2.3) with the following parameters

$$
\begin{aligned}
a_1 &= 0, \\
b_1 &= mgk_0^{-1}T_0^{-1}, \\
N_0 &= \exp(-c_1) = \mu P_0 T_0^{-1}.
\end{aligned}
\tag{2.7}
$$

Here m is the effective molecular weight of atmospheric gas, k_0 is the Boltzmann constant, g is gravitational acceleration, P_0 is the near-surface pressure, T_0 is the mean temperature of the atmosphere, and μ is a coefficient dependent on the gas composition of the atmosphere. Simple expressions (2.3) and (2.7) allow the refraction angle to be found analytically. When $N(h)$ is an exponential function, we can derive from (2.2) for $a_1 = 0$ and $b_1 a N_0 \ll 1$ the following formula

$$\xi = 2b_1 a N_0 \int_0^\infty \frac{\exp(-x)dx}{\sqrt{x(x + 2b_1 a)}}. \tag{2.8}$$

Here we took into account (2.3) and introduced a new variable $x = b_1(r - a)$. From (2.8) it follows that

$$\xi = 2b_1 a N_0 \exp(-b_1 a)I_0(b_1 a), \tag{2.9}$$

where I_0 is the Bessel function of an imaginary argument. Most commonly, $b_1 a \gg 1$; therefore, one makes use of the asymptotic form of function I_0. Then from (2.9) follows the formula

$$\xi = N_0 \sqrt{2\pi b_1 a} \,. \tag{2.10}$$

This refraction angle corresponds to the ray that touches the planet's surface. At an arbitrary ray altitude, the angle of refraction is given by

$$\xi = N_0 \sqrt{2\pi b_1 (a + h_1)} \exp(-b_1 h_1) \,.$$

Taking into account that $h_1 \ll a$, we arrive at

$$\xi = N_0 \sqrt{2\pi b_1 a} \, \exp(-b_1 h_1) \,. \tag{2.11}$$

Below the theoretical analysis of the refraction angle ξ is given with reference to the atmospheres of Venus, Mars, and Jupiter.

For the numerical analysis of refractive phenomena, one needs an atmospheric model; in other words, it is necessary to define the parameters a_1, b_1, and c_1 in formula (2.3). The Venusian and Martian atmospheres are composed primarily of carbon dioxide, with trace amounts of nitrogen. According to [6], the refractivity N of such an atmosphere is given by

$$N = \mu P T^{-1} \,. \tag{2.12}$$

Here

$$\mu = k_0^{-1} (1.8 e_2 + 1.1 e_3) \cdot 10^{-17} \,,$$

P is pressure expressed in atmospheres, T is temperature expressed in degrees Kelvin, k_0 is the Boltzmann constant, e_2 and e_3 are the fractional contents of carbon dioxide and nitrogen in the atmosphere.

Taking the CO_2 and N_2 contents of the Venusian atmosphere to be, respectively, 96 and 4% and neglecting the weak effect of other atmospheric constituents on the refractive index, we find that parameter μ is equal to 0.135 deg/atm. Using experimental data on the height dependence of pressure and temperature and taking into account formula (2.3), an approximate height dependence of refractivity can be derived. Numerical analysis performed in [3] showed that the altitude profile of the refractive index in the Venusian atmosphere is satisfactorily approximated by expression (2.3) at the following values of the component parameters:

$$a_1 = 5.79 \cdot 10^{-4}\,\mathrm{km}^{-2},$$
$$b_1 = 4.40 \cdot 10^{-2}\,\mathrm{km}^{-1}, \qquad (2.7A)$$
$$c_1 = 4.11.$$

This approximation corresponds to an arbitrary zero altitude corresponding to a pressure P equal to 92 atm. Using expressions (2.3), (2.4), and parameters (2.7A), we can find the refraction angle ξ.

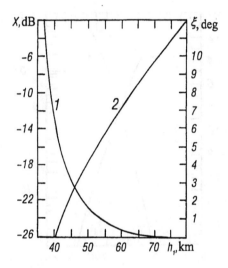

Fig. 2.2. Refraction angle ξ (curve 1) and refractive attenuation X (curve 2) as functions of the minimal ray height h_l in the Venusian atmosphere.

Curve 1 in Fig. 2.2 represents the calculated dependence of the refraction angle on the minimal ray altitude in the Venusian atmosphere. At altitudes $h_1 > 90$ km, the refraction angle is insignificant, but it steeply rises at $h_1 \le 60$ km. For instance, at so close altitudes as $h_1 = 50$ and 40 km, the refraction angle ξ is equal to 1.6 and 6°, respectively. At $h_1 = 34$ km, the Venusian atmosphere captures radio waves; this altitude corresponds to so-called critical refraction.

The Jovian atmosphere is composed of hydrogen and helium, with trace amounts of methane and ammonia. Analysis of the altitude profile of the refractive index and some other refraction effects in this atmosphere showed that the relevant parameters in formula (2.3) are as follows:

$$a_1 = 4.45 \cdot 10^{-4} \, \text{km}^{-2},$$
$$b_1 = 3.22 \cdot 10^{-2} \, \text{km}^{-1}, \qquad\qquad (2.7\text{B})$$
$$c_1 = 7.75.$$

In this case the altitude h_1 is measured from an arbitrary zero level corresponding to $P = 0.5$ atm. Curve 1 in Fig. 2.3 illustrates the theoretical dependence of angle ξ on the ray altitude relative to the arbitrary zero level in the Jovian atmosphere. At $h_1 > 100$ km, the refraction angle ξ is insignificant, but it rises steeply in the region $h_1 < 80$ km to reach a value of $1°$ at $h_1 = 35$ km. Critical refraction in the Jovian atmosphere occurs at $h_1 = 0$, i.e., in the region where the pressure is about 0.5 atm.

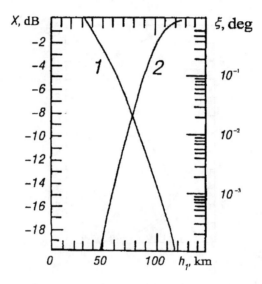

Fig. 2.3. Refraction angle ξ (curve 1) and the refractive attenuation X (curve 2) as functions of the minimal ray height h_1 in the Jovian atmosphere.

The refraction angle in the Martian atmosphere can be calculated by formula (2.11), taking that the planet radius $a = 3393$ km, parameter b_1 is about 0.1 km^{-1}, and N_0 averages $4.7 \cdot 10^{-6}$. The refraction angle ξ in the Martian atmosphere is small: even for a ray touching the planet's surface, it averages $6 \cdot 10^{-4}$ radians.

The results of this quantitative analysis are used below to estimate the refractive attenuation and frequency shift of radio waves. In occultation by dense atmospheres, knowledge of the refraction angle ξ is essential, since highly directional spacecraft-borne antennae oriented earthward emit radio signals at an angle ξ to the direction of its main lobe, and this should be taken into account.

For further analysis, one has to find a ray with the impact parameter p that passes through point B with polar coordinates R and θ (Fig. 2.1). The relationship between the ray element dl and the angle ψ is given by

$$dl = \frac{dr}{\cos\psi} = \frac{rndr}{\sqrt{r^2 n^2 - p^2}} \, . \tag{2.13}$$

In this expression, $\cos\psi$ was substituted allowing for (2.1). Going over to the polar coordinate system and integrating from r_1 to R, we arrive at

$$\Theta = \frac{\pi - \xi}{2} - p \int_{r_1}^{R} \frac{dr}{r\sqrt{r^2 n^2 - p^2}} \, . \tag{2.14}$$

Recall that R and θ are the coordinates of the spacecraft located at point B (Fig. 2.1). Using (2.14) and (2.2), after appropriate rearrangements, we obtain

$$\Theta = \frac{\pi}{2} + p \int_{r_1}^{R} \frac{\left(r\dfrac{dn}{dr} - n\right)dr}{rn\sqrt{r^2 n^2 - p^2}} \, . \tag{2.15}$$

This expression relates coordinates R and θ to the impact parameter p, and formula (2.1) relates p to h_1.

The above relations describe the atmospheric refraction effects that give rise to the refractive attenuation of radio waves. To estimate the refractive attenuation, consider two close rays that have impact parameters p and $p + dp$ (Fig. 2.1). Beyond the atmosphere, the annular tube bounded by these rays has the cross section $S_0 = 2\pi p dp$. The refractive deformation of this tube leads to an increase in its cross section at point B, given by

$$S_1 = \frac{2\pi R^2 \sin\Theta d\Theta}{\cos(\Theta + \xi)} \, . \tag{2.16}$$

The refractive attenuation of radio waves is expressed by the ratio

$$X = \frac{p\cos(\Theta + \xi)}{R^2 \sin\Theta\left(\dfrac{d\Theta}{dp}\right)} \, . \tag{2.17}$$

From the geometry given in Fig. 2.1 it follows:

$$\cos(\Theta + \xi) = \frac{\sqrt{R^2 - p^2}}{R}, \tag{2.18}$$

and hence

$$\Theta = \frac{\pi}{2} - \xi - \arctan \frac{\sqrt{R^2 - p^2}}{p}; \tag{2.19}$$

therefore,

$$\frac{d\Theta}{dp} = -\frac{d\xi}{dp} + \frac{1}{\sqrt{R^2 - p^2}}. \tag{2.20}$$

From (2.17), (2.18), and (2.20) one can derive the following formula for the refractive attenuation of radio waves:

$$X = \frac{p}{R \sin \Theta \left[1 - \left(\dfrac{d\xi}{dp} \right) \sqrt{R^2 - p^2} \right]}. \tag{2.21}$$

If the refraction angle is small, then $p = R \sin\theta$ and $\sqrt{R^2 - p^2} = L$. In such a case the following approximate expression can be written for thin atmospheres:

$$X = \left[1 - L \left(\frac{d\xi}{dp} \right) \right]^{-1}, \tag{2.22}$$

where $L = DB$ is the distance from the spacecraft to the planet's limb. From (2.21) and (2.22) it follows that the refractive attenuation of radio waves is determined by the distance L and the derivative of the refraction angle with respect to the impact parameter. Using (2.11) and (2.22), we can get the following approximate formula for the refractive attenuation of radio waves:

$$X = \left[1 + b_1 L \sqrt{2\pi b_1 a} N_0 \exp(-b_1 h_1) \right]^{-1}. \tag{2.23}$$

For dense atmospheres, one has to use (2.2) in addition to (2.21). Then,

$$
\frac{d\xi}{dp} = 2 \int_{r_1}^{\infty} \frac{\left(\dfrac{2p^2}{r^2 n^2} - 1\right)\left[n\dfrac{dn}{dr} + 5r\left(\dfrac{dn}{dr}\right)^2 - rn\dfrac{d^2 n}{dr^2} + \dfrac{2r^2}{n}\left(\dfrac{dn}{dr}\right)^3 \right] dr}{\left(n + r\dfrac{dn}{dr}\right)^2 \sqrt{r^2 n^2 - p^2}} .
$$

(2.24)

Expressions (2.21) and (2.24) give the refractive attenuation of radio waves in planetary atmospheres.

From these expressions it follows that the energy flux of radio signals vanishes when the term $(n + rdn/dr) \to 0$; therefore, the following expression is valid for the critical refraction level:

$$
a + h_c = r_c = -n\left(\frac{dn}{dr}\right)^{-1} .
$$

(2.25)

This simple formula corresponds to the beam which is bent so severely that it is captured by the atmosphere and cannot escape it. Therefore, atmospheres can be investigated by the radio occultation technique only if $r > r_c$. Parameters r_c and h_c refer to a critical level of the radio wave refraction. From (2.25) and (2.3) follows the expression describing the critical refraction in dense atmospheres:

$$
(a + h_c)(2a_1 h + b_1)\exp(-a_1 h_c^2 - b_1 h_c - c_1) = 1 + \exp(-a_1 h_c^2 - b_1 h_c - c_1) .
$$

(2.26)

Let us turn now to the numerical analysis of the refractive attenuation of radio waves. Curve *2* in Fig. 2.2 represents the theoretical dependence of the refractive attenuation of radio waves, *X*, on the minimal ray altitude in the Venusian atmosphere. This dependence was found by invoking relations (2.21) and (2.24) for the function $N(h)$ of the type (2.3) and using the parameter values shown in (2.7A) and distance $L = 10\,220$ km. It is evident from this figure that attenuation is equal to 3 and 14.5 dB at h_1 of about 75 and 55 km, respectively. At lower ray altitudes, refractive attenuation abruptly increases to -26 dB at $h_1 = 40$ km. From (2.26) we can find that h_c is 32 km; this corresponds to a pressure of about 6.8 atm.

Curve *2* in Fig. 2.3 represents the refractive attenuation of radio waves in the Jovian atmosphere for a spacecraft located 10^5 km from the planet's center. The altitude h_1 is measured from an arbitrary level corresponding to a pressure of 0.5 atm. It is obvious from this curve that the attenuation is approximately 3 dB at a ray altitude of about 100 km and is as strong as 20 dB at a ray altitude of 50 km. Estimation by formula (2.26) yields $h_c = -10$ km; in other words, this altitude corresponds to the top of the cloud cover of the planet.

Stratified irregularities, which are always present in atmospheres, affect the smooth refractivity profile $N(h)$. Let us consider how such irregularities may influence the refractive attenuation of radio waves. Refractivity in this case can be approximated by the expression that differs from (2.3) by having an additional term accounting for the effect of a bell-shaped layer:

$$N(h) = \exp(-a_1 h^2 - b_1 h - c_1) + M_3 \exp\left[-m_4 (h - h_i)^2\right].$$ (2.27)

Here h_i is the altitude of the layer's midpoint, M_3 is a deviation of the refractive index at this altitude from the mean profile, and parameter m_4 measures the layer thickness. The aforementioned feature of the refractive index is typical of, for example, temperature inversions in the Venusian atmosphere. Taking into account (2.12) and differentiating with respect to T, we arrive at

$$M_3 = -\mu P T^{-2} \Delta T .$$ (2.28)

Here ΔT is the temperature deviation at the center of a stratified temperature inversion. The refractive attenuation of radio waves in the presence of such an inversion can be calculated by (2.21), (2.24), and (2.27).

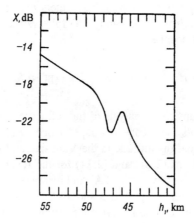

Fig. 2.4. Effect of stratified inhomogeneities on the refractive attenuation of radio waves, X.

The curve in Fig. 2.4 is drawn based on the results of such calculations for the Venusian atmosphere with an inverse layer. The calculations were performed by taking the following values of the relevant parameters: $h_i = 46$ km, $m_4 = 1$ km^{-2}, $M_3 = 2 \cdot 10^{-5}$, and $R = 10\ 220$ km. The indicated value of M_3 corresponds to the temperature difference ΔT being equal to 12 K. As follows from Fig. 2.4, the stratified inhomogeneities that affect the refractive index must manifest themselves as peculiarities in the altitude dependence of the refractive attenuation of radio waves.

In occultation by thin atmospheres, such as that of Mars, the radio ray touches the planet's surface, so that the first Fresnel zone is comparable in size with the arbitrary tropospheric thickness. In such a case the results obtained (e.g., formula (2.23)) need a correction because of diffraction phenomena that occur at the light–shadow boundary. The role of diffraction effects in radio occultation studies of thin atmospheres was analyzed in [7, 8]. Since the quantitative analysis of the effects of the atmosphere and the planet's surface on the diffraction of radio waves is sophisticated, we shall give only a qualitative description of the diffraction phenomena. If the planetary atmosphere is absent, the diffraction of radio waves is given by

$$\left(\frac{E}{E_0}\right)^2 = \left[\frac{\exp\left(\frac{i\pi}{4}\right)}{\sqrt{\pi}}\int_{\tau_0}^{\infty}\exp(iy^2)dy\right]^2.$$

(2.29)

Here $(E/E_0)^2$ expresses a decrease in the radio wave energy flux, and $\tau_0 = \psi_1\sqrt{\pi L\lambda^{-1}}$ is a geometrical parameter dependent on the spacecraft–planet's limb distance L and the diffraction angle ψ_1 (this angle, which is made by the geometrical shadow line and the direction towards the spacecraft, has the vertex located at the point where the ray touches the planetary surface). Relation (2.29) can be given in terms of the following tabulated Fresnel integrals

$$\left(\frac{E}{E_0}\right)^2 = \frac{1}{2}\left[\left(\frac{1}{2}-\text{Si}(\tau_0)\right)^2 + \left(\frac{1}{2}-\text{Co}(\tau_0)\right)^2\right],$$

$$\text{Si}(\tau_0) = \int_0^{\tau_0}\sin\left(\frac{\pi t^2}{2}\right)dt,$$

(2.30)

$$\text{Co}(\tau_0) = \int_0^{\tau_0}\cos\left(\frac{\pi t^2}{2}\right)dt.$$

In the illuminated region, where the angle $\psi_1 > 0$, the following asymptotic formula is valid:

$$\left(\frac{E}{E_0}\right)^2 = \left[1 + \frac{\sin\left(\tau_0^2 - \frac{\pi}{4}\right)}{\sqrt{\pi}|\tau_0|}\right]^2.$$

(2.31)

This formula accounts for the oscillatory pattern of the field strength in the illuminated area.

When a planetary atmosphere is present, the diffraction pattern becomes more complex, and one has to treat the problem of the diffraction of radio waves from a sphere surrounded with an atmosphere. Such problem was solved in [7, 9] for the case of the dependence $N(r) \sim r^{-2}$, which only roughly expresses the altitude profile of refractivity in the atmosphere but, surprisingly, allows this problem to be solved exactly. The field strength in this case is given, according to [9], by

$$\left(\frac{E}{E_0}\right)^2 = X \left[1 + \frac{\sin\left(\tau_0^2 - \frac{\pi}{4}\right)}{\sqrt{\pi}|\tau_0|} \right]^2 . \tag{2.32}$$

Here parameter X represents the refractive attenuation of radio waves found in terms of rays, and τ_0 depends on the refraction angle according to the formula

$$\tau_0 = (\psi_1 + \xi)\sqrt{\pi L \lambda^{-1}} . \tag{2.33}$$

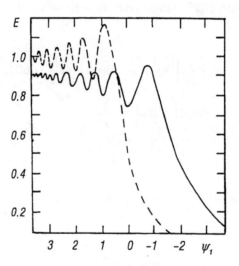

Fig. 2.5. Effect of the Martian atmosphere on diffractive variations in the field strength.

The authors of [7] approximately solved the diffraction problem for the case of an exponential decrease in the refractivity. Such an approximation allows a good description of real atmospheres but does not allow an exact solution of the diffraction problem. As was shown in [7], if the dependence $N(h)$ is exponential, the diffraction of radio waves can be well described by expression (2.32) in which the refractive attenuation X and the refraction angle ξ are substituted according to (2.23) and (2.11).

Figure 2.5 presents the results of the calculation of the field strength of 30-cm radio waves during the radio occultation studies of the Martian atmosphere (it was taken in the calculation that $N_0 = 4.7 \cdot 10^{-6}$, $b_1 = 0.1$ km^{-1}, and $L = 3 \cdot 10^4$ km). In this figure, the dashed line corresponds to the diffraction of radio waves from a planet that lacks an atmosphere, the solid line was calculated allowing for the effect of an atmosphere, and the angle ψ_1 is measured in 10^{-4} radian. Figure 2.5 and the relations presented above indicate that the effect of the thin Martian atmosphere on the diffraction of radio waves manifests itself as (1) a small decrease in the field strength due to the refractive attenuation of radio waves, (2) a shift of the diffraction curve by the refraction angle ξ, and (3) small changes in the shape of the oscillations of the field strength.

Let us now consider changes in the frequency and phase of radio waves induced by the atmosphere. General expressions for the frequency and phase of radio signals have the form

$$\varphi = \frac{2\pi}{\lambda} \int n dl,$$

$$f = \frac{1}{2\pi} \frac{d\varphi}{dt}.$$

(2.34)

Taking into account equation (2.13), we can derive from (2.34) the following phase formula

$$\varphi = \frac{2\pi}{\lambda} \int \frac{n^2 r dr}{\sqrt{r^2 n^2 - p^2}},$$

(2.35)

where λ is the wavelength, and integration is performed along the ray AB (Fig. 2.1). Now decompose the integral (2.35) into three parts:

$$\varphi = \frac{2\pi}{\lambda} \left[2 \int_{r_1}^{R_a} \frac{n^2 r dr}{\sqrt{r^2 n^2 - p^2}} + \int_{R_a}^{R} \frac{r dr}{\sqrt{r^2 - p^2}} + \int_{R_a}^{R_3} \frac{r dr}{\sqrt{r^2 - p^2}} \right],$$

(2.36)

where R_3 is the distance from the planet's center to the Earth, and R_a is an arbitrary radius corresponding to the upper boundary of the ionosphere (Fig. 2.1). Let us rearrange the formula (2.36) by invoking the equality

$$\int_{r_1}^{R_a} \frac{n^2 r dr}{\sqrt{r^2 n^2 - p^2}} = -\frac{p\xi}{2} + \int_{r_1}^{R_a} \frac{1}{n}\frac{dn}{dr}\sqrt{r^2 n^2 - p^2}\,dr\,, \tag{2.37}$$

where ξ is defined by integral (2.2). From (2.36) and (2.37) we obtain the formula for the phase of radio waves traveling through the planet's atmosphere:

$$\varphi = \frac{2\pi}{\lambda}\left[p\xi - 2\int_{r_1}^{R_a} \frac{1}{n}\frac{dn}{dr}\sqrt{r^2 n^2 - p^2}\,dr + \sqrt{R^2 - p^2} + \sqrt{R_3^{\,2} - p^2}\,\right]. \tag{2.38}$$

Taking into account that

$$R_3 \gg R,$$

$$R_3 \gg p, \tag{2.39}$$

$$\sqrt{R^2 - p^2} = R\cos(\Theta + \xi),$$

one can derive from (2.38) the following formula

$$\varphi = \frac{2\pi}{\lambda}\left[p\xi - 2\int_{r_1}^{R_a} \frac{1}{n}\frac{dn}{dr}\sqrt{n^2 r^2 - p^2}\,dr + R\cos(\Theta + \xi) + R_3\right]. \tag{2.40}$$

If there is no atmosphere, from (2.40) and the geometry in Fig. 2.1 it follows:

$$\varphi_0 = \frac{2\pi}{\lambda}\left(R\cos\Theta + R_3\right).$$

Atmospheric changes in the phase of radio waves are given by

$$\Delta\varphi = \varphi - \varphi_0 = \frac{2\pi}{\lambda}\left[p\xi - 2\int_{r_1}^{R_a} \frac{1}{n}\frac{dn}{dr}\sqrt{n^2 r^2 - p^2}\,dr + R\cos(\Theta + \xi) - R\cos\Theta\right]. \tag{2.41}$$

Taking into account (2.34) and (2.41), atmospheric changes in the frequency of radio waves are given by

$$\Delta F = \frac{1}{\lambda}\frac{d}{dt}\left[p\xi - 2\int_{r_1}^{R_a}\frac{1}{n}\frac{dn}{dr}\sqrt{n^2 r^2 - p^2}\,dr + R\cos(\Theta + \xi) - R\cos\Theta \right]. \qquad (2.42)$$

By differentiating (2.42) and taking into account (2.2), we get

$$\Delta F = \frac{1}{\lambda}\left\{ [\cos(\Theta + \xi) - \cos\Theta]\frac{dR}{dt} - R[\sin(\Theta + \xi) - \sin\Theta]\frac{d\Theta}{dt} \right\}. \qquad (2.43)$$

Expression (2.43) is a solution to the problem of frequency shifts in radio occultation studies of atmospheres. Let us go over from the parameters of the spacecraft trajectory, dR/dt and $d\Theta/dt$, to the projection of its velocity vector onto the earthward direction, V_2, and the normal velocity component, V_1 (Fig. 2.1), which are related as

$$\frac{dR}{dt} = -V_1\sin(\Theta + \xi) + V_2\cos(\Theta + \xi),$$

$$\qquad (2.44)$$

$$R\frac{d\Theta}{dt} = -V_1\cos(\Theta + \xi) - V_2\sin(\Theta + \xi).$$

These expressions allow the formula (2.43) to be rearranged to the form

$$\Delta F = \frac{1}{\lambda}\left[V_1\sin\xi + V_2(1 - \cos\xi) \right]. \qquad (2.45)$$

This exact formula relates the frequency shift ΔF induced by the atmosphere or ionosphere to the refraction angle ξ. According to the expression (2.2), ξ is defined by the altitude profile of the refractive index, $n(h)$; therefore, formula (2.45) also gives a relationship between ΔF_0 and n. It should be taken into account that the motion of a spacecraft always causes a significant Doppler shift in frequency, Δf_0, which is proportional to V_2 and does not depend on the properties of the planetary atmosphere. The frequency shift ΔF is considerably less than the Doppler shift in frequency, Δf_0. If $\xi \ll 1$, formula (2.45) is simplified to

$$\Delta F = \frac{V_1\xi}{\lambda}. \qquad (2.46)$$

Expressions (2.45) and (2.46) show that ΔF depends only on the refraction angle ξ and spacecraft velocity projections. The approximate formula (2.46) immediately follows from the formula expressing the Doppler effect

$$\Delta F = \frac{V_3 - V_2}{\lambda} .$$
(2.47)

Here V_3 and V_2 are the velocity vector projections onto, respectively, the refracted ray path and the line of sight (that is, on the ray path that would occur in the absence of the refractive effect of the atmosphere) (Fig. 2.1). Taking into account the geometry in Fig. 2.1, one can easily derive formula (2.46) from formula (2.47). Using expression (2.11), which is valid for an isothermal atmosphere, and formula (2.46), we obtain the following approximate formula

$$\Delta F = \frac{V_1 N_0}{\lambda} \sqrt{2\pi b_1 a} \exp(-b_1 h_1) .$$
(2.48)

In radio occultation experiments, the ray altitude $h_1 \approx V_1 t$; therefore, ΔF must change exponentially. As has already been mentioned, radio occultation observations of thin atmospheres are accompanied by diffraction phenomena. Based on the approximate solution to the diffraction problem, it was shown that formula (2.46) is valid even in the penumbra region [8]. Formula (2.46) and curves *1* in Figs. 2.2 and 2.3 allow a simple estimation of the frequency shift in the Venusian and Jovian atmospheres from radio occultation data. The spacecraft velocity component V_1 varies from 1 to 7 km s^{-1}; therefore, ΔF in the Venusian atmosphere may reach hundreds of Hz. At the same time, ΔF in the Martian atmosphere is as small as a few Hz.

Consider now the specificity of the radio occultation studies of planetary ionospheres. Basically, the above formulae, which were derived for atmospheres, are equally valid for ionospheres provided that the refractive index is expressed through the electron density. Atmospheres and ionospheres differ in only having qualitatively different altitude dependences of the refractive index. Namely, the altitude profile of refractivity in planetary atmospheres, $N(h)$, is primarily determined by an inverse relationship between pressure and altitude, while in the case of planetary ionospheres this profile is determined by the complex altitude dependence of the electron density, N_e. According to (1.42), the refractive index in plasma depends on the frequency of radio signals. Let us derive expressions for the refraction angle, refractive attenuation, and ionospheric frequency shift. With allowance for (2.2) and (1.42), the refraction angle in occultation by planetary ionospheres is given by

$$\xi = 2p\gamma f^{-2} \int_{r_1}^{\infty} \frac{\dfrac{dN_e}{dr} dr}{(1 - \gamma f^{-2} N_e)\sqrt{r^2(1 - \gamma f^{-2} N_e)^2 - p^2}} .$$
(2.49)

Here integration is carried out from the minimum ray altitude, which, according to (1.42) and (2.1), can be found from the relation

$$p = (a + h_1)\left[1 - \gamma f^{-2} N_e(h_1)\right].$$ (2.50)

In calculations, it is convenient to specify the minimal ray altitude h_1 and calculate the impact parameter p by formula (2.50). Refractive changes in the radio wave energy flux are given by the approximate formula (2.22); this yields

$$X = \left[1 - 2\gamma f^{-2} L \frac{d}{dp}\left(p \int_{r_1}^{\infty} \frac{\frac{dN_e}{dr} dr}{(1 - \gamma f^{-2} N_e)\sqrt{r^2(1 - \gamma f^{-2} N_e)^2 - p^2}}\right)\right]^{-1}.$$ (2.51)

The ionosphere-induced frequency shift is given by (2.46), where the refraction angle is defined by formula (2.49). To determine ξ, X, and ΔF, one has to know the vertical profile of the electron density, $N_e(h)$, which can be approximated by formulae (1.43) and (1.44). Recall that in this case the ionosphere is characterized by four parameters: N_m, peak electron density; h_m, the altitude of the peak electron density; d_i, the ionosphere thickness beneath the peak electron density; and b_i, an index expressing the rate of decrease of electron density in the upper ionosphere. From (1.43) and (1.44) it follows that the derivative dN_e/dh is negative at $h > h_m$ and positive at $h < h_m$. Therefore, in occultation by the upper ionosphere, rays are bent outwards from the planet and thus may give rise to the so-called caustics (ray intersections), for which the above expressions for ΔF and X are not valid. At $h < h_m$, rays in the ionosphere are bent towards the planet's surface. Restricting ourselves to the case of high frequencies, which is of interest for most practical applications, we can obtain simple approximations for the refraction angle, atmospheric Doppler shift, and refractive changes in the wave energy flux.

Let us find the refraction angle for occultation by planetary ionospheres. Electron density in the upper ionosphere decreases with altitude; therefore, by analogy with expression (2.11), we can write

$$\xi = -\gamma f^{-2} N_m \sqrt{2\pi b_1 (a + h_1)} \exp\left[-b_i (h_1 - h_m)\right].$$ (2.52)

The first minus sign in (2.52) indicates that the ray deviates outwards from the planet. To calculate the refraction angle for $(h_m - d_i) < h_1 < h_m$, let us divide the refraction integral into two parts

$$\xi = \xi_1 + \xi_2.$$ (2.53)

The quantity ξ_1 corresponds to integrating between the limits h_m and infinity and can be found by formula (2.52), assuming that $h_1 = h_m$:

$$\xi_1 = -\gamma f^{-2} N_m \sqrt{2\pi b_i (a + h_m)}.$$ (2.54)

The quantity ξ_2 corresponds to integrating between the limits h_1 and h_m. Recall that we assume the radio frequency be high; if so, it is possible to make the following approximations in (2.2):

$$\frac{1}{n}\frac{dn}{dr} = \frac{dn}{dh},$$

(2.55)

$$r^2n^2 - p^2 = 2(a + h_m)(h - h_1).$$

From (2.2) and (2.55), it follows:

$$\xi_2 = \frac{4\gamma N_m(a + h_1)}{f^2 d_i^2 \sqrt{2(a + h_m)}} \int_{h_1}^{h_m} \frac{(h_m - h)dh}{\sqrt{h - h_1}}.$$

(2.56)

Taking into account that

$$a + h_1 \approx a + h_m \approx a,$$

we can find from (2.56)

$$\xi_2 = \frac{8\gamma N_m \sqrt{2a}(h_m - h_1)^{3/2}}{3 f^2 d_i^2}.$$

(2.57)

Expressions (2.54) and (2.57) yield the following formula for the refraction angle in the region $(h_m - d_i) < h_1 < h_m$:

$$\xi = \gamma f^{-2} N_m \sqrt{2a}\left[\frac{8(h_m - h_1)^{3/2}}{3d_i^2} - \sqrt{\pi b_i}\right].$$

(2.58)

At $h_i < h_m - d_i$, the refraction angle is given by the formula

$$\xi = \gamma f^{-2} N_m \sqrt{2a}\left[\frac{8}{3\sqrt{d_i}} - \sqrt{\pi b_i}\right].$$

(2.59)

According to (2.46), the ionospheric frequency shift is proportional to the refraction angle. Allowing for (2.54), (2.58), and (2.59), we arrive at the following formulae for the frequency shift in the ionosphere:

At $h_1 > h_m$,

$$\Delta F = -\gamma V_1 N_m c^{-1} f^{-1} \sqrt{2\pi b_i (a + h_1)} \exp[-b_i (h_1 - h_m)];$$

at $(h_m - d_i) < h_1 < h_m$,

$$\Delta F = \gamma V_1 N_m c^{-1} f^{-1} \sqrt{2a} \left[\frac{8(h_m - h_1)^{3/2}}{3 d_i^2} - \sqrt{\pi b_i} \right]; \tag{2.60}$$

and at $h_1 < (h_m - d_i)$,

$$\Delta F = \gamma V_1 N_m c^{-1} f^{-1} \sqrt{2a} \left[\frac{8}{3\sqrt{d_i}} - \sqrt{\pi b_i} \right].$$

According to (2.22), to determine the refractive attenuation of radio signals upon occultation by ionospheres, one has to find the derivative $d\xi/dp \approx d\xi/dh$. Then the following formulae for the refractive attenuation of radio waves in the ionosphere are valid:

At $h_1 > h_m$,

$$X = \left[1 - L\gamma f^{-2} b_i N_m \sqrt{2\pi b_i a} \exp[-b_i (h_1 - h_m)] \right]^{-1};$$

at $(h_m - d_i) < h_1 < h_m$,

$$X = \left[1 + 4 L\gamma f^{-2} N_m d_i^{-2} \sqrt{2a(h_m - h_1)} \right]^{-1}; \tag{2.61}$$

and at $h_1 < (h_m - d_i)$,

$$X = 1.$$

Expressions (2.60) and (2.61) are approximate because the function $N_e(h)$ and relations (2.55) are approximate. From (2.60) it follows that in occultation by the upper ionosphere the frequency shift ΔF changes exponentially, undergoing an abrupt increase and the reversal of its sign in the region $(h_m - d_i) < h_1 < h_m$. At $h_1 < (h_m - d_i)$, ΔF changes insignificantly; this allows one to separate the contributions from the troposphere and ionosphere provided that the ray altitudes are low. At $h_1 > h_m$, refractive changes in the wave energy flux given by expressions (2.61) are small. With the minimal ray altitude h_1 approaching the altitude of the maximum electron density, h_m, the field strength slightly rises and then decreases.

As has already been mentioned, the effect of the ionosphere on frequency, ΔF, is small when compared with the Doppler effect Δf_0. To exclude the interfering effect of Δf_0, the dual-frequency technique was proposed [2]. It should be noted that this method is appropriate only for thin ionospheres or high frequencies, when the refractive bending of rays can be neglected. Let us consider the principle of the dual-frequency phase measurements. Neglecting the ionospheric nonstationarity, we can derive, similarly to the derivation of (1.126), the following formula:

$$\Delta F = 2K_1 \frac{d}{dt} \int_{h_1}^{\infty} N_e(h) dl \,, \tag{2.62}$$

where the ray element dl corresponds to the line AB in Fig. 2.1. By integrating (2.62) with respect to time, we can find the differential phase for the dual-frequency radio occultation method:

$$\varphi = 4\pi K_1 \int_{h_1}^{\infty} N_e(h) dl + \varphi_0 \,. \tag{2.63}$$

The initial differential phase difference φ_0 can readily be determined, since $\varphi = \varphi_0$ just before radio occultation. According to (2.63), the differential phase difference φ is proportional to the electron column density along the ray provided that the electron density is low and radio frequency is high.

When summing up this section, we might reason that the refraction angle, refractive changes in the radio wave energy flux, and frequency shifts in the known atmospheres and ionospheres can be found by solving the direct radio occultation problem. Conversely, experimental investigations of unknown atmospheres and ionospheres require solving the respective inverse problem.

2.2 Inverse problem of the radio occultation investigations of planetary atmospheres and ionospheres

The solution to the inverse radio occultation problem lies in the determination of unknown atmospheric and ionospheric parameters from experimentally measured changes in the frequency and amplitude of radio waves. The inverse radio occultation problem was theoretically analyzed in [10–13]. Let us treat this problem as it has been considered in these publications.

Ballistic data provide the time dependences of the spacecraft coordinates R and θ, the velocity components V_1 and V_2, and the Doppler frequency shift, Δf_0. Functions $\Delta F(t)$ and $X(t)$ are also sought experimentally. We can now rewrite formulae (2.21) and (2.45) in the form

$$\frac{d\xi}{dp} = \frac{1}{\sqrt{R^2 - p^2}}\left(1 - \frac{p}{XR\sin\Theta}\right), \tag{2.64}$$

$$\xi = 2\arctan\left(\frac{\dfrac{\Delta F\lambda}{V_1}}{1 + \sqrt{1 - 2\dfrac{\Delta F V_2\lambda}{V_1^2} - \dfrac{\Delta F^2\lambda^2}{V_1^2}}}\right). \tag{2.65}$$

In the case of thin planetary atmospheres and ionospheres, expressions (2.64) and (2.65) can be simplified:

$$\frac{d\xi}{dp} = \frac{X-1}{LX}, \tag{2.66}$$

$$\xi = \frac{\Delta F\lambda}{V_1}\left(1 + \frac{\Delta F\lambda V_2}{2V_1^2}\right). \tag{2.67}$$

In some cases, formula (2.67) can be further simplified by neglecting the second parenthesized term. Formulae (2.64)–(2.67), together with (2.1), allow the dependences of the derivative $d\xi/dp$ and angle ξ on the impact parameter p to be found provided that the refractive attenuation X and the atmospheric frequency shift ΔF have been estimated experimentally. It should be noted that $d\xi/dp$ is determined from amplitude data, while ξ is sought from independent frequency data.

Turn now to the general formula for the refraction angle in a spherically symmetric medium (2.2). Let us introduce new variables

$$y = r^2 n^2 - R_a^2,$$

$$z = p^2 - R_a^2,$$

(2.68)

where R_a is the radius of the conventional boundary of the atmosphere. Then from (2.2) and (2.68) it follows:

$$\xi = -2p \int_z^0 (y - z)^{-1/2} \frac{d \ln n(r)}{dy} dy.$$

(2.69)

This formula can be rearranged using the inverse Abel transforms,

$$F_1(z) = -A \int_0^z (y - z)^{-1/2} \frac{dF_2(y)}{dy} dy,$$

$$F_2(y) = \frac{1}{A\pi} \int_0^y (z - y)^{-1/2} F_1(z) dz,$$

(2.70)

where $F_{1,2}$ are arbitrary functions, into

$$\ln n(r) = \frac{1}{\pi} \int_{rn}^\infty \frac{\xi(p) dp}{\sqrt{p^2 - r^2 n^2}}.$$

(2.71)

Formula (2.71), which relates the refraction angle $\xi(p)$ and the altitude profile of the refractive index of radio waves, $n(r)$, can be simplified by taking into account the following approximations:

$$\ln n(r) \approx N(r),$$

$$rn \approx r.$$

(2.72)

The first approximation is always correct, but the second approximation is correct only for thin planetary atmospheres and ionospheres. From expressions (2.71) and (2.72) it follows:

$$N(r) = \frac{1}{\pi} \int_r^\infty \frac{\xi(p) dp}{\sqrt{p^2 - r^2}}.$$

(2.73)

This expression allows the altitude profile of the refractivity, $N(r)$, to be found provided that the dependence of the refraction angle on the parameter p, $\xi(p)$, is known from the measurements of the frequency shift ΔF.

Consider now the inverse radio occultation problem in relation to the refractive attenuation of radio waves. From (2.71) and (2.72) it follows:

$$N(r) = \frac{1}{\pi} \int\limits_{rn}^{\infty} \frac{\xi(p)dp}{\sqrt{p^2 - r^2 n^2}} .$$ (2.74)

By integrating this expression by parts, we have

$$N(r) = -\frac{1}{\pi} \int\limits_{rn}^{\infty} \text{arcch}\left[\frac{p}{rn(r)}\right] \frac{d\xi}{dp} dp .$$ (2.75)

Formulae (2.75) and (2.64) allow the altitude profile of the refractivity to be found from amplitude data, that is, from the experimental values for $d\xi/dp$.

The function $N(h)$ thus obtained allows one to find the altitude profiles of the gas density $n_1(h)$ in the atmosphere and electron density $N_e(h)$ in the ionosphere, since the following relations are valid:

$$\begin{aligned} N &= \mu\, k_0 n_1, \\ N &= -\gamma\, f^{-2} N_e. \end{aligned}$$ (2.76)

Parameter μ depends on the gas composition of the atmosphere. Knowing the gas density $n_1(h)$ and using the equations of an ideal gas and hydrostatic equilibrium along with the formula (2.12), it is possible to find the dependence of the pressure P and temperature T on the altitude h [12]:

$$\begin{aligned} P &= k_0 n_1 T, \\ dP &= -mg n_1 dh, \\ N &= \mu P T^{-1} = \mu k_0 n_1. \end{aligned}$$ (2.77)

Here g is the acceleration due to gravity and m is the molecular mass of the major gaseous component of the atmosphere. From (2.77) it follows:

$$T(h) = \frac{T_2 N_2}{N(h)} - \frac{m}{k_0 N(h)} \int\limits_{h_2}^{h} g(h)N(h)dh ,$$ (2.78)

where T_2 and N_2 are the temperature and refractivity at the height h_2 above the planet's surface corresponding to the upper atmosphere (in this case, as the authors of [12] show, the inaccuracy of the determination of T_2 does not significantly affect the desired temperature profile). The dependences $P(h)$ and $T(h)$ can be found from expressions (2.77) and (2.78).

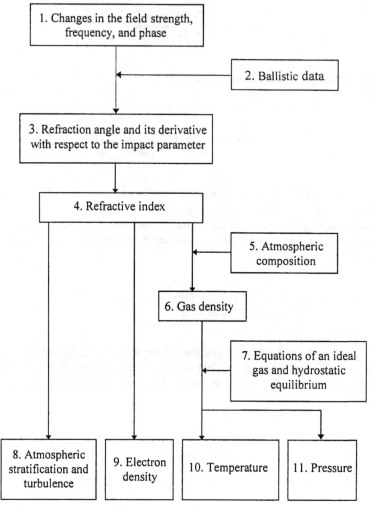

Fig 2.6. Block diagram showing the principle of the evaluation of atmospheric and ionospheric parameters by the radio occultation technique.

In solving the inverse problem of the radio occultation observations of atmospheres and ionospheres, it has become a common practice to use a multi-step algorithm for processing experimental data (Fig. 2.6).

Block 1 represents primary experimental data on changes in the frequency and amplitude of radio waves. Ballistic data (block 2), together with formula (2.1), allow the time dependence of the impact parameter to be found. The dependences of ξ and $d\xi/dp$ on the impact parameter are sought by formulae (2.66) and (2.67) (block 3). Block 4 corresponds to the Abel transform in terms of (2.74) or (2.75), which allows the altitude profile of the refractive index to be found from two independent sets of experimental data. According to (2.76), block 4 can yield the electron density in the ionosphere (block 9) and, by taking into account the gas composition of the atmosphere, the altitude profile of the gas density (block 6). The dependences of temperature and pressure on height above the planet's surface can be retrieved using formulae (2.77) and (2.78). Amplitude data are sensitive to stratified and random inhomogeneities in the atmosphere; therefore, block 1 can yield additional information about atmospheric stratification and turbulence (block 8). If radio signals are strongly absorbed in the atmosphere, the altitude distribution of the absorbing substance can be found from amplitude data by eliminating the refractive attenuation.

It is evident from Fig. 2.6 that the inverse radio occultation problem for atmospheres can be solved with great difficulty. It is especially difficult to perform its first step, namely, to measure precisely changes in the field strength and the frequency of radio signals emitted from a low-power source moving near the planet's surface. Considerable progress along this research line became possible only due to great advances in the instrumental complex for deep-space radio communications, which included transmitters with a high frequency stability, high-gain antennae, low-noise amplifiers, and efficient facilities for self-tuning and frequency measurements.

The radio occultation technique was used for exploring the atmospheres and ionospheres of Mars, Venus, Jupiter, Saturn, Titan, and Uranus. This technique was also shown to be highly efficient for the global monitoring of Earth's atmosphere and ionosphere and the exploration of plasma around the minor bodies of the solar system, such as the Moon, Mercury, Io, and Halley's Comet.

2.3 Radio occultation investigations of the Martian and Venusian atmospheres and ionospheres

Pioneering radio occultation investigations of the planetary atmosphere date back to 1965 when the U.S. *Mariner 4* space probe flew past Mars [14, 15]. Before that flight, the ground-based observations of the carbon dioxide spectral lines have yielded uncertain values for the near-surface pressure in the Martian atmosphere (between 10 and 40 mbar). As for the information about the Martian ionosphere, it was absent. *Mariner 4* disappeared behind the Martian illuminated side at a solar zenith angle of 67° and reappeared from behind the Martian nightside. During this occultation session, *Mariner 4* transmitted

2300-MHz radio signals that were received by a few Earth-based stations. As was noted in section 2.1, the amplitude and frequency of radio waves depend not only on the properties of the atmosphere, but also on the normal component of the spacecraft velocity, V_1, and the planetary limb-to-spacecraft distance L. At the time of the spacecraft's immersion, these parameters were 2.07 km/s and 25 570 km, respectively. Amplitude measurements during the occultation period showed that radio waves underwent diffraction (Fig. 2.5), implying that they touched the Martian surface. The refractive changes in the field strength were too small (less than 2 dB) to be used for the reliable determination of the atmospheric parameters. However, analysis of variations in the frequency shift ΔF allowed the parameters of the Martian atmosphere and ionosphere to be estimated. Experimental ΔF values recorded at a few ground-based stations showed that the neutral Martian atmosphere caused a frequency shift of 5.5 Hz, i.e., about 30 complete phase periods. The Martian ionosphere caused a phase shift corresponding to 11 periods (in these experiments, the accuracy of the phase and frequency measurements was ±0.3 period and ±0.1 Hz, respectively). As was shown in Section 2.2, to retrieve the atmospheric pressure and temperature from the frequency data, one has to know the gas composition of the planetary atmosphere. Based on the spectroscopic data, the authors of [14, 15] assumed that the Martian atmosphere contains 90–100% of carbon dioxide. The solution of the inverse problem with respect to frequency data showed that the surface temperature and pressure on the Martian dayside average 180±20 K and 5±1 mbar, respectively. The temperature was almost constant at heights of 0÷20 km, which corresponded to an atmospheric height scale of 9±1 km. On the Martian nightside, the pressure was 8 mbar, and the temperature decreased with height. The Martian ionosphere was detected only on the dayside at a solar zenith angle of 67°. The height profile of the ionospheric electron density recorded between 90 and 230 km displayed a peak of the electron density equal to $(9.5\pm1) \cdot 10^4$ cm^{-3} and located at a height of 120 km. Above its maximum, the electron density decreased exponentially with an ionospheric scale height equal to 24±3 km. Measurements performed during the *Mariner 4* mission showed the efficiency of the radio occultation technique for studying planetary atmospheres.

Investigations of the Martian atmosphere were resumed in 1969 by *Mariner 6* and *Mariner 7*, which flew past Mars along the trajectories that allowed radio occultation measurements to be carried out in four new regions of Mars [16–21]. The equipment, radio frequency, and data processing schedule were the same as during the mission of *Mariner 4*. *Mariner 6* and *Mariner 7* revealed the same effects of the Martian atmosphere on the frequency and amplitude of probing radio waves as *Mariner 4*. The novelty of the experiments discussed was that experimental errors were given a more careful consideration than previously [14, 15] and that four new Martian regions were investigated.

Figure 2.7 shows the dynamics of the frequency shift, $\Delta F(t)$, during the immersion of *Mariner 6* behind the Martian dayside (it should be noted that this figure gives a smooth curve, since unprocessed frequency data showed a wide scatter). As is evident from Fig. 2.7, the frequency variations caused by the Martian ionosphere preceded those induced by the thin Martian troposphere (a rapid change in ΔF induced by the troposphere occurred immediately before the signal fadeout). The thorough analysis of frequency data made it possible to derive the dependence of temperature *T* on height *h*.

Figure 2.8 illustrates the temperature lapses *T(h)* derived from the radio occultation data provided by *Mariner 6* and *Mariner 7*. Numerals along the curves indicate the Martian latitudes in degrees; temperature is expressed in degrees Kelvin; and height is expressed in km. The surface temperature was estimated to an accuracy of about ±10 K. The error of the temperature determination increased with altitude and was ±20 K at a height *h* = 20 km.

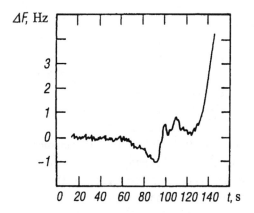

Fig. 2.7. A representative frequency shift *ΔF* derived from the *Mariner 6* radio occultation investigation of the Martian ionosphere and atmosphere [17].

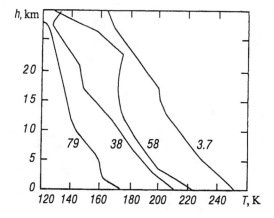

Fig. 2.8. Dependence of temperature *T* on height *h* derived from the results of the radio occultation investigation of four Martian regions [19].

The temperature lapses presented in Fig. 2.8 show the effectiveness of the radio occultation technique in studying the temperature state of planetary atmospheres. Analysis of the temperature lapses $T(h)$ allowed the height at which CO_2 is condensed to be estimated as being equal to 15 km in polar regions and about 30 km in equatorial regions. Surface pressure ranged from 4 to 8 mbar, depending on the planetary region. Ionospheric studies yielded the altitude profiles of the electron density, $N_e(h)$, for a solar zenith angle of 56°. The maximum electron density ($N_m = 1.7 \cdot 10^5$ cm^{-3}) was observed at an altitude of 135 km. Above this altitude, the electron density showed an exponential decline with a height scale equal to 27 km (data from *Mariner 7*) or 38 km (data from *Mariner 6*); this corresponds to a plasma temperature ranging between 400 and 450 K at heights of 150–200 km.

Radio waves sent out from a spacecraft flying behind a planet penetrate the planetary atmosphere twice: during the phases of spacecraft's immersion into and emersion from the occultation. Therefore, radio occultation measurements can yield information about the planet's atmosphere in just two regions. Radio occultation measurements from orbiting satellites are more informative. Thus, the first artificial extraterrestrial satellite *Mars 2*, launched by the Soviet Union in 1971, allowed the Martian atmosphere and ionosphere to be studied in several regions [22, 23]. Investigations were carried out in the 32-cm wavelength band. To improve the accuracy of estimation of the ionospheric parameters, an auxiliary low-power generator of centimeter radio waves was used. *Mars 2* explored the Martian dayside atmosphere in equatorial regions. Its distance from the planet's limb, L, ranged from 4300 to 6400 km; the spacecraft velocity component normal to the radio-beam was equal to 0.8–1.6 km/s. The solar zenith angle during the measurements was about 50°.

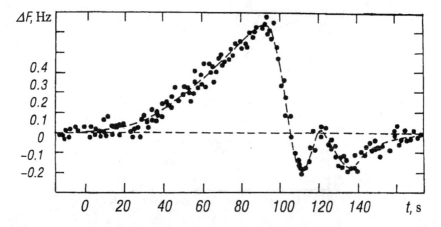

Fig. 2.9. A representative time dependence of the ionospheric frequency shift ΔF derived from the radio occultation data obtained during the mission of *Mars 2* [22].

When entering into occultation, *Mars 2* first probed the Martian ionosphere (a typical dynamics of the ionosphere-induced frequency shift ΔF observed in this case is shown in Fig. 2.9). The initial exponential increase in the frequency shift ΔF corresponds to the propagation of radio waves through the upper Martian ionosphere. The following steep decline in the frequency shift ΔF to negative values with two characteristic minima corresponds to the propagation of probing radio signals through the main ionospheric layer and the lower Martian ionosphere. Since the frequency shift ΔF is proportional to the refraction angle ξ, the positive values of ΔF in the upper Martian ionosphere imply that the ray path is bent outwards from the planet, and, conversely, the negative values of ΔF in the lower Martian ionosphere and troposphere imply that the ray is refracted towards the planet.

As the ray altitude decreased, radio signals began to probe the Martian troposphere (a typical time dependence of the tropospheric frequency shift is shown in Fig. 2.10). The time of spacecraft's disappearance behind the planet (it is indicated in the figure by the arrow) was determined from diffractive changes in the field strength, that is, from the amplitude data. Since the solar zenith angle in the *Mars 2* experiments did not significantly change, measurements gave close values of the ionospheric parameters: the lower border of the Martian ionosphere corresponded to a height of 60 km, the electron density in the lower ionospheric maximum lying at a height of 110 km was equal to $7.5 \cdot 10^4$ cm^{-3}, the electron density in the main ionospheric maximum lying at a height of 138 km was $1.7 \cdot 10^5$ cm^{-3}, the electron density at an altitude of 200 km was $3.4 \cdot 10^4$ cm^{-3}, and the ionospheric scale height in the region $h = 150$–250 km averaged 35 km. The surface pressure in the Martian equatorial regions varied from 6 to 9 mbar.

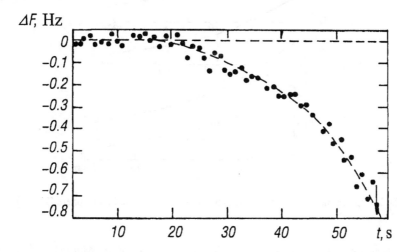

Fig. 2.10. An example of the effect of the Martian atmosphere on the radio frequency shift (data from *Mars 2*) [22].

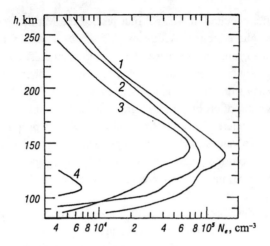

Fig. 2.11. Height profiles of the electron density in the Martian ionosphere [24].

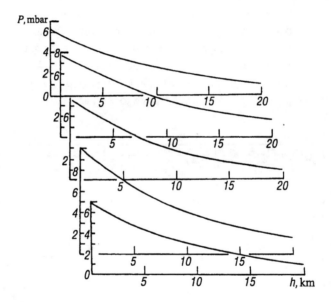

Fig. 2.12. Representative vertical pressure profiles in the Martian atmosphere [22, 24].

Similar investigations were performed during the missions of *Mars 4* and *Mars 6*, which probed the Martian atmosphere in three other regions and obtained the first evidence for the existence of an ionosphere on the Martian nightside [24–26]. Together with

Mars 2, the planetary probes *Mars 4* and *Mars 6* provided the altitude profiles of electron density for the Martian day, evening, and night. Figure 2.11 illustrates a decrease in the ionospheric electron density with increasing solar zenith angle. In this figure, altitude is given in km, N_e is expressed in cm^{-3}; curves *1*, *2*, *3*, and *4* correspond to solar zenith angles of 50, 72, 82°, and the Martian night, respectively. As can be seen from Fig. 2.11, the Martian nightside ionosphere is very thin, as a result of which its height profiling is difficult. Using a highly sensitive dual-frequency radio occultation method, the authors of [26] succeeded in obtaining an accurate profile of the Martian nightside ionosphere, $N_e(h)$. A clear-cut maximum of electron density equal to $4.8 \cdot 10^3$ cm^{-3} lay at a height of 120 km above the planet's surface. The planetary probes of the *Mars* series also provided vertical pressure profiles for six Martian regions (Fig. 2.12).

Table 2.1. Parameters of the Martian atmosphere and ionosphere evaluated in the first radio occultation experiments

Space-craft	Longi-tude, deg	Lati-tude, deg	Solar zenith angle, deg	Surface pressure, mbar	Surface tempe-rature, deg	Peak electron density, cm^{-3}	Altitude of peak electron density, km	Iono-spheric scale height, km
Mariner 4	177	−55	67	5	180	$9.5 \cdot 10^4$	120	24
	34	66	night	8	200			
Mariner 6	355	4	57	5.5	250	$1.7 \cdot 10^5$	135	38
	84	79	night	7	164			
Mariner 7	30	−58	56	4.2	220	$1.7 \cdot 10^5$	135	27
	211	38	night	7.5	205			
Mars 2	98	10.8	50	6.1	220	$1.7 \cdot 10^5$	138	35
	360	12	50	7.8	220	$1.7 \cdot 10^5$	138	35
	76	11.4	50	7.5	220	$1.7 \cdot 10^5$	138	35
	23	9	50	10	220	$1.7 \cdot 10^5$	138	35
Mars 4	17	−52	82	5.5	183	$6.5 \cdot 10^4$	142	24
	236	−9	night	4.1	205	$5.0 \cdot 10^3$	108	30
Mars 6	14	−35	72			$8.7 \cdot 10^4$	130	32

Altogether, the Soviet *Marses 2, 4,* and *6* and the U.S. *Mariners 4, 6,* and *7* have explored the Martian atmosphere in thirteen planet's regions (data are summarized in Table 2.1). Different values of surface pressures obtained in various experiments are due to the effect of the Martian terrain. Pressure in these experiments was measured to an accuracy of ±0.5 mbar, which corresponded to an accuracy of the terrain height determination of about ±1 km.

Mariner 9, a long-lived U.S. planetary probe, conducted numerous radio occultation observations of the Martian atmosphere, ionosphere, and terrain [27–30]. With this probe, circumstantial data on the main parameters of the Martian atmosphere, including the surface temperature, pressure, and temperature lapse, were obtained for 218 regions of the planet. The altitude profiles of the electron density were obtained for the Martian dayside and nightside ionosphere. The experimental dependence of the peak electron density on the solar zenith angle is shown in Fig. 2.13. The electron density N_e in this figure is expressed in 10^4 cm^{-3}; points correspond to data from *Mariner 9*; squares represent data from *Mars*; and triangles correspond to the observations from *Mariners 4, 6,* and *7*. This figure shows the decisive role of photoionization in the formation of the Martian dayside ionosphere. The radio occultation observations from *Mariner 9* yielded valuable information that advanced our knowledge of the Martian atmosphere and its global-scale dynamics.

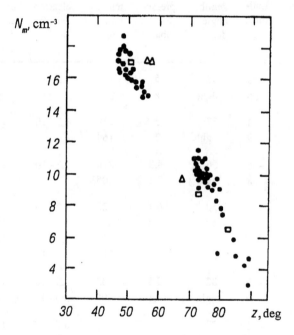

Fig. 2.13. Peak electron density in the Martian ionosphere as a function of the solar zenith angle.

The *Viking* orbiter greatly contributed to the radio occultation investigations of the Martian atmosphere and terrain [31]. In the work cited, the surface temperature was determined for various latitudes and the results obtained were processed to retrieve the planet's topography. To date, the Martian atmosphere has been studied in sufficient detail. Relevant data obtained by various methods were reviewed in [32].

Consider now the results of the radio occultation investigations of the Venusian atmosphere and ionosphere. Investigations from the Soviet landers of the *Venera* series provided the first reliable data on the lower atmosphere of Venus. In particular, these landers carried out the direct measurements of pressure, temperature, chemical composition, illuminance, and wind speed in the lower atmosphere (from 0 to 40 km above the Venusian surface) and obtained information about the upper Venusian atmosphere and ionosphere in a height range between 40 and 400 km. It should be noted that pioneering radio occultation investigations of the Venusian atmosphere were conducted from the U.S. *Mariner 5* and *Mariner 10* flying past this planet [33–38]. Complex triple-frequency observations were performed from *Mariner 5* with the aid of a master transmitter sending out 2297-MHz radio signals towards the Earth. A ground-based station received these signals and analyzed changes in their frequency and amplitude. In turn, *Mariner 5* received two auxiliary 49.8- and 423.3-MHz coherent radio signals from the Earth and sent telemetry information about changes in the amplitude and differential frequency of the auxiliary signals back to the Earth. 49.8-MHz radio waves turned out to be unsuitable for the radio occultation studies of the Venusian ionosphere and atmosphere because they underwent a strong refraction in the upper Venusian ionosphere and, hence, could not penetrate deeper. At the same time, the amplitude data obtained in the 423.3-MHz frequency band and the frequency data obtained in the 2297-MHz frequency band yielded valuable information.

Figure 2.14 presents the first data on the refractive attenuation of radio waves in the Venusian nightside atmosphere at $L = 10\ 220$ km obtained from *Mariner 5* [35]. The dashed line in this figure corresponds to $\lambda = 13$ cm, and the broken solid line to $\lambda = 71$ cm. A comparison of Figs. 2.14 and 2.2 shows that the approximation for $N(h)$ obtained in Section 2.1 is in good agreement with experimental data on the refractive attenuation of radio waves at heights $h_1 > 50$ km. At $h_1 < 50$ km, the experiment and theory slightly disagree; in particular, the refractive attenuation was found to be sensitive to stratified inhomogeneities. To make the results more reliable, the authors of [35, 36] retrieved the height profiles $T(h)$ and $P(h)$ from the independent amplitude and frequency measurements and found that the temperature lapses $T(h)$ derived from amplitude and frequency data are very close. The temperature lapse appeared to have a fine structure at heights of 46 and 48.5 km above the planet's surface, suggesting the occurrence of small temperature inversions. Temperature was profiled for both Venusian sides, day and night; however, the single atmospheric probing did not allow an unambiguous inference as to whether the temperatures on the night- and daysides of the planet are similar or different. Analysis of the attenuation of the 2297-MHz and 423.3-MHz radio waves during the radio occultation experiments showed that radio waves are attenuated because of their refraction and absorption by the atmospheric gases [35, 36]; in this case the absorption of radio waves, unlike their refraction, depends on the frequency of radio signals. Therefore, the dual-frequency radio occultation method must allow the absorption of radio waves to be measured. Estimations showed that the attenuation of 13-cm radio waves at altitudes of 40–45 km amounted to $5 \cdot 10^{-3}$ dB km^{-1}, which was somewhat greater that would be expected for an atmosphere composed of only CO_2.

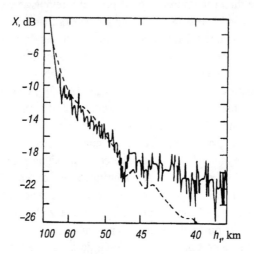

Fig. 2.14. First observations of the refractive attenuation of radio waves in the Venusian atmosphere [35].

Mariner 10 provided information on the Venusian atmosphere in two other planet's regions [37, 38]. This planetary probe, which carried two transmitters emitting 2295-MHz and 8415-MHz radio signals, immersed behind the Venusian nightside at a solar zenith angle of 118° and emerged from behind the dayside at a solar zenith angle of 67°. The temperature lapse obtained in these experiments was found to be very close to that retrieved from the *Mariner 5* data. *Mariner 10* refined the data concerning the temperature inversions at heights of 54–63 km. The mean rate of the temperature lapse at heights between 46 and 54 km above the Venusian surface was found to be 9 K/km, i.e., close to the dry adiabatic lapse rate typical of a carbon dioxide atmosphere. Below 46 km, the temperature lapse was not so steep, probably because of the boundary effect of the cloud cover of the planet [37]. Analysis of the amplitude data from *Mariner 5* and *Mariner 10* revealed rapid fluctuations in radio waves, which were due to the random refractive index inhomogeneities in the turbulent Venusian atmosphere. These data prompted the authors of [60–64] to treat the problem of atmospheric turbulence in relation to its effect on fluctuations in the amplitude of radio waves. Analysis of such fluctuations in two radio occultation experiments showed that atmospheric turbulence tends to vary with height above the planet's surface. The most intense turbulence was observed at a height of 60 km; the external turbulence scale was found to be equal to about 5 km; the structural constant of refractive index was close to $c_n = 3.5 \cdot 10^{-8}$ cm^{-3}; and that of temperature fluctuations was $2.8 \cdot 10^{-2}$ deg cm^{-3}. The last value corresponded to temperature fluctuations of about 0.1 K for two observation points 1 m apart.

The first information on the Venusian ionosphere was obtained from *Mariner 5* flying past Venus. The spacecraft probed the Venusian ionosphere by the triple-frequency method. The employment of relatively low frequencies of 49.8 and 423.3 MHz in combination with a high frequency of 2297 MHz made it possible to estimate the basic parameters of the Venusian ionosphere.

Let us now consider how the amplitude of different radio frequencies changes as they propagate in the ionosphere. Figure 2.15 illustrates the dependence of the refractive changes in the field strength on the minimal ray altitude (data are taken from [35]). Curve *2*, corresponding to 49.8-MHz radio waves, shows that the low-frequency radio waves incident on the ionosphere at grazing angles will be reflected from it when h_1 is about 400 km. Curve *1*, corresponding to 423.3-MHz radio waves, shows a steep decline in the field strength at h_1 within 120÷140 km.

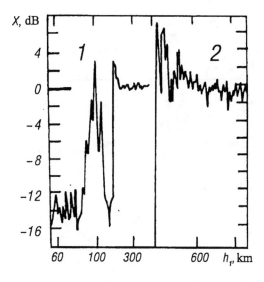

Fig. 2.15. Refractive variations in the radio field strength in the Venusian ionosphere [35].

Investigation of the Venusian ionosphere from *Mariner 5* showed that, at a solar zenith angle of 33°, the main maximum of the electron density equal to $5 \cdot 10^5$ cm^{-3} lies 138 km above the planet's surface. At a height of 124 km, there was one more indistinct ionospheric maximum with an electron density of $2 \cdot 10^5$ cm^{-3}. The ionospheric scale height immediately above the main ionization maximum was 13 km. Electron density in the day-side ionosphere exhibited a steep decline above a height of about 500 km, the effect being related to plasma 'sweeping' by the solar wind. The highest electron density in the Venusian nightside ionosphere was estimated to be 10^4 cm^{-3}.

Investigations of the Venusian ionosphere were resumed by *Mariner 10* equipped with two transmitters emitting 2295-MHz and 8415-MHz coherent radio waves [37, 38]. This spacecraft, when going behind the Venusian nightside, provided the vertical profiling of the electron density in the nightside ionosphere. The profile exhibited two narrow maxima, one with $N_m = 9 \cdot 10^3$ cm^{-3}, which occurred 143 km above the planet's surface, and the other, with an electron density of $7 \cdot 10^3$ cm^{-3}, which occurred at a height of 123 km. The radio occultation probing of the evening ionosphere during the emersion of *Mariner 10* from behind the planet (the solar zenith angle in this experiment was equal to 67°) showed that the main maximum of the electron density in the evening ionosphere was equal to $3 \cdot 10^5$ cm^{-3} at a height of 145 km.

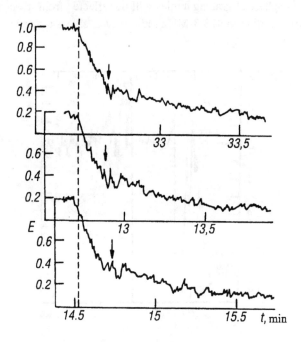

Fig. 2.16. Decline in the radio field strength during the immersion of *Venera 9* behind the nightside of Venus [39].

The artificial satellites of Venus have widened the potential of the radio occultation technique for studying the atmosphere and ionosphere of this planet. A series of experiments was carried out in 1975 from the two first artificial satellites of Venus, *Venera 9* and *Venera 10*, which conducted radio occultation measurements in 50 Venusian regions. The second and third sets of experiments were carried out in 1978 from *Pioneer Venus 1* and in 1984 from *Venera 15* and *Venera 16*. These planetary probes allowed the extensive radio occultation studies of various Venusian regions and provided global-scale data on the Venusian atmosphere and ionosphere. Let us briefly describe some of these experiments to illustrate the potential of the radio occultation technique, without going into experimental details.

The radio occultation investigations of Venus from *Venera 9* and *Venera 10* can be outlined as follows. During the immersion of spacecraft behind the Venusian nightside and its emergence from behind the Venusian dayside, the spacecraft-borne transmitters sent out 32-cm radio waves towards the Earth. To improve the accuracy of measurements, an auxiliary low-power transmitter emitted 8-cm coherent radio waves. These signals were received at two ground-based stations. The main difficulty in these studies was to reliably record the field strength and frequency of low-intensity signals. As has already been mentioned, in the case of the radio occultation by dense atmospheres, amplitude data are no less informative than frequency data.

Figure 2.16 presents three traces of the field strength E recorded during the immersion of *Venera 9* behind the planet's nightside (the field strength in the absence of the effect of the Venusian atmosphere is taken to be unity). This figure, in which the ordinate indicates the field strength E and the abscissa indicates time in min, shows the dependences $E(t)$ recorded on November 17, 19, and 21, 1975, at 48-h intervals and matched to the instants in time when the ray altitude above the planetary surface was close to 80 km (these instances are indicated in Fig. 2.16 by the vertical dashed line). The good reproducibility of the results of the amplitude measurements (especially of the maxima and minima marked in Fig. 2.16 by the arrows) was due to stable stratified formations in the Venusian atmosphere concentrated between 55 and 66 km above the planet's surface (the arrows in Fig. 2.16 correspond to an altitude of 62 km).

Figure 2.17 presents the time dependences of the field strength obtained during the emersion of *Venera 9* from behind the illuminated Venusian side on November 18, 20, and 22, 1975. In this figure, time expressed in min is plotted as the abscissa, the experimental curves are matched in such a way that the dashed line corresponds to $h_1 = 138$ km. It should be noted that due to the strong refraction of radio waves in the Venusian atmosphere, the reception of the satellite signals began before the satellite's exit from behind the planet's limb. Initially, field strength exhibited characteristic maxima and minima (they are marked in Fig. 2.17 by the arrows) produced by stratified formations in the Venusian atmosphere. Some time afterwards the effect of the troposphere tended to diminish, while that of the ionosphere tended to rise (this is evident from the appearance of sharp maxima and minima followed by a substantial decline in the field strength). Then the field strength E increased up to 1.2–1.3 but finally decreased to unity (recall that the field strength equal to unity corresponds to the free propagation of radio waves).

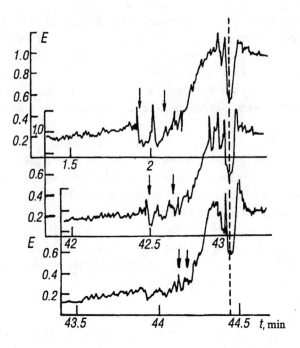

Fig. 2.17. Variations in the radio field strength during the exit of *Venera 9* from behind the dayside of Venus [39].

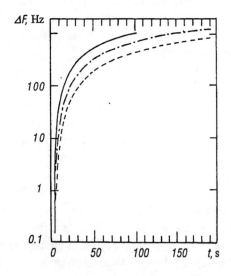

Fig. 2.18. Effect of the Venusian atmosphere on the radio frequency shift [39].

The entry of *Venera 9* and *Venera 10* into occultation was accompanied by abrupt changes in the frequency shift ΔF induced by the dense Venusian atmosphere. Figure 2.18 shows three $\Delta F(t)$ traces recorded on November 19, 21, and 27, 1975, during radio occultation observations of the Venusian nightside atmosphere. In this figure, the frequency is expressed in Hz, and time is given in sec. Some differences in the curves $\Delta F(t)$ recorded in the aforementioned days were due to different speeds at which the satellites immersed behind the planet rather than the effect of the atmosphere. The emergence of the satellites from behind the planet was accompanied by high-magnitude frequency shifts induced by the Venusian atmosphere followed by smaller shifts ΔF induced by the ionosphere. Figure 2.19 illustrates typical frequency shifts induced by the Venusian dayside ionosphere (ΔF is expressed in Hz, and time is given in sec). Now, again, the differences in the ΔF curves recorded on various days were most likely due to the different speeds at which the satellites appeared from behind the planet but not to changes in the electron density, since the solar zenith angles corresponding to these curves were almost the same.

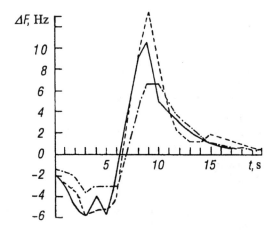

Fig. 2.19. Effect of the dayside Venusian ionosphere on the radio frequency shift [39].

Radio occultation measurements from *Venera 9* and *Venera 10* made it possible to derive the dependences $E(t)$ and $\Delta F(t)$ for 50 Venusian equatorial and mid-latitude regions. The solution of the inverse radio occultation problem with allowing for orbital data allowed the height dependences of the refractive index, density, pressure, temperature, and electron density to be derived.

During the missions of *Venera 15* and *Venera 16*, the Venusian atmosphere was studied by the radio occultation method at 300 sites, primarily in the subpolar and polar regions. The authors of [46, 47] analyzed the height profile of the refractive index, $N(h)$, and the dependence of the refractive index on the minimal ray altitude, $\xi(h_1)$. The refraction angle was determined from data on ΔF by the formulae (2.65) or (2.67), and $N(h)$ was obtained through the solution of the inverse problem, by invoking the Abel transform (2.73).

Table 2.2 summarizes relevant experimental data. The first column lists the values of the height h measured from a level of $a = 6051$ km; the second column gives experimental data on the refractivity N_2 in the polar regions of Venus; the third column, which lists the differences between N_1 in equatorial regions and N_2 in polar regions, gives an idea of how the refractive index changes in various Venusian regions; and the fourth column lists experimental values for the refraction angle in the polar Venusian atmosphere. More detailed experimental data and analytical expressions approximating the height profile of the refractive index for radio waves propagating in the Venusian atmosphere can be found in [46, 47].

Table 2.2. Refractive index and refraction angle in the middle Venusian atmosphere

h, km	$N_2 \cdot 10^6$	$(N_1-N_2) \cdot 10^6$	$\xi \cdot 10^4$ rad	h, km	$N_2 \cdot 10^6$	$(N_1-N_2) \cdot 10^6$	$\xi \cdot 10^4$ rad
84	1.1	0.4	1.0	62	75.1	22.9	66.5
82	1.6	0.7	1.5	60	112	21.0	104
80	2.5	0.8	2.1	58	158	21.0	138
78	3.5	1.8	3.1	56	210	20	171
76	5.3	2.9	4.6	54	273	17	218
74	7.7	4.3	6.8	52	347	11	275
72	11.2	6.8	9.8	50	438	−3.0	351
70	16.5	8.5	14.5	48	540	−3.0	438
68	23.9	11.1	21.0	46	670	−10	540
66	34.8	15.2	30.3	44	820	−20	700
64	51.0	19.0	44.3	42	995	−30	880

Radio occultation studies showed that, in addition to the refractive attenuation, radio waves undergo a weak absorption [40–45]. The absorption of radio waves by carbon dioxide can be neglected when the wavelength λ is longer than 3 cm and when the minimal ray altitude h_1 is greater than 40 km. Analysis showed that in this case radio waves are primarily absorbed by the sulfuric acid vapor. To determine the integral absorption coefficient Y, one has to eliminate the refractive attenuation of radio waves, X, whose value is much greater than that of Y. This can be most reliably done by using the dual-frequency method, in which the refractive attenuation X is measured in the decimeter wavelength band, whereas the sum of the absorption (X) and refractive attenuation (Y) is measured in the centimeter wavelength band. With this technique, the authors of [44, 45] succeeded in measuring the absorption of 5-cm radio waves.

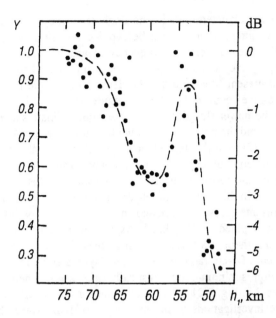

Fig. 2.20. Integral absorption of 5-cm radio waves in the Venusian atmosphere vs. the minimal ray altitude [44].

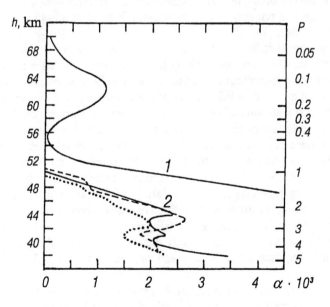

Fig. 2.21. Altitude profiles of the absorption coefficient in two wavelength bands [44, 45].

Alternatively, Y can be determined by the single-frequency method. In this case the effect of the refractive attenuation X can be avoided by using experimental data on the refractive index N retrieved from the frequency measurements. Such an approach was used, for example, by the authors of [43] to determine the absorption of 13-cm radio waves. Figure 2.20 presents the experimental dependence of the integral absorption Y of 5-cm radio waves on the height h_1 [44]. In this figure, the left ordinate represents the integral absorption of the power flux density (Y), the right ordinate shows the same quantity but expressed in dB, and the abscissa shows the height h_1 given in km. The application of the integral Abel transform to the dependence $Y(h_1)$ allows the absorption coefficient α to be determined as a function of height above the Venusian surface [43–45].

Figure 2.21 shows the dependences $\alpha(h)$ for two wavelengths, $\lambda = 5$ cm (curve *1*) and $\lambda = 13$ cm (curves *2*). In this figure, the left ordinate shows the height h given in km, the right ordinate represents pressure expressed in bars, and the abscissa represents the absorption coefficient α given in 10^{-3} km^{-1}. As is evident from Figs. 2.20 and 2.21, at heights of 60–64 km, there is a layer with an increased absorption of radio waves. The absorption coefficient of 5-cm radio waves in this layer is equal to 10^{-3} km^{-1}. At a height $h = 54$ km, absorption is minimum but drastically rises at lower heights. A comparison of curve *1* with curves *2* in Fig. 2.21 shows that the absorption coefficient strongly depends on the wavelength. Investigations of the absorption of radio waves by the sulfuric acid vapor under laboratory conditions allowed the author of [42] to derive the dependence of the absorption on pressure and wavelength. The comparison of these data with those obtained with the *Venera* satellites showed that, at heights of 52–36 km, radio waves are primarily absorbed by the sulfuric acid vapor and that the concentration of sulfuric acid vapor in the Venusian atmosphere rises approximately linearly with the height decreasing from 52 to 45 km and remains almost constant (at a level of 35–40 ppm) at the heights between 44 and 38 km.

The propagation of radio waves in the Venusian atmosphere is accompanied by amplitude fluctuations, which are due not only to atmospheric turbulence but also to the fine stratified atmospheric formations related to temperature inversions and other factors. The authors of [60–70] gave a detailed analysis of the influence of the atmospheric turbulence on the amplitude fluctuations of radio waves propagating through the Venusian atmosphere and attempted to estimate this turbulence. The greatest fluctuations in the amplitude of radio waves were observed between 57 and 64 km above the Venusian surface, suggesting that this is the region of the most intense turbulence of the Venusian atmosphere. The atmospheric turbulence depended on Venusian latitude, so that it was higher in the mid-latitude than in equatorial regions. Analysis of the amplitude fluctuation spectra showed that the spatial spectrum of the refractive index inhomogeneities is largely determined by temperature fluctuations and corresponds to a developed atmospheric turbulence. It should be noted that one can hardly distinguish between the fluctuations of radio waves induced by atmospheric turbulence and those caused by stratified formations in the atmosphere.

The height profiling of the temperature and pressure for different Venusian regions is one of the most important results of the investigation of the Venusian atmosphere by the

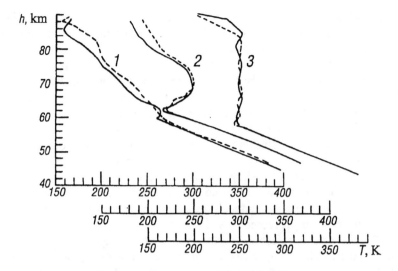

Fig. 2.22. Representative temperature lapses in the Venusian atmosphere at various latitudes [59].

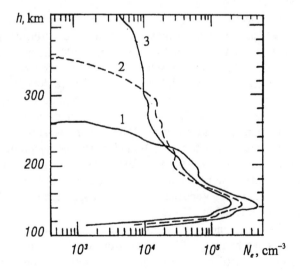

Fig. 2.23. Height profiles of electron density in the Venusian ionosphere at different solar zenith angles [83, 86].

radio occultation technique [48–59, 75]. Figure 2.22 presents the typical temperature lapses for subequatorial, subpolar, and polar regions (they are labeled by numerals *1*, *2*, and *3*, respectively). In this figure, the height *h* and temperature are expressed in km and degrees Kelvin, respectively. This figure illustrates the latitude dependence of the temperature lapses in the Venusian atmosphere.

The temperature lapses $T(h)$ derived for various latitudes made it possible to determine the distribution of winds in the Venusian atmosphere. The highest wind speed, averaging 100 m s^{-1}, was observed at altitudes of 62–65 km. More data on the height distribution of winds in various Venusian regions can be found in [71–74].

Radio occultation measurements from *Venera 9*, *Venera 10*, *Pioneer Venus*, *Venera 15*, and *Venera 16* provided valuable information on the Venusian ionosphere and greatly contributed to our knowledge of its origin. In the *Venera* experiments, the master 1000-MHz transmitter and an auxiliary generator of coherent centimeter radio waves were used. 1000-MHz radio waves turned out to be convenient for investigating the Venusian ionosphere, since the ionosphere-induced changes in their frequency and phase were sufficiently high to be reliably recorded (see Fig. 2.19).

Figure 2.23 presents the electron density profiles for the dayside ionosphere of Venus. In this figure, the height *h* is given in km and the electron density N_e is given in cm^{-3}; curves *1*, *2*, and *3* correspond to solar zenith angles of 14, 71, and 82°. In 1975, the highest electron density ($N_e = (3.8–4) \cdot 10^5$ cm^{-3} at a solar zenith angle of 14–16°) was recorded at a height of 145 km. At solar zenith angles z_0 equal to 54 and 62°, N_m was found to be $3.2 \cdot 10^5$ and $2.9 \cdot 10^5$ cm^{-3}, respectively. In the nightside ionosphere at $z_0 = 74$ and 81°, N_m was equal to $2.5 \cdot 10^5$ and $1.6 \cdot 10^5$ cm^{-3}, respectively. Above the main ionospheric maximum, the electron density changed in a complex manner. At small zenith angles ($z_0 < 16°$), the upper ionization maximum with an electron density of $6 \cdot 10^4$ to $8 \cdot 10^4$ cm^{-3} was observed at a height of 190 km. At $z_0 > 40°$, the upper ionization maximum was not observed. The dayside ionosphere of Venus was found to possess a sizable plasmopause resulting from the solar wind. At heights of more than 300 km, the electron density in the nightside ionosphere was higher than in the dayside ionosphere.

The experimental dependences $N_e(h)$ for the Venusian dayside ionosphere have a strong theoretical underpinning. Indeed, at $h < 160$ km, there is a region of photochemical equilibrium, where the predominant ion is O_2^+ and where the quadratic law of recombination is satisfied. Above 170 km, vertical diffusion becomes noticeable; the region of diffusive equilibrium, where the predominant ion is O^+, extends from $h > 220$ km to the ionopause. Detailed data on the Venusian dayside ionosphere obtained by the radio occultation technique can be found in [76–86].

The presence of an ionosphere on the nightside of Venus was reported in [87–97]. Figure 2.24 illustrates the height profiles of electron density in the Venusian nightside ionosphere (N_e is given in 10^4 cm^{-3}). At a low solar activity, two narrow ionospheric maxima 20 km apart could occasionally be observed on the nightside of Venus. At a high solar activity, the nightside ionosphere often exhibited one narrow peak of electron density. In general, the height profiles of electron density, $N_e(h)$, in the nightside ionosphere are variable.

Although the radio occultation technique provided vast information on the Venusian nightside ionosphere, the occurrence of a few narrow maxima of electron density in this

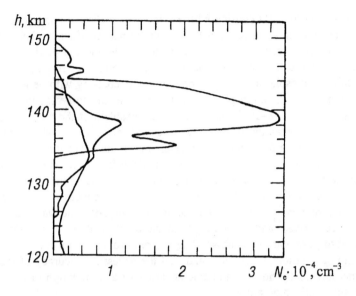

Fig. 2.24. Some height profiles of the electron density in the nightside Venusian ionosphere [87, 91].

ionosphere still remains to be understood. Presumably the formation of the Venusian nightside ionosphere is due to two factors, namely, plasma leakage from the dayside of Venus and the effect of an unknown ionizing agent.

Experimental data on the Venusian polar ionosphere near the terminator obtained by the radio occultation method can be found in [98–99].

2.4 Investigations of the Earth's atmosphere and ionosphere by the radio occultation technique

The radio occultation technique has proved to be promising for studying the Martian and Venusian atmospheres. The high efficiency of this method encouraged the researchers [100–123] to apply it to the monitoring of the Earth's atmosphere and ionosphere. It should be noted that approaches to the investigation of unknown planetary atmospheres and the well-studied Earth atmosphere must be different. Indeed, in the former case any information is useful, whereas in the latter case the radio occultation technique must demonstrate its advantage over conventional methods used for the acquisition of meteorological data. The radio occultation studies of planetary atmospheres and the monitoring of the Earth's atmosphere and ionosphere require two satellites, one of which operates as an emitter of radio waves and the other operates as their receiver.

The authors of [100, 101] considered the effect of the Earth's atmosphere on the satellite-to-satellite radio links in relation to the determination of atmospheric density and pressure. They derived formulae for the refraction angle and the phase of radio waves propagating in a spherically symmetric medium and performed corresponding quantitative estimations of the atmospheric density exponentially decreasing with height. The inverse Abel transform allowed the atmospheric density and pressure to be found as functions of height. The authors of [102–108] treated the direct problem of the radio occultation observations of the Earth's atmosphere, derived general formulae for estimating the atmosphere-induced changes in the frequency, phase, amplitude, refraction angle, and absorption of radio waves, and analyzed some ionospheric effects. The expected atmospheric and ionospheric effects were theoretically treated for the case of the geostationary satellite-to-orbital spacecraft radio links. The theoretical dependences of the refractive loss, refraction angle, frequency, and the water-vapor absorption of radio waves on the minimal ray altitude were derived, and the necessity of the determination of the atmospheric humidity from absorption data was shown. The authors proposed to use more than two satellites in radio occultation studies and analyzed the necessary accuracy of the frequency and phase measurements and the attainable accuracy of determination of the atmospheric density, pressure, and temperature.

Preliminary radio occultation experiments showed that the Earth's atmosphere and ionosphere affect the frequency and amplitude of radio waves in a complex manner [109–111]. In view of this, the comprehensive investigations of radio waves propagating along the satellite-to-satellite paths were undertaken using the radio links between the orbital station *Mir* and a geostationary satellite in the decimeter ($\lambda = 32$ cm) and centimeter ($\lambda = 2$ cm) wavelength bands [112–117].

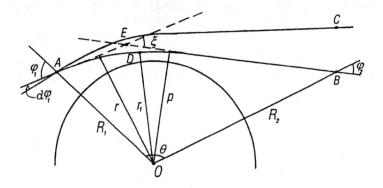

Fig. 2.25. Diagram illustrating the radio occultation observations of the Earth's atmosphere and ionosphere.

Let us treat, after the authors of [102–117], the theory of the remote sensing of the Earth's atmosphere and ionosphere. The geometry of such remote sensing is presented in

Fig. 2.25. For definiteness, assume that an emitting satellite is situated at point A and a receiving satellite is situated at point B. The satellites occur at distances R_1 and R_2 from the Earth's center designated by O. Point D on the ray ADB corresponds to the minimal ray altitude above the Earth's surface, $h_1 = r_1 - a$. The tangents to the ray, represented in Fig. 2.25 by the dashed lines, form the refraction angle, which is dependent on the altitude profile of the refractive index, $n(r)$, and the impact parameter p according to formula (2.2).

The expression for the refractive attenuation of radio waves can be derived by introducing the angles φ_1, φ_2, and θ and considering the refractive bending of a ray tube. As is evident from Fig. 2.25, $\pi - \varphi_1$ is the angle made by the radius-vector OA and the ray ADB at point A; φ_2 is the angle made by the radius-vector OB and the ray ADB at point B; and θ is the central angle between OA and OB. At point A, the ray tube has an angular size $d\varphi_1$ in the plane of the figure. At point B, the linear size of the ray tube in the plane of the figure is given by

$$BC = R_2 \cos\varphi_2 d\theta.$$

The interfacial angle $d\chi$ between the plane of the figure and the plane that crosses it along the line OA determines the angular size of the ray tube normal to the figure's plane. At point B, the linear size of the ray tube normal to the plane of the figure is given by

$$BB' = R_2 \sin\theta d\chi .$$

At point B, the cross-sectional area of the ray tube, S, is equal to the product of BC and BB', that is,

$$S_1 = R_2^2 \sin\Theta \cos\varphi_2 d\theta d\chi . \qquad (2.79)$$

If the refraction were absent, the ray tube with the angular sizes $d\varphi_1$ and $d\chi$ at point A would have the following cross-sectional area at point B:

$$S_0 = L^2 \sin\varphi_1 d\varphi_1 d\chi ,$$

where $L^2 = R_1^2 + R_2^2 - 2R_1R_2\cos\theta$ is the distance between points A and B. The refractive attenuation of radio waves with respect to their power, X, which is equal to the ratio S_0/S_1, is given by the formula

$$X = \frac{L^2 \sin\varphi_1 d\varphi_1}{R_2^2 \sin\Theta \cos\varphi_2 d\Theta} . \qquad (2.80)$$

This expression can be transformed by eliminating the angles φ_1 and φ_2 and introducing the impact parameter p and the refractive indices n_1 and n_2 at points A and B, respectively. In order to perform this transformation, Snell's law should be written in the form of expression (2.1):

$$p = n_1 R_1 \sin \varphi_1 = n_2 R_2 \sin \varphi_2 , \tag{2.81}$$

from which it follows:

$$d\varphi_1 = \frac{dp}{\sqrt{n_1^2 R_1^2 - p^2}} ,$$

$$\sin \varphi_1 = \frac{p}{n_1 R_1} , \tag{2.82}$$

$$\cos \varphi_2 = \frac{\sqrt{n_2^2 R_2^2 - p^2}}{n_2 R_2} .$$

In formula (2.80), the derivative $d\theta / d\varphi_1$ should be substituted by the derivative $d\theta / dp$. From the consideration of the quadrangle $AOBE$, we may write the following formula for the angle θ:

$$\Theta = \pi - \varphi_1 - \varphi_2 + \xi , \tag{2.83}$$

from whence, allowing for (2.81), we obtain

$$\Theta = \pi - \arcsin\left(\frac{p}{n_1 R_1}\right) - \arcsin\left(\frac{p}{n_2 R_2}\right) + \xi . \tag{2.84}$$

Differentiating (2.84) with respect to p, we obtain

$$\frac{d\Theta}{dp} = \frac{d\xi}{dp} - \frac{1}{\sqrt{n_1^2 R_1^2 - p^2}} - \frac{1}{\sqrt{n_2^2 R_2^2 - p^2}} . \tag{2.85}$$

Substituting expressions (2.82) and (2.85) into (2.80), we arrive at the following general formula for the refractive attenuation of radio waves:

$$X = \frac{n_2 p (R_1^2 + R_2^2 - 2R_1 R_2 \cos\Theta)}{n_1 R_1 R_2 \sin\Theta \left[\sqrt{n_1^2 R_1^2 - p^2} + \sqrt{n_2^2 R_2^2 - p^2} - \frac{d\xi}{dp} \sqrt{\left(n_1^2 R_1^2 - p^2\right)\left(n_2^2 R_2^2 - p^2\right)} \right]}. \tag{2.86}$$

For atmospheres, $n_1 = n_2 = 1$; this approximation is also valid for ionospheres provided that the frequency of radio signals is sufficiently high. Then we have with good accuracy

$$X = \frac{p (R_1^2 + R_2^2 - 2R_1 R_2 \cos\Theta)}{R_1 R_2 \sin\Theta \left[\sqrt{R_1^2 - p^2} + \sqrt{R_2^2 - p^2} - \frac{d\xi}{dp} \sqrt{\left(R_1^2 - p^2\right)\left(R_2^2 - p^2\right)} \right]}. \tag{2.87}$$

The resultant formula (2.87) makes it possible to analyze the refractive attenuation of radio waves for any dependence of the refractive index on height, $N(h)$. For radio occultation by the upper atmosphere, the refraction angle is small; therefore, we can write the approximate formulae

$$R_1^2 + R_2^2 - 2R_1 R_2 \cos\Theta = \left(L_1 + L_2\right)^2,$$

$$\sqrt{R_1^2 - p^2} = L_1, \tag{2.88}$$

$$\sqrt{R_2^2 - p^2} = L_2.$$

where $L_1 = AD$ and $L_2 = DB$. Taking into account (2.88), we have from (2.87)

$$X = \frac{p\left(L_1 + L_2\right)^2}{R_1 R_2 \sin\Theta \left(L_1 + L_2 - L_1 L_2 \frac{d\xi}{dp} \right)}. \tag{2.89}$$

If at least one of the distances to points A and B is much greater than the planet's radius, then (2.89) transforms into formula (2.22).

The refractive attenuation of radio waves in the radio occultation by the Earth's atmosphere was analyzed in [103, 112, 113]. To determine the dependence of X on the minimum ray altitude h_1, the altitude dependence of the refractivity should be written in the form

$$N(h) = N_0 \exp(-\beta h) \quad \text{if} \quad h \le h_t,$$

$$N(h) = N_1 \exp(-\beta_1 h) \quad \text{if} \quad h \ge h_t. \tag{2.90}$$

Here $N_1 = N_0 \exp(-\beta h_t)$ is the refractivity near the tropopause, h_t is the tropopause height, and N_0 is the near-surface value of N.

Figure 2.26 shows the refractive attenuation of the radio wave strength $E = X^{1.2}$ calculated by formula (2.87) with allowance for (2.90). It was assumed in calculations that $b_1 = 0.149 \ \mathrm{km}^{-1}$, $b = -0.1 \ \log(93/N_0) \ \mathrm{km}^{-1}$, and that satellite A moves along a 300-km circular orbit, whereas B is a geostationary satellite. Curves *1*, *2*, and *3* correspond to the same height $h_t = 12$ km but different N_0 values of 275, 325, and 375 N units, respectively (recall that one N unit is 10^{-6}).

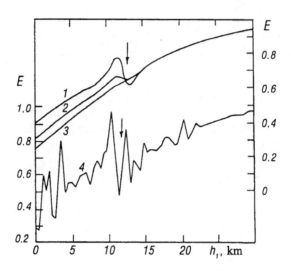

Fig. 2.26. Theoretical dependences of the refractive attenuation of the radio field strength on the minimal ray height h_1 [112].

In Fig. 2.26, the dependences $E(h_1)$ labeled by numerals *1*, *2*, and *3* correspond to the right ordinate, the minimal ray altitude h_1, given in km, is plotted as the abscissa. It is evident from Fig. 2.26 that the refractive attenuation of radio waves is significant at $h_1 < 30$ km; thus, at $h_1 = 20$ and 5 km, E is equal to 0.82 and 0.53–0.41, respectively. It should be noted that the dependence $N(h)$ of the type (2.90) does not reflect the finely stratified structure of real atmospheres. Curve *4* in Fig. 2.26 (this curve corresponds to the left ordinate) represents the dependence $E(h_1)$ that was calculated allowing for the actual altitude profile $N(h)$ derived from meteorological data. Curve *4* illustrates the effect of the atmospheric stratification caused by the temperature and humidity inversions on the refractive attenuation of radio waves. The calculated dependences $E(h_1)$ suggest the occurrence of specific changes in the field strength at $h_1 \approx h_t$ (in Fig. 2.26, these changes are indicated by the arrows).

Experimental evidence on the refractive attenuation of radio waves is presented in [112, 113]. Figure 2.27 illustrates two experimental dependences of the refractive attenuation of decimeter ($\lambda = 32$ cm) radio waves (estimated from the field strength E) on

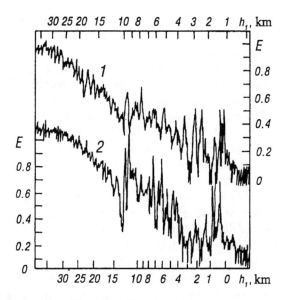

Fig. 2.27. Experimental dependences of the refractive attenuation of the radio field strength on the minimal ray height h_1 in the radio occultation by the Earth's atmosphere [112].

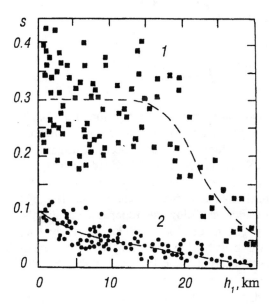

Fig. 2.28. The r.m.s. amplitude fluctuations of centimeter and decimeter radio waves versus the minimum ray height [114].

the minimal ray altitude h_1 expressed in km. Trace *1*, which corresponds to the right ordinate, illustrates the dependence $E(h_1)$ for a water area in the Indian Ocean in December 1990. Trace *2*, which corresponds to the left ordinate, illustrates the dependence $E(h_1)$ in a continental Kazakhstan region in June of the same year. At $h_1 > 15$ km, experimental dependences well fit theoretical ones. Near the tropopause, specific changes in the field strength are clearly seen. The irregular changes in the field strength recorded at altitudes of 1–8 km are induced by clouds and the temperature inversions. The dependences $E(h_1)$ are sensitive to stratified formations in the atmosphere, since these formations are characterized by steep vertical gradients of refractivity leading to sizable changes in the amplitude of radio waves. This makes possible the monitoring of the tropopause, temperature inversions, and cloud boundaries.

Random refractive index inhomogeneities, which are primarily due to atmospheric turbulence, cause irregular amplitude fluctuations. Analysis of such fluctuations in the radio occultation studies of the Earth's atmosphere was performed in [113, 114]. Experimentally, only the dependence of the standard deviation of amplitude fluctuations, s, on the minimum ray altitude h_1 can be obtained (in Fig. 2.28, curves *1* and *2* correspond to $\lambda = 2$ and 32 cm, respectively). Analysis of these data together with the temporal spectra of amplitude fluctuations showed that, at $h_1 < 15$ km, fluctuations are due to atmospheric turbulence. It follows from the results presented in [113, 114] that the turbulence of the Earth's atmosphere can be studied by the radio occultation method.

The absorption of radio waves in radio occultation experiments was analyzed in [103, 107, 115]. Centimeter and decimeter radio waves can be absorbed by water vapor and oxygen. The attenuation of the power flux density due to the absorption of radio waves can be described by the formula

$$Y(h_1) = 2 \int_{h_1}^{\infty} \frac{\gamma_1(h,\lambda)n(h)(a+h)dh}{\sqrt{n^2(a+h)^2 + p^2}}. \tag{2.91}$$

Here $\gamma_1(h,\lambda)$ is the absorption coefficient as a function of the height h and wavelength λ. The authors of [107, 115] calculated the total absorption of radio waves for mid-latitude regions in terms of a model for a standard atmosphere, using the following near-surface values of meteorological parameters: $T = 278$ K and humidity $e = 11.1$ g m^{-3} in July, $T = 257$ K and $e = 2.1$ g m^{-3} in January. The altitude profile of the water vapor content corresponded to a model of a humid stratosphere characterized by the humidity that monotonically decreases with height below the tropopause but slightly increases with height above the tropopause to an altitude of about 30 km. Similarly, temperature was assumed to decrease linearly with height below the tropopause and increase linearly with height above the tropopause. The results of calculation of the absorption of 13.5-mm radio waves by water vapor are presented in Fig. 2.29, where curve *1* illustrates January data and curve *2* corresponds to July data. As is evident from this figure, at $h_1 = 8$–15 km, $Y \approx$ 2.5 dB in wintertime and ≈ 7 dB in summertime; at $h_1 = 5$ km, $Y \approx 7$ dB in wintertime and ≈ 20 dB in summertime. It should be noted that Fig. 2.29 gives only a general idea of the

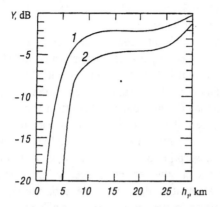

Fig. 2.29. Absorption of 13.5-mm radio waves by water vapor vs. the minimal ray height h_1 [107].

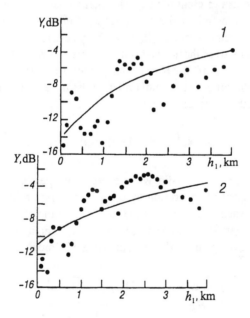

Fig. 2.30. Experimental dependences of the absorptivity of 2-cm radio wave on h_1 [115].

effect of humidity on the absorption of radio waves: actually, this absorption strongly depends on meteorological conditions. Analysis of the absorption of millimeter radio waves in low-loss transmission windows [107] showed that 8.6-mm radio waves are similarly absorbed by water vapor and oxygen. At $h_1 > 15$ km, their absorption is insignificant; at

h_1 = 8 km, the absorption is 4 dB; and at h_1 = 2 km, the absorption may be as high as 18-20 dB, depending on the air humidity.

Millimeter and shorter-wavelength centimeter radio waves undergo appreciable attenuation in clouds. Indeed, theoretical analysis performed in [107] showed that a stratus with a typical water content of 0.3 g cm^{-3} can considerably attenuate centimeter radio waves. Thus, the attenuation of 2-cm radio waves in the maxima of strati may reach 3-5 dB. Thicker altostrati may produce even higher attenuation.

The authors of [115] described their experiments with decimeter (λ_1 = 32 cm) and centimeter (λ_2 = 2 cm) radio waves linking the orbital station *Mir* with a geostationary satellite. To estimate the absorption of radio waves, one has to measure a decrease in the field strength E and cancel out the refractive attenuation of radio waves. It should be noted that the refractive attenuation does not depend on the wavelength of radio signals and that the absorption of centimeter radio waves is much stronger than that of decimeter radio waves. To cancel out the refractive attenuation of radio waves, the absorption of centimeter radio waves, Y, was determined as the ratio of the experimental values of the field strength of the wavelength λ_2 to that of λ_1.

In Fig. 2.30, solid dots represent the experimental values of the integral absorption of 2-cm radio waves versus the minimal ray altitude h_1. Curve *1* corresponds to wintertime experiments above the Aleutian Islands (the Pacific Ocean) and curve *2* corresponds to summertime observations over the Kazakhstan territory. At h_1 = 0, Y = –14 dB for the Aleutian Islands and Y = –12 dB for Kazakhstan. When analyzing experimental data, the authors of [115] used the following formula relating the differential absorption γ_1 and the height h_1 above the Earth's surface:

$$\gamma_1(h) = \gamma_0 \exp\left(-hH_0^{-1}\right) + \gamma_w \exp\left(-hH_w^{-1}\right). \tag{2.92}$$

Here γ_1 is the attenuation of the radio wave energy flux per unit length, expressed in dB/km; γ_0 is the coefficient of absorption by oxygen in dB/km at h = 0; γ_w is the coefficient of absorption by water vapor at h = 0; H_0 and H_w are the effective altitudes of radio wave absorption by oxygen and water vapor, respectively. The integration of the differential absorption along the ray path yields the approximate formula

$$Y(h_1) = \sqrt{2\pi a}\left[\gamma_0 \sqrt{H_0} \exp\left(-h_1 H_0^{-1}\right) + \gamma_w \sqrt{H_w} \exp\left(-h_1 H_w^{-1}\right)\right], \tag{2.93}$$

where a is the Earth's radius and h_1 is the minimal ray altitude. Parameters γ_0, γ_w, H_0, and H_w can be determined from the experimental dependences $Y(h_1)$ obtained in the decimeter and centimeter wavelength bands using formula (2.93). According to the data presented in [115], for λ = 2 cm, γ_w = 3 · 10^{-2} dB/km and H_w = 1.5 km in wintertime above the sea near the Aleutian Islands and γ_w = 1.6 · 10^{-2} dB/km and H_w = 1.5 km in summertime above the continental region of Kazakhstan. Oxygen parameters appeared to be equal in both regions: $\gamma_0 = 1.5 \cdot 10^{-2}$ dB/km and $H_0 = 5.3$ km. Curves in Fig. 2.30 were calculated by formula (2.93), using the indicated values of the relevant parameters. It can be seen that the calculated curves agree well with experimental data. The authors of [115] compared γ_w with meteorological data on humidity e and found that the normalized absorption of 2-cm radio waves by water vapor is $\gamma_w/e = 2.2 \cdot 10^{-3}$ dB/km g. The data presented in [107, 115] suggest that the radio occultation method can be used to monitor the distribution of water vapor in the atmosphere.

Now, after the authors of [103, 105, 116], let us consider frequency changes in radio occultation measurements. Using the phase equation in the form (2.35) with designations as in Fig. 2.25, we get

$$\varphi = \frac{2\pi}{\lambda} \left[\int_{r_1}^{R_1} \frac{n^2 r dr}{\sqrt{r^2 n^2 - p^2}} + \int_{r_1}^{R_2} \frac{n^2 r dr}{\sqrt{n^2 r^2 - p^2}} \right]. \tag{2.94}$$

Integrating (2.94) by parts, we obtain

$$\varphi = \frac{2\pi}{\lambda} \left[\sqrt{n_1^2 R_1^2 - p^2} + \sqrt{n_2^2 R_2^2 - p^2} - \int_{r_1}^{R_1} \frac{n r^2 \frac{dn}{dr} dr}{\sqrt{r^2 n^2 - p^2}} - \int_{r_1}^{R_2} \frac{n r^2 \frac{dn}{dr} dr}{\sqrt{n^2 r^2 - p^2}} \right]. \tag{2.95}$$

This expression can be transformed into

$$\varphi = \frac{2\pi}{\lambda} \left[\sqrt{n_1^2 R_1^2 - p^2} + \sqrt{n_2^2 R_2^2 - p^2} - \int_{r_1}^{R_1} \frac{1}{n} \frac{dn}{dr} \sqrt{n^2 r^2 - p^2} dr - \right.$$
$$\left. - \int_{r_1}^{R_2} \frac{1}{n} \frac{dn}{dr} \sqrt{n^2 r^2 - p^2} dr + p\xi \right], \tag{2.96}$$

where n_1 and n_2 are the refractive indices at points A and B, respectively. The frequency shift Δf can be found by differentiating (2.96) with respect to time and taking into account that $\Delta f = 1/2\pi \, d\varphi/dt$:

$$\Delta f = \frac{1}{\lambda} \left[\frac{R_1 \frac{dR_1}{dt}}{\sqrt{R_1^2 - p^2}} + \frac{R_2 \frac{dR_2}{dt}}{\sqrt{R_2^2 - p^2}} - \frac{p \frac{dp}{dt}}{\sqrt{R_1^2 - p^2}} - \frac{p \frac{dp}{dt}}{\sqrt{R_2^2 - p^2}} + p \frac{d\xi}{dt} \right]. \tag{2.97}$$

When deriving formula (2.97), we assumed that $n_1 = n_2 = 1$ and that $dn/dt = 0$ at points A and B. Differentiating expression (2.84) with respect to time, one can get the auxiliary relation

$$\frac{d\xi}{dt} = \frac{d\Theta}{dt} + \frac{\dfrac{dp}{dt}}{\sqrt{R_1^2 - p^2}} + \frac{\dfrac{dp}{dt}}{\sqrt{R_2^2 - p^2}} - \frac{p\dfrac{dp}{dt}}{R_1\sqrt{R_1^2 - p^2}} - \frac{p\dfrac{dp}{dt}}{R_2\sqrt{R_2^2 - p^2}}, \qquad (2.98)$$

which allows formula (2.97) to be transformed into

$$\Delta f = \frac{1}{\lambda}\left(p\frac{d\Theta}{dt} + \frac{1}{R_1}\frac{dR_1}{dt}\sqrt{R_1^2 - p^2} + \frac{1}{R_2}\frac{dR_2}{dt}\sqrt{R_2^2 - p^2} \right). \qquad (2.99)$$

In order to express the refraction angle ξ through the impact parameter p and angle θ, let us rewrite (2.83) in the form

$$p^2 - \sqrt{\left(R_1^2 - p^2\right)\left(R_2^2 - p^2\right)} = R_1 R_2 \cos(\Theta - \xi), \qquad (2.100)$$

from whence we have

$$p = \frac{R_1 R_2 \sin(\Theta - \xi)}{\sqrt{R_1^2 + R_2^2 - 2R_1 R_2 \cos(\Theta - \xi)}}. \qquad (2.101)$$

Equations (2.99) and (2.101) yield the expression for the frequency shift induced by both the atmosphere and the Doppler effect:

$$\Delta f = \frac{[R_1 - R_2\cos(\Theta - \xi)]\dfrac{dR_1}{dt} + [R_2 - R_1\cos(\Theta - \xi)]\dfrac{dR_2}{dt} + R_1 R_2\sin(\Theta - \xi)\dfrac{d\Theta}{dt}}{\lambda\sqrt{R_1^2 + R_2^2 - 2R_1 R_2\cos(\Theta - \xi)}}. \qquad (2.102)$$

To determine the frequency shift produced by the atmosphere alone (ΔF), one has to subtract from the total frequency shift Δf the frequency shift Δf_0 that would occur in the absence of the atmosphere:

$$\Delta F = \Delta f - \Delta f_0. \qquad (2.103)$$

In so far as

$$\Delta f_0 = \frac{1}{\lambda}\frac{d}{dt}\left(\sqrt{R_1^2 + R_2^2 - 2R_1R_2\cos\Theta}\right),$$

the Doppler effect in the absence of the atmosphere is given by

$$\Delta f_0 = \frac{\left(R_1 - R_2\cos\Theta\right)\dfrac{dR_1}{dt} + \left(R_2 - R_1\cos\Theta\right)\dfrac{dR_2}{dt} + R_1R_2\sin\Theta\,\dfrac{d\Theta}{dt}}{\lambda\sqrt{R_1^2 + R_2^2 - 2R_1R_2\cos\Theta}}. \tag{2.104}$$

Formulae (2.102)–(2.104) allow the determination of the atmospheric and ionospheric frequency shifts provided that the satellite trajectories and the altitude profile of the refractive index are known. In the radio occultation experiments, the refraction angle ξ can be taken to be small; if so, we can easily derive from formulae (2.102)–(2.104) the final formula

$$\Delta F = -\frac{\xi}{2\lambda}\left(\frac{R_2\dfrac{dR_1}{dt}\sin\Theta + R_1\dfrac{dR_2}{dt}\sin\Theta + R_1R_2\cos\Theta}{\sqrt{R_1^2 + R_2^2 - 2R_1R_2\cos\Theta}}\right) \tag{2.105}$$

or

$$\Delta F = -\frac{\xi R_1 R_2\cos\Theta}{2\lambda L}\left[\frac{d\Theta}{dt} + \left(\frac{1}{R_1}\frac{dR_1}{dt} + \frac{1}{R_2}\frac{dR_2}{dt}\right)\tan\Theta\right], \tag{2.106}$$

where L is the distance between the satellites. It should be noted that expression (2.106) transforms into expression (2.46) at $R_1 \gg R_2$; in this case, the frequency shift ΔF is proportional to the refraction angle ξ. Formula (2.105) or (2.106), together with (2.2), allow one to determine the frequency shifts induced by an atmosphere or ionosphere, that is, to solve the direct radio occultation problem.

To solve the inverse radio occultation problem, it would be more useful to use another approach based on the analysis of the Doppler effect. As follows from the geometry given in Fig. 2.31, V_1 is the projection of the velocity vector of a satellite confined in the plane of the figure onto the normal to the dashed line AB, and V_2 is the projection of that vector onto the line AB. The Doppler effect in the presence of an atmosphere is given by

$$\Delta f = \lambda^{-1}(V_2 \cos\alpha + V_1 \sin\alpha) . \tag{2.107}$$

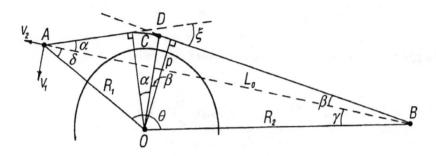

Fig. 2.31. Diagram illustrating the analysis of variations in radio frequency.

If the atmosphere were absent, the frequency shift induced by the motion of the satellite would be

$$\Delta f_0 = \lambda^{-1} V_2 . \tag{2.108}$$

The frequency shift induced by the planetary atmosphere or ionosphere is equal to the difference $\Delta F = \Delta f - \Delta f_0$; therefore

$$\Delta F = \lambda^{-1}\left[V_2(\cos\alpha - 1) + V_1 \sin\alpha\right]. \tag{2.109}$$

Comparing this formula with (2.45) shows that (2.109) contains the angle α instead of the refraction angle ξ. If $R_2 \gg R_1$, then $\xi = \alpha$ and formula (2.109) transforms into formula (2.45). If satellites A and B are at equal altitudes (i.e., $R_1 = R_2$), then $\xi = 2\alpha$. From (2.109) we can derive the following expression for the angle α:

$$\alpha = \arcsin\left[\frac{V_1(\lambda\Delta F + V_2) - V_2(V_1^2 - \lambda^2\Delta F^2 - 2\lambda V_2\Delta F)}{V_1^2 + V_2^2}\right]^{1/2}. \tag{2.110}$$

Formula (2.110) allows the angle α to be found from experimental values for the Doppler shift in frequency, ΔF. However, to solve the inverse radio occultation problem, one has to find the refraction angle ξ rather than the angle α. Bearing this in mind, we can write the relations that follow from the analysis of the geometry of the problem given in Fig. 2.31:

$$\delta = \arcsin\left(\frac{R_2 \sin \Theta}{L}\right),$$

$$\gamma = \arcsin\left(\frac{R_1 \sin \Theta}{L}\right),$$

$$p = R_1 \sin(\alpha + \delta), \tag{2.111}$$

$$\beta = \arcsin\left(\frac{p}{R_2}\right) - \gamma,$$

$$\xi = \alpha + \beta.$$

Formulae (2.110) and (2.111) make it possible to find the experimental dependence of the refraction angle on the impact parameter, $\xi(p)$, and, consequently, the altitude profile of the refractive index, $N(h)$.

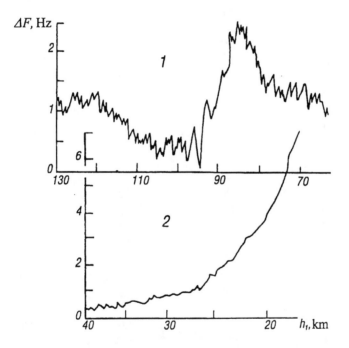

Fig. 2.32. Representative variations in the radio frequency during the radio occultation monitoring of the Earth's ionosphere and stratosphere [116].

Experimental investigations of frequency shifts during radio occultation are described in [111, 116]. In these experiments, the station *Mir* and a geostationary satellite communicated in the 32-cm wavelength band. Figure 2.32 shows typical changes in frequency,

ΔF, during radio occultation by the ionosphere (curve *1*) and stratosphere (curve *2*). In the ionosphere, at $h_1 > 120$ km, ΔF slightly changes with height, but at $h_1 < 100$ km, ΔF shows the reverse of its sign and a steep decline in magnitude. In the stratosphere ($h_1 = 40 \div 20$ km), ΔF rises almost exponentially. In the troposphere ($h_1 < 15$ km), the frequency shift ΔF rises steeply to reach 140–170 Hz at $h_1 = 0$. Radio occultation experiments described in [111, 116] have also demonstrated the efficiency of this method for obtaining the altitude profiles of the refraction angle ξ, phase delay ΔL, and refractivity N. Relevant experimental data obtained in December over the Indian Ocean are presented in Table 2.3 [116].

Table 2.3. Representative altitude profiles of the refraction angle, phase delay, and refractivity over the Indian Ocean.

H_0, km	ξ, s	ΔL, m	N
29	65	1.9	4.2
28	84	2.4	5.5
27	112	3.1	6.8
26	125	3.9	8.2
25	154	5.0	9.8
24	185	6.1	11.0
23	220	7.5	13.8
22	273	9.6	15.2
21	325	12	19.9
20	387	15	23.4
19	490	19	28.0
18	578	24	33.4
17	633	30	39.0
16	745	36	45.5
15	796	43	52.0
14	935	53	59.4
13	1038	64	67.7
12	1190	76	76.0
11	1300	90	86.8
10	1410	97	94.0
9	1615	128	109.0
8	1770	153	121.0
7	1920	182	136.0
6	2270	212	152.0
5	2350	246	169.0
4	2560	295	190.0
3	3100	363	212.0
2	3390	464	239.0

According to our data, radio occultation experiments make it possible to retrieve the dependence $N(h)$ and, consequently, the height profiles of the ionospheric electron density and the atmospheric temperature and pressure from the frequency shift ΔF. It should be noted that the function $N(h)$ depends on temperature, pressure, and humidity (see (1.30)). At $h = 40 \div 8$ km, the effect of humidity can be neglected, and the altitude profiles of pressure and temperature can be determined according to formulae (2.77) and (2.78). In the troposphere ($h < 5$ km), where the effect of humidity cannot be neglected, it can be accounted for by measuring, for example, the absorption of radio waves.

The accuracy of the retrieval of the altitude profiles of pressure and temperature depends on the interfering effects of the ionosphere, frequency fluctuations induced by atmospheric turbulence, and the horizontal gradients of the atmospheric parameters. The role of frequency and phase fluctuations in the radio occultation experiments was analyzed in [117], and that of the horizontal gradients was considered in [118]. The interfering effect of the ionosphere can be minimized by using the dual-frequency technique or sufficiently high radio frequencies.

The efficiency of the radio occultation monitoring of the atmosphere and ionosphere can be enhanced by collecting relevant data from several Earth-orbiting satellites flying above various regions of the Earth. In [119–121], the authors attempted to analyze the efficiency of the radio occultation system including the GPS satellites as transmitters of radio waves and some auxiliary satellites as their receivers. The radio occultation experiments with the GPS satellites and a small *Microlab* satellite have demonstrated that the radio occultation technique can be used to obtain the temperature lapses at $h = 2 \div 30$ km and for vertical profiling of the ionospheric electron density [122, 123, 165–187].

The authors of [166–184] analyzed the accuracy of the determination of the vertical profiles of temperature and electron density by the radio occultation technique. The effects of the multibeam propagation and the oblateness of the Earth's atmosphere were considered in terms of the wave approach.

Theoretically, the radio occultation technique rests on a ray approximation, which implies that there is only one ray trajectory between the points of the emission and reception of radio waves. However, as was shown in [174, 175], actually several ray trajectories between these points may exist in the troposphere due to the effects of the temperature inversions and the reflection of waves from the Earth's surface. This may essentially increase errors in the temperature determination at altitudes lower than 6 km. Furthermore, the theory implies a spherical symmetry of the refractive index function $N(h)$ and, consequently, of the atmospheric parameters. The real atmosphere is not, however, ideally symmetrical. The influence of a deviation from the spherical symmetry of $N(h)$ on the accuracy of the determination of the altitude profile $T(h)$ has been considered in [176].

In order to solve the problem of the radio wave propagation through the atmosphere for an arbitrary dependence $N(h)$, the respective wave equation should be solved exactly, which has not yet been accomplished. In an attempt to solve the wave problem of the radio occultation monitoring of the atmosphere approximately, the authors of [169] applied the Fresnel diffraction theory, the authors of [175] used a phase screen model, and the authors of [183, 191] used a radioholographic approach. The development of these approaches may increase the accuracy of the determination of temperature and electron density and the vertical resolution of the radio occultation method.

To validate the efficiency of the radio occultation technique, the radio occultation data should be compared with the results of determination of the atmospheric and ionospheric characteristics by conventional methods. Convincing evidence for the high accuracy of determination of the altitude temperature profiles from the radio occultation data was presented by the authors of [177, 185, 187, 189], who compared the dependences $T(h)$ obtained by the radio occultation method and by means of meteorological radars and lidars and found that in the first case the temperature determination error was ~ 2 K for altitudes $h = 16 \div 30$ km and ~ 1 K for altitudes $h = 8 \div 16$ km. At the same time, at low altitudes $h < 5$ km, the error may reach 6 K. The high accuracy of determination of $T(h)$ within the altitude interval $h = 8 \div 30$ km by the radio occultation technique allowed the authors of [178] to suggest that this technique can be efficiently used for studying the atmospheric gravitational waves. As for the measurements of electron density in the ionosphere, the radio occultation technique makes it with a high vertical resolution and accuracy at altitudes between 70 and 400 km [123, 172, 178, 182]. Further evidence for the efficiency of the radio occultation technique in monitoring the Earth's atmosphere and ionosphere was presented by the authors of [178–187].

The efficiency of the radio occultation technique for monitoring the Earth's atmosphere and ionosphere was confirmed in experiments with the radio links between the orbital station *Mir* and a geostationary satellite in Russia [111–117, 172, 186] and between the navigational GPS satellites and the *Microlab* satellite in the United States [122, 123, 165–185]. It is planned to create the Constellation Observing System for the Meteorology of Ionosphere and Climate (COSMIC) including eight small satellites, which will be able to perform daily about 4000 radio occultation measurements of the vertical profiles of temperature in the atmosphere and of electron density in the ionosphere using radio signals from the GPS satellites [187–190]. Of great potential must be a multi-frequency radio occultation system operated on coherent radio waves of the milli-, centi-, and decimeter wavelength bands of an extremely high frequency stability. Millimeter radio waves will make it possible to control the atmospheric humidity, ozone content, and turbulence from amplitude data; changes in the frequency of centimeter radio waves can yield the temperature and pressure profiles; and variations in the differential frequency of centi- and decimeter radio waves can yield the distribution of electron density. Such a system should allow the Earth's atmosphere and ionosphere to be globally monitored.

2.5 Radio occultation investigations of giant planets

Preliminary theoretical analysis of the applicability of the radio occultation method for investigating the Jovian atmosphere was performed in [4, 5] using various atmospheric models and calculating anticipated changes in the frequency and amplitude of radio waves. It was found that the radio occultation method is of limited applicability to the investigation of the Jovian atmosphere, since the critical level of refraction in this atmosphere lies near the top of the cloud cover of Jupiter, where the pressure is about 1 atm. Nevertheless, analysis showed that the effects of Jupiter's atmosphere on the frequency and amplitude of radio waves may be sufficiently high to be recorded.

The first radio occultation observations of the Jovian atmosphere date back to 1974, when the spacecraft *Pioneer 10* and *Pioneer 11* flying past the planet conducted three sets of radio occultation experiments in the 13-cm wavelength band. Analysis of relevant data showed that the Abel transform, which gives correct results for a spherically symmetric medium, yielded overestimated values for the temperature of the Jovian atmosphere and, therefore, the oblateness of Jupiter should be taken into account [124–127]. This allowed the first reliable data on the temperature and pressure in the upper Jovian atmosphere to be obtained [128, 129].

Radio occultation data from *Voyager 1* and *Voyager 2* appeared to be of particular interest. These interplanetary probes were launched to the outer planets of the solar system in 1977 and flew past Jupiter in March and July 1979 and past Saturn in November 1980 and August 1981. *Voyager 2* flew past Uranus in January 1986 and past Neptune in August 1989. The *Voyagers* conducted the radio occultation observations of the atmospheres and ionospheres of four planets in the 13-cm and 3.6-cm wavelength bands. To enhance the accuracy of measurements, the power and frequency characteristics of the transmitters installed on board these spacecraft were substantially improved.

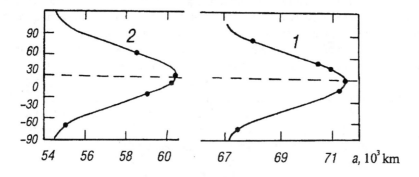

Fig. 2.33. Shapes of Jupiter and Saturn at a 100-mbar pressure level [131, 139].

Voyager 1 and *Voyager 2* conducted radio occultation measurements over three regions of Jupiter [130, 131]. Comparing the results of the six sets of the radio occultation observations of the Jovian atmosphere (three from *Pioneer 10* and *Pioneer 11* and three from the *Voyagers*) made it possible to elucidate the shape of this planet. In Fig. 2.33, curve *1* illustrates the latitude dependence of Jupiter's radius at a 100-mbar pressure level (dots represent experimental data). The altitude profiles of temperature were obtained for

the region where pressure varied from 2 mbar to 1 bar. The atmospheric profiles somewhat differ for different latitudes; so that curve *1* in Fig. 2.34 is the representative dependence of the atmospheric parameters of Jupiter. According to the radio occultation
data, the temperature lapse at altitudes where the pressure exceeds 100 mbar is adiabatic.
At altitudes corresponding to pressures of 100–10 mbar, the temperature is almost constant and equals 110 K. At higher altitudes, the temperature decreases with height. Analysis of radio occultation data and those obtained by infrared spectrometry showed that the
upper Jovian atmosphere contains 89% hydrogen and 11% helium. Radio occultation
measurements from the *Pioneers* and *Voyagers* provided the first information on the Jovian ionosphere, which extends to a height of 3000 km and has a maximum electron density of $3 \cdot 10^5$ cm^{-3} [132–136]. The height scale in the upper ionosphere, which is characterized by the inversely related electron density and height, is 380 and 800 km at solar
zenith angles of 81 and 95°, respectively. There are three additional maxima of ionization
below the main ionospheric maximum. In general, the Jovian ionosphere was found to
differ greatly from the ionospheres of terrestrial planets.

The first radio occultation observations of Saturn were carried out by *Pioneer 11* flying past this planet. More detailed information on the Saturnian atmosphere was gained by
Voyager 1 and *Voyager 2* flying past Saturn [137–140]. Radio occultation measurements
from these spacecraft were conducted in five regions to a depth corresponding to an atmospheric pressure of 1.5 bar. The shape of Saturn was found to be oblate: at a 100-mbar
pressure level, the equatorial radius of this planet was 60500 km, whereas only 54900 km
at a latitude of 75°. The latitude dependence of the Saturnian radius retrieved from radio
occultation data is presented in Fig. 2.33 by curve *2* (dots represent experimental data).
Analysis of the radio occultation and infrared spectrometry data showed that the upper
Saturnian atmosphere consists of 94% hydrogen and 6% helium. The atmospheric parameters were measured at a height of about 400 km, where the pressure varied from 1.4
bar to 0.2 mbar. The abrupt increase in the refractive attenuation of radio signals that was
revealed in these experiments suggests that critical refraction in the Saturnian atmosphere
takes place at a depth corresponding to a pressure of 1.6 bar. The atmospheric profile of
Saturn in the pressure–temperature coordinates is represented by curve *2* in Fig. 2.34. The
Saturnian ionosphere has a thickness of about 2500 km and a maximal electron density of
$2.3 \cdot 10^4$ and $1.5 \cdot 10^4$ cm^{-3}, as measured in duplicate experiments. Experiments revealed
a fine stratified structure of the lower Saturnian ionosphere.

Voyager 1 traversed the satellite system of Saturn in a way allowing the radio occultation observations of the atmosphere of Titan [141, 142]. Analysis of the data obtained
showed that the near-surface pressure and temperature are equal to 1.5 bar and 94 K, respectively. The tropopause lies at a height of about 42 km, where the temperature is 71 K
and the pressure is 130 mbar. Above the tropopause, the temperature rises and reaches
170 K at a height of about 200 km. The upper atmosphere of Titan is ionized to a peak
electron density of $(3–5) \cdot 10^3$ cm^{-3}. It is assumed that the atmosphere of Titan is composed primarily of nitrogen, with traces of CH_4 [141, 142]. According to radio occultation
data, Titan has a radius $a = 2575$ km. Following Saturn, *Voyager 2* passed by the immedi-
ate vicinity of Uranus and probed its atmosphere [143, 144]. The altitude profiles of temperature and pressure retrieved from the radio occultation data corresponded to a region

of about 120 km in height, where the pressure varied from 10 to 2000 mbar, and the temperature was from 52 to 78 K. The atmosphere of Uranus was found to possess a sizable tropopause with the temperature and pressure equal to 52 K and 110 mbar, respectively. The extended ionosphere of Uranus has two clear-cut maxima of the electron density, lying 2000 and 3500 km above a 100-mbar pressure level. The lower ionosphere has some additional maxima of electron density. Analysis of the radio occultation and optical data showed that the upper atmosphere of Uranus is composed of 84% hydrogen and 16% helium.

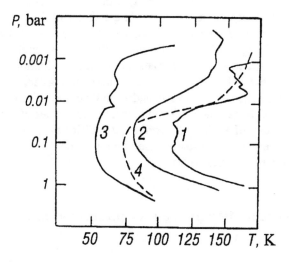

Fig. 2.34. Temperature–pressure plots for Jupiter (curve *1*), Saturn (curve *2*), Uranus (curve *3*), and Neptune (curve *4*) [131, 139, 146, 156].

Voyager 2 also performed the radio occultation observations of the Neptunian atmosphere [145, 146]. The temperature lapses were obtained for pressures ranging from 1 mbar to 6 bar. The Neptunian radius was found to be equal to 24764 km at a 1-bar pressure level and 24340 km at a probable level of the solid-state surface. Analysis of radio occultation and infrared spectrometry data showed that the Neptunian atmosphere contains 77–85% hydrogen.

Voyager 2 moving behind Saturn and Uranus performed the radio occultation observations of the planets' rings. Relevant experimental data and the theoretical substantiation of the phenomena observed can be found in [147–156]. The Saturnian rings were studied from *Voyager 2* flying 215 000 km behind the planet, the spacecraft's speed being 21 km/s relative to the planetary surface. In this case the projection of its velocity normal to the ring's plane was equal to 8.8 km/s. Radio occultation measurements were carried out using 3.6-cm and 13-cm radio signals. The spatial resolution in these experiments was very high, since the size of the first Fresnel zone normal to the ray path was as great as 2 km (for λ = 3.6 cm). Analysis of the data obtained for two wavelengths yielded the following three characteristics of radio waves: the attenuation and the phase shifts of the coherent component of the signal and the energy spectrum of its incoherent component. The results were interpreted in terms of an approximate model accounting for the composition of the ring's particles, their distribution in sizes and velocities, etc. The most reliable information was obtained for the spatial structure of the Saturnian rings, which was due to a high spatial resolution of the method used. The most important inference drawn from these experiments was that the structure of the Saturnian rings is miscellaneous. The attenuation of the radio signals traveling through various regions of the rings suggested the occurrence of waves of density and bending in the rings' structure. It was found that the density waves appear at three distances from the center of Saturn, where the orbital period of the ring's particles is comparable with the rotation period of Mimas, a Saturnian satellite. The waves of ring bending are also due to the commensurability of the rotation periods of the ring's particles and Mimas. Data on the size distribution of the ring's particles turned out to be less reliable. Most likely, the Saturnian rings are composed of a great number of small particles about 1 cm in size and a small number of large aggregates.

Of great interest are the results of the radio occultation studies of the rings of Uranus [155, 156]. The trajectory of *Voyager 2* permitted two different regions of the rings to be explored. In these experiments, the spatial resolution was about 1 km, which made possible a detailed investigation of these rings.

2.6 Some radio occultation data on the plasmas of minor bodies in the solar system

Since the mass of minor bodies of the solar system is insignificant, it could be anticipated that residual gases around them are very rarefied and ionized on the illuminated side. Furthermore, the surface of such bodies can emit plasma due to photoionization. As a result, the electron density near these bodies may appear sufficiently high to exert measurable effects on radio signals. The parameter of radio waves that is most susceptible to the encountered plasma is their phase. To avoid the phase shifts induced by the motion of transmitters (i.e., to the Doppler effect) and the instability of their frequency, the dual-frequency method should preferably be used. This method allows the increase in the electron column density along the spacecraft–Earth ray path to be measured. Knowing the increase, one can derive the dependence of the plasma density on the distance to the body's center under the supposition of a spherically symmetric distribution of electrons. This method was used to explore plasmas in the vicinities of the Moon, Mercury, Halley's Comet, and Io, a small satellite of Jupiter.

The results of the investigation of the circumlunar plasma by the dual-frequency radio occultation method with transmitters installed on board lunar satellites are described in [157, 158]. The 32-cm and 8-cm radio signals emitted by the satellite transmitters were received at a ground-based station, which measured the differential phase of these signals. Because of variations in the electron density of the Earth's ionosphere, it was difficult to distinguish the weak effect produced by the circumlunar plasma. Fortunately, analysis of the interfering effect of the Earth's ionosphere showed the existence of time intervals during which the effect of the ionosphere is sufficiently weak to allow the effect of the circumlunar plasma to be registered. Multiple radio occultation measurements on the day- and nightsides of the Moon revealed the presence of a plasma near the dayside Moon surface, which had an electron density of about $6 \cdot 10^2$ cm^{-3} [157, 158].

The investigation of Mercury by the dual-frequency radio occultation method from *Mariner 10* flying past the planet [159, 160] demonstrated the presence of only an ionosphere, whereas the neutral atmosphere of Mercury was probably too thin to affect the parameters of the probing radio signal. Peak electron density on the dayside of Mercury was equal to $1.5 \cdot 10^3$ cm^{-3}. Surprisingly, electron density on the nightside of Mercury appeared to be relatively high ($4 \cdot 10^3$ cm^{-3}). The authors of [159, 160] explained these unexpected results as the interfering effect of the Earth's ionosphere and the instability of the transmitter frequency. It is of note that the circumlunar plasma and the dayside ionosphere of Mercury have close electron densities; this was likely due to the fact that they both result from the interaction of the solar wind with the planet's surface rather than from the photoionization of gases. The very thin atmosphere of Mercury made possible the precise timing of the signal fade-out and fade-in when *Mariner 10* disappeared behind the planet and reappeared from behind it. Together with the trajectory data, this allowed the radius of Mercury to be estimated as 2439±1 km.

The results of the radio occultation investigations of the atmosphere of Io, a Jovian satellite, from *Pioneer 10* flying past Jupiter are described in [161, 162]. The conditions of these experiments greatly differed from those under which Mars or Venus were explored. The spacecraft passed behind Io at a high speed and great distance ($5 \cdot 10^5$ km from the Io's limb to the spacecraft). This resulted in that the size of the first Fresnel zone, whose value determines the resolution in height, was about 8 km. The projection of the spacecraft velocity normal to the ray was equal to 40 km/s, so that the time of measurement of the atmospheric parameters was short. Unexpectedly, these experiments detected a relatively dense ionosphere around Io. Even at a solar zenith angle as high as 81°, the peak electron density at a height of 100 km was $6 \cdot 10^4$ cm^{-3}. Above this ionospheric maximum, the electron density decreased almost exponentially, with the ionospheric height scale equal to 93 km. On the dayside, the ionosphere extended to a height of 700 km. On the night side, the electron density peak ($9 \cdot 10^3$ cm^{-3}) occurred at a height of 50 km. The nightside ionosphere had a smaller thickness (200 km), and its electron density decreased with height more steeply (the nightside ionospheric scale was equal to 30 km) that the dayside ionosphere. The great difference in the electron densities of the dayside and nightside ionospheres implies that, even at a distance of as great as 750 million km from the Sun, photoionization is the primary source of the dayside ionosphere. The authors were unable to obtain direct evidence for the existence of the neutral Io's atmosphere, probably because it was too thin to affect the amplitude of radio signals. At the same time, the thin atmosphere of Io allowed the determination of its radius from the duration of the radio signal fade-out and fade-in. The most likely radius of Io was estimated to be 1875 km.

The interplanetary probes *Vega 1* and *Vega 2* conducted the radio occultation investigations of plasma near Halley's Comet [163, 164]. These spacecraft flew 8500 km from the comet's core. Of great importance is the fact that *Vega 1* and *Vega 2* passed through the medium investigated. Measurements were carried out using 32-cm and 5-cm radio signals. A ground-based station received these signals and generated differential frequency proportionally to the rate of change of the electron column density along the spacecraft–ground station ray path. The phase shift of the decimeter radio signals induced by the plasma of Halley's Comet was determined by integrating the differential frequency with respect to time. The two sets of the radio occultation measurements showed that the electron density 8500 km far from the comet's core was equal to $(2–3) \cdot 10^3$ cm^{-3} and changed proportionally to r^{-2}.

Table 2.4. Celestial bodies investigated by the radio occultation technique

Object	Spacecraft	Years	Ref.
Mars	*Mariner 4, 6, 7*	1965, 1969	[12–31]
	Mars 2, 4, 6	1971, 1974	
	Mariner 9	1971	
	Viking	1976	
Venus	*Mariner 5, 10*	1967, 1974	[33–99]
	Venera 9, 10	1975	
	Pioneer Venus	1978	
	Venera 15, 16	1983	
Earth	*Mir* – Geostationary satellite	1990	[100–123,
	GPS – *Microlab*	1995	165–187]
Jupiter	*Pioneer 10, 11*	1974	[124–136]
	Voyager 1, 2	1979	
Saturn	*Pioneer 11*	1976	[137–140]
	Voyager 1, 2	1980	
Titan	*Voyager 1*	1980	[141, 142]
Uranus	*Voyager 2*	1986	[143, 144]
Neptune	*Voyager 2*	1988	[145, 146]
Saturnian and Uranium rings	*Voyager 2*	1980	[147–156]
Moon	*Luna 19, 22*	1973, 1974	[157, 158]
Mercury	*Mariner 10*	1974	[159, 160]
Io	*Pioneer 10*	1974	[161, 162]
Halley's Comet	*Vega 1, 2*	1986	[163,164]

The radio occultation studies of the small bodies of the solar system were performed at the threshold sensitivity of the method, which is determined by the interfering effects of the Earth's ionosphere and interplanetary plasma. The minimum electron density level detectable by the radio occultation method is $(5-10) \cdot 10^2 \, cm^{-3}$.

Overall, fourteen bodies of the solar system have been investigated by the radio occultation technique between the years 1965 and 1995 (Table 2.4). The first radio occultation observations of the atmospheres and ionospheres of Mars and Venus were performed from the planetary probes *Mariner 4* and *Mariner 5* flying past these planets. These experiments convincingly demonstrated the efficiency of the radio occultation method for investigating the planetary atmospheres and ionospheres. It became clear that artificial satellites are a powerful tool for the global-scale observations of atmospheres and ionospheres at various planetary latitudes and under various illumination conditions. The extensive radio occultation observations of Mars and Venus were conducted from the satellites *Mars 2*, *Venera 9*, and *Venera 10*.

The Earth's atmosphere and ionosphere can be efficiently studied by the radio occultation technique employing two satellites, one of which emits radio signals and the other receives them. The first large-scale studies of the Earth's atmosphere and ionosphere with the *Mir* orbital station and a geostationary satellite already demonstrated the potential of the multi-satellite radio occultation method for the global monitoring of the Earth's atmosphere and ionosphere. Such a monitoring was later accomplished using the *Microlab* satellite that received signals from the multi-satellite navigation system, GPS. The radio occultation monitoring of the Earth's atmosphere and ionosphere is still at a stage of refinement, since considerable research efforts are required to demonstrate its advantages over the conventional methods of acquisition of meteorological data.

The radio occultation studies of the atmospheres of Jupiter, Saturn, Uranus, and Neptune by the interplanetary probes of the *Pioneer* and *Voyager* series are the outstanding achievements of space science. These experiments have demonstrated the existence of ionospheres around giant planets and allowed the parameters of their upper atmospheres to be determined. In particular, the relative hydrogen and helium contents of these atmospheres were estimated. Radio occultation observations have greatly enlarged our knowledge of Saturn and Uranus.

Furthermore, the remote sensing of the minor bodies of the solar system, such as the Moon, Mercury, Io, and Halley's Comet, has shown the efficiency of the radio occultation technique for studying very rarefied plasmas.

References

1. Fjeldbo, G. and Eshleman, V.R. (1965) *J. Geoph. Res.*, **70**, 13: 3217.
2. Fjeldbo, G., Eshleman, V.R., Carriott, O.K., and Smith, F.L. (1965) *J. Geoph. Res.*, **70**, 15: 3701.
3. Yakovleva, G.D. and Yakovlev, O.I. (1971) *Radiotekhnika i Elektronika*, **16**, 5: 665 (in Russian).

4. Yakovleva, G.D., Yakovlev, O.I., and Timofeeva T.S. (1973) *Kosmicheskie Issledovaniya*, **11**, 5: 774 (in Russian).

5. Ungar, S.G. (1974) *J. Geoph. Res.*, **79**, 13: 1969.

6. Tyler, G.L. and Howard, H.T. (1969) *Radio Sci.*, **4**, 10: 899.

7. Andrianov, V.A. (1969) *Radiotekhnika i Elektronika*, **14**, 8: 1355 (in Russian).

8. Andrianov, V.A. and Yakovleva, G.D. (1975) *Radiotekhnika i Elektronika* (Radio Engineering and Electronic Physics), **20**, 1: 123 (in Russian)*.

9. Andrianov, V.A. (1968) *Radiotekhnika i Elektronika*, **13**, 8: 1374 (in Russian).

10. Phinnely, R.A. and Anderson, D.L. (1968) *J. Geoph. Res.*, **73**, 5: 1819.

11. Hays, P.B. and Roble, R.G. (1968) *Planet. Space Sci.*, **16**, 9: 1197.

12. Fjeldbo, G. and Eshleman, V.R (1968) *Planet. Space Sci.*, **16**, 8: 1035.

13. Eshleman, V.R. (1973) *Planet. Space. Sci.*, **21**, 9: 1521.

14. Kliore, A., Gain, D.L., Levy, S., *et al.* (1965) *Science*, **149**, 3689: 1243.

15. Fjeldbo, G., Fjeldbo, W., and Eshleman, V.R. (1966) *J. Geophys. Res.*, **71**, 9: 2307.

16. Kliore, A., Fjeldbo, G., Seidel, B., and Rasool, S. (1969) *Science*, **166**, 3911: 1393.

17. Kliore, A., Fjeldbo, G., and Seidel, B. (1970) *Radio Sci.*, **5**, 2: 373.

18. Fjeldbo, G., Kliore, A., and Seidel, B. (1970) *Radio Sci.*, **5**, 2: 381.

19. Rasool, S.I., Hogan, I.S., Stewart, R.W., and Russell, L.H. (1970) *J. Atmos. Sci.*, **27**, 5: 841.

20. Hogan, I.S., Stewar, R.W., and Rasool, S.I. (1972) *Radio Sci.*, **7**, 5: 525.

21. Stewart, R.W. and Hogan, I.S. (1973) *Radio Sci.*, **8**, 2: 109.

22. Kolosov, M.A., Yakovlev, O.I., Kruglov, Yu. M., *et al.* (1972) *Radiotekhnika i Elektronika*, **17**, 12: 2483 (in Russian).

23. Kolosov, M.A., Savich, N.A., Azarkh, S.L., *et al.* (1973) *Radiotekhnika i Elektronika*, **18**, 10: 2009 (in Russian).

24. Kolosov, M.A., Yakovlev, O.I, Yakovleva, G.D., *et al.* (1975) *Kosmicheskie Issledovaniya* (Cosmic. Res.), **13**, 1: 118 (in Russian)*.

25. Vasyl'ev, M.B., Kolosov, M.A., *et al.* (1975) *Kosmicheskie Issledovaniya* (Cosmic Res.), **13**, 1: 48 (in Russian)*.

26. Savich, N.A., Samovol, V.A., Vasilyev, M.B., *et al.* (1975) in: *Solar-Wind Interaction with Planets*, Ness N.F. (Ed.), NASA, p. 41.

27. Kliore, A.I., Gain, D.L., Fjeldbo, G., *et al.* (1972) *Icarus*, **17**, 2: 484.

28. Kliore, A., Fjeldbo, G., Seidil, B., *et al.* (1973) *J. Geophys. Res.*, **78**, 20: 4331.

29. Woiceshyn, P.M. (1974) *Icarus*, **22**, 3: 325.

30. Gain, D.L., Kliore, A.I., Seidel, B.L., and Sykes, M.I. (1972) *Icarus*, **17**, 2: 517.

31. Fjeldbo, G., Sweetnam, D., Brenkle, E., *et al.* (1977) *J. Geophys. Res.*, **82**, 28: 4317.

32. Moroz, V.I. (1976) *Space Sci. Rev.*, **19**, 6: 763.

33. Kliore, A., Levy, G., Gain, D., *et al.* (1967) *Science*, **158**, 3809: 1683.

34. Kliore, A., Gain, L., Levy, G., *et al.* (1969) *COSPAR Space Research*, Amsterdam, **9**, 712.

35. Fjeldbo, G. and Eshleman, V.R. (1969) *Radio Sci.*, **4**, 10: 879.

36. Fjeldbo, G., Kliore, A., and Eshleman, V.R. (1971) *Astronom. J.*, **76**, 2: 123.

37. Howard, H., Tyler, G., Fjeldbo, G., *et al.* (1974) *Science*, **183**, 4131: 1297.

38. Fjeldbo, G., Seidel, B., Sweetnam, D., and Howard, T. (1975) *J. Atmos. Sci.*, **32**, 6: 1232.

39. Kolosov, M.A., Yakovlev, O.Y., Efimov, A.I., *et al.* (1979) *Radio Sci.*, **14**, 1: 163.

40. Cimno, I.B., Elachi, C., Kliore, A.I., *et al.* (1980) *J. Geophys. Res.*, **85**, A13: 8082.

41. Steffes, P.G. and Eshleman, V.R. (1982) *Icarus*, **51**, 2: 322.

42. Steffes, P.G. (1985) *Icarus*, **64**, 3: 575.

43. Jenkins, I.M. and Steefes, P.G. (1991) *Icarus*, **90**, 1: 129.

44. Gubenko, V.N., Yakovlev, O.I., Matyugov, S.S., *et al.* (1989) *Radiotekhnika i Elektronika* (Soviet J. of Communications Technology and Electronics), **34**, 11: 2278 (in Russian)*.

45. Matyugov, S.S., Yakovlev, O.I., and Gubenko, V.N. (1990) *Kosmicheskie Issledovaniya* (Cosmic Res.), **28**, 2: 277 (in Russian)*.

46. Yakovlev, O.I., Matyugov, S.S., and Yakovleva, G.D. (1984) *Radiotekhnika i Elektronika* (Radio Engineering and Electronic Physics), **29**, 11: 2081 (in Russian)*.

47. Yakovlev, O.I., Gubenko, V.N., and Matyugov, S.S. (1990) *Radiotekhnika i Elektronika* (Soviet J. of Communications Technology and Electronics), **35**, 1: 21 (in Russian)*.

48. Kolosov, M.A., Yakovlev, O.I., Trusov, B.P., *et al.* (1976) *Radiotekhnika i Elektronika* (Radio Engineering and Electronic Physics), **21**, 8: 1585 (in Russian)*.

49. Yakovlev, O.I., Efimov, A.I., Matyugov, S.S., *et al.* (1978) *Kosmicheskie Issledovaniya* (Cosmic Res.), **16**, 1: 413 (in Russian)*.

50. Kolosov, M.A., Yakovlev, O.I., Matyugov, S.S., *et al.* (1978) *Kosmicheskie Issledovaniya* (Cosmic Res.), **16**, 2: 278 (in Russian)*.

51. Kolosov, M.A., Yakovlev, O.I., Efimov, A.I., *et al.* (1980) *Acta Astronautica*, **7**, 219.

52. Kliore, A.I. and Patel, I.R. (1982) *Icarus*, **52**, 2: 320.

53. Kliore, A.I. (1985) *Adv. Space Res.*, **5**, 9: 41.

54. Yakovlev, O.I., Gubenko, V.N., Matyugov, S.S., and Yakovleva, G.D. (1987) *Kosmicheskie Issledovaniya* (Cosmic Res.), **25**, 2: 213 (in Russian)*.

55. Yakovlev, O.I., Gubenko, V.N., Matyugov, S.S., *et al.* (1987) *Kosmicheskie Issledovaniya* (Cosmic Res.), **25**, 2: 206 (in Russian)*.

56. Yakovlev, O.I., Matyugov, S.S., Efimov, A.I., *et al.* (1987) *Kosmicheskie Issledovaniya* (Cosmic Res.), **25**, 2: 219 (in Russian)*.

57. Matyugov, S.S., Gubenko, V.N., Yakovlev, O.I., *et al.* (1988) *Kosmicheskie Issledovaniya* (Cosmic Res.), **26**, 4: 500 (in Russian)*.

58. Yakovlev, O.I., Matyugov, S.S., and Gubenko, V.N. (1988) *Kosmicheskie Issledovaniya* (Cosmic Res.), **26**, 5: 655 (in Russian)*.

59. Yakovlev, O.I., Matyugov, S.S., and Gubenko, V.N. (1991) *Icarus*, **94**, 493.

60. Golitsyn, G.S. and Gurvich, A.S. (1971) *J. Atmos. Sci.*, **28**, 1: 138.

61. Woo, R. and Ichimary, A. (1974) *IEEE Trans. on Ant. and Prop.*, **22**, 4: 566.

62. Woo, R. and Ishimaru, A. (1973) *Radio Sci.*, **8**, 2: 103.

63. Woo, R., Ishimary, A., and Kendall, W. (1974) *J. Atmos. Sci.*, **31**, 9: 1698.

64. Woo, R. (1975) *J. Atmos. Sci.*, **32**, 6: 1084.

65. Woo, R., Armstrong, I., and Kliore, A. (1982) *Icarus*, **52**, 2: 335.

66. Eshleman, R. and Haugstad, S. (1978) *Icarus*, **34**, 2: 396.

67. Haugstad B.S. (1978) *Icarus*, **35**, 1: 121 and 3: 410.

68. Timofeeva, T.S., Yakovlev, O.I., and Efimov, A.I. (1978) *Kosmicheskie Issledovaniya* (Cosmic Res.), **16**, 2: 226 (in Russian)*.

69. Timofeeva, T.S., Efimov, A.I., and Yakovlev, O.I. (1980) *Radiotekhnika i Elektronika* (Radio Engineering and Electronic Rhysics), **25**, 3: 449 (in Russian)*.

70. Timofeeva, T.S., Efimov, A.I., and Yakovlev, O.I. (1980) *Kosmicheskie Issledovaniya* (Cosmic Res.), **18**, 5: 775 (in Russian)*.

71. Chub, E. and Yakovlev, O.I. (1980) *Kosmicheskie Issledovaniya* (Cosmic Res.), **18**, 3: 435 (in Russian)*.

72. Newman, M., Schubert, G., Kliore, A., *et al.* (1984) *J. Atmos. Sci.*, **41**, 12: 1901.

73. Vaganov, I.R., Yakovlev, O.I., Matyugov, S.S., and Gubenko, V.N. (1992) *Kosmicheskie Issledovaniya* (Cosmic Res.), **30**, 5: 695 (in Russian)*.

74. Gubenko, V.N., Matyugov, S.S., Yakovlev, O.I., and Vaganov, I.R. (1992) *Kosmicheskie Issledovaniya* (Cosmic Res.), **30**, 3: 390 (in Russian)*.

75. Matyugov, S.S., Gubenko, V.N., Yakovlev, O.I., and Vaganov, I.R. (1994) *Kosmicheskie Issledovaniya* (Cosmic Res.), **32**, 3: 108 (in Russian)*.

76. Ivanov, G.S., Kolosov, M.A., Savich, N.A., *et al.* (1979) *Icarus*, **39**, 2: 209.
77. Cravens, T.E., Kliore, A.J., Kozyra, J.U., and Nagy, A.F. (1981) *J. Geoph. Res.,* **86**, A13: 11323.
78. Kalashnikov, I.E., Matyugov, S.S., Yakovleva, G.D., and Yakovlev O.I. (1983) *Radiotekhnika i Elektronika* (Radio Engineering and Electronic Physics), **28**, 8: 1457 (in Russian)*.
79. Brace, L.H., Taylor, H.A., Gombosi, T.I., *et al.* (1983) *Venus,* Hunten D.M. (Ed.), University of Arizona Press, p. 779.
80. Gavrik, A.L. and Samoznaev, L.N. (1985) *Kosmicheskie Issledovaniya* (Cosmic Res.), **23**, 1: 148 (in Russian)*.
81. Bauer, S.I., Brace, L.M., Taylor, H.A., *et al.* (1985) *Adv. Space Res.,* **5**, 11: 233 and 269.
82. Savich, N.A., Andreev, ., Vyshlov, A.S., and Gavrik, A.L. (1986) *Radiotekhnika i Elektronika* (Soviet J. of Communications Technology and Electronics), **31**, 11: 2113 (in Russian)*.
83. Gavrik, A.L. and Samoznaev, L.N. (1987) *Kosmicheskie Issledovaniya* (Cosmic Res.), **25**, 2: 228 (in Russian)*.
84. Kliore, A.J. and Mullen, L.F. (1989) *J. Geoph. Res.,* **94**, A10: 13339.
85. Aleksandrov, Yu. N., Vasyl'ev, M.B., Vyshlov, A.S., *et al.* (1978) *Radiotekhnika i Elektronika* (Radio Engineering and Electronic Physics), **23**, 9: 1840 (in Russian)*.
86. Samoznaev, L.N. (1991) *Kosmicheskie Issledovaniya* (Cosmic Res.), **29**, 1: 104 (in Russian)*.
87. Aleksandrov, Yu. N., Vasyl'ev, M.B., Vyshlov, A.S., *et al.* (1976) *Kosmicheskie Issledovaniya* (Cosmic Res.), **14**, 6: 812 (in Russian)*.
88. Kliore, A.J., Patel, I.R., Nagy, A.F., *et al.* (1979) *Science*, **205**, 4401: 99.
89. Osmolovskii, I.K., Savich, N.A., and Samoznaev, L.N. (1984) *Radiotekhnika i Elektronika* (Radio Engineering and Electronic Physics), **29**, 12: 2302 (in Russian)*.
90. Kalashnikov, I.E., Matyugov, S.S., Yakovlev, O.I., and Yakovleva, G.D. (1981) *Radiotekhnika i Elektronika* (Radio Engineering and Electronic Physics), **26**, 2: 319 (in Russian)*.
91. Savich, N.A., Andreev, ., Vyshlov, A.S., *et al.* (1986) *Radiotekhnika i Elektronika* (Soviet J. of Communications Technology and Electronics), **31**, 3: 433 (in Russian)*.
92. Gavrik, A.L., Osmolovskij, I.K., and Samoznaev, L.N. (1986) *Kosmicheskie Issledovaniya* (Cosmic Res.), **24**, 4: 620 (in Russian)*.
93. Knudsen, W.C., Kliore, A.J., and Whitten, R.C. (1987) *J. Geoph. Res.,* **92**, A12: 13391.
94. Osmolovskii, I.K. and Samoznaev, L.N. (1987) *Kosmicheskie Issledovaniya* (Cosmic Res.), **25**, 2: 233 (in Russian)*.
95. Zhang, M.H., Luhman, J.G., and Kliore, A.J. (1990) *J. Geoph. Res.,* **95**, A10: 17095.
96. Samoznaev, L.N. (1990) *Kosmicheskie Issledovaniya* (Cosmic Res.), **28**, 2: 282 (in Russian)*.
97. Kliore, A.J., Luhman, J.G., and Zhang, M.H. (1991) *J. Geoph. Res.,* **96**, A7: 11065.
98. Kliore, A.J., Woo, R., Armstrong, J.W., *et al.* (1979) *Science,* **205**, 4382: 765.
99. Savich, N.A., Andreev, V.Y., Vyshlov, A.S., Gavrik, A.L., *et al.* (1986) *Kosmicheskie Issledovaniya* (Cosmic Res.), **24**, 3: 448 (in Russian)*.
100. Kliore, A. (1969) *Space Res.,* **9**: 590.
101. Lusignan, B., Modrell, G., Morrison, A., *et al.* (1969) *Proc. IEEE*, **57**, 4: 458.
102. Ungar, S.G. and Lusignan, B.B. (1973) *J. Applied Meteorology,* **12**, 2: 396.
103. Kalashnikov, I.E. and Yakovlev, O.I. (1978) *Kosmicheskie Issledovaniya* (Cosmic Res.), **16**, 6: 943 (in Russian)*.
104. Kalashnikov, I.E. and Yakovlev, O.I. (1981) in: *Proceeding of the Fifth Conference of USSR on Radiometeorology,* Moscow: Gidrometeoizdat, p. 184 (in Russian).
105. Kalashnikov, I.E., Matyugov, S.S., and Yakovlev, O.I. (1986) *Radiotekhnika i Elektronika* (Soviet J. of Communications Technology and Electronics), **31**, 1: 56 (in Russian)*.

106. Kalashnikov, I.E., Matyugov, S.S. Pavelyev, A.G., and Yakovlev, O.I. (1986) in: *Electromagnetic Waves in the Atmosphere and Space*, Sokolov, A.V. and Semenov, A.A. (Eds.), Moscow: Nauka, p. 208 (in Russian).

107. Eliseev, S.D. and Yakovlev, O.I. (1989) *Radiofizika* (Radiophysics and Quantum Electronics), **32**, 1: 3 (in Russian)*.

108. Sokolovskii, S. (1986) *Issledovanie Zemli iz Kosmosa*, 3: 13 (in Russian).

109. Rangaswamy, S. (1976) *Geophys. Res. Letters*, **3**, 8: 483.

110. Liu, A.S. (1978) *Radio Sci.*, **13**, 4: 709.

111. Yakovlev, O.I., Grishmanovskij, V.A., Eliseev, S.D., *et al.* (1990) *Doklady Akademii Nauk SSSR*, **315**, 1: 101 (in Russian).

112. Yakovlev, O.I., Vilkov, I.A., Grishmanovskij, V.A., *et al.* (1992) *Radiotekhnika i Elektronika* (J. Communications Technology and Electronics), **37**, 1: 42 (in Russian)*.

113. Yakovlev, O.I., Matyugov, S.S., and Vilkov, I.A. (1995) *Radio Sci.*, **30**, 3: 591.

114. Vilkov, I.A., Matyugov, S.S., and Yakovlev, O.I. (1993) *Radiotekhnika i Elektronika* (J. Communications Technology and Electronics), **38**, 5: 795.

115. Matyugov, S.S., Yakovlev, O.I., and Vilkov, I.A. (1994) *Radiotekhnika i Elektronika* (J. Communications Technology and Electronics), **39**, 8/9: 1251 (in Russian)*.

116. Yakovlev, O.I., Vilkov, I.A., Zakharov, A.I., *et al.* (1995) *Radiotekhnika i Elektronika* (J. Communications Technology and Electronics), **40**, 12: 73 (in Russian)*.

117. Yakovlev, O.I., Matyugov, S.S., Vilkov, I.A., *et al.* (1996) *Radiotekhnika i Elektronika* (J. Communications Technology and Electronics), **41**, 9: 1088 (in Russian)*.

118. Sokolovskii, S. (1994) *Doklady RAN*, **333**, 5: 650 (in Russian).

119. Hardy, K.R., Hajj, G.A., and Kursinski, E.R. (1994) *International J. Satellite Communications*, **12**: 463.

120. Yuan, L.L., Anthes, B.A., Ware, R.H., Rocken, C., *et al.* (1993) *J. Geoph. Res.*, **98**, D8: 14925.

121. Kursinski, E.R., Hajj, G.A., Hardy, K.R., *et al.* (1995) *Geophys. Res. Lett.*, **22**, 17: 2365.

122. Ware, R., Exner, M., Feng, D., *et al.* (1996) *Bulletin of the Amer. Meteorological Soc.*, **77**, 1: 19.

123. Hajj, G.A. and Romans, L.I. (1998) *Radio Sci.*, **33**, 1: 175.

124. Eshleman, R. (1975) *Science*, **189**, 4206: 876.

125. Kliore, A., Fjeldbo, G., Seidel, B., *et al.* (1975) *Science*, **188**, 4187: 474.

126. Hubbard, W.B., Hunten, D.M., and Kliore, A. (1975) *Geophys. Res. Lett.*, **2**, 7: 265.

127. Hogan, I.S. and Gess, R.D. (1975) *J. Atmos. Sci.*, **32**, 4: 860.

128. Kliore, A.D., Woiceshyn, P.M., and Hubbard, W.B. (1976) *Geophys. Res. Lett.*, **3**, 3: 113.

129. Kliore, A.D. and Woiceshyn, O.M. (1976) The Atmosphere of Jupiter from the Pioneer 10/11 Radio Occultation Measurements, in: *Jupiter*, Gehrels, T. (Ed.), The University Arizona Press.

130. Eshleman, V.R., Tyler, G.L., Wood, G.E., *et al.* (1979) *Science*, **204**, 4396: 976.

131. Lindal, G,F., Wood, G.E., Levy, G.S., *et al.* (1981) *J. Geophys. Res.*, **86**, A10: 8721.

132. Fjeldbo, G., Kliore, A., Seidel, B., *et al.* (1975) *Astron. and Astrophys.*, **39**, 1: 91.

133. Woo, R. and Yang, F. (1976) *J. Geophys. Res.*, **81**, 19: 3417.

134. Fjeldbo, G., Kliore, A., Seidel, B., and Suitnem, D. (1976) The Ionosphere of Jupiter from Pioneer Radio Occultation, in: *Jupiter*, Gehrels, T. (Ed.), The University Arizona Press.

135. Atreya, S.K., Donahue, T.M., and Waite, I.H. (1979) *Nature*, **280**, 5725: 795.

136. Hinson, D.P. and Tyler, G.L. (1982) *J. Geoph. Res.*, **87**, A7: 5275 .

137. Tyler, G.L., Eshleman, V.R., Anderson, I.D., *et al.* (1981) *Science*, **212**, 4491: 201.

138. Hinson, D.P. (1984) *J. Geoph. Res.*, **89**, A1: 65.

139. Lindal, G.F., Sweetnam, D.N., and Eshleman, V.R. (1985) *Astronom. J.*, **90**, 6: 1136.

140. Gehrels,T. and Matthews, M. (Eds.) (1984) *Saturn,* Tucson: The University Arizona Press.
141. Lindal, G.F., Wood, G.E., Hotz, H., *et al.* (1983) *Icarus,* **53**, 2: 348.
142. Hinson, D.P. and Tyler, G.L. (1983) *Icarus,* **54**, 2: 337.
143. Tyler, G.L., Sweetnam, D.N., Anderson, I.D., *et al.* (1986) *Science,* **233**, 4759: 79.
144. Lindal, G.F., Lyons, I.R., Sweetnam, D.N., *et al.* (1987) *J. Geoph. Res.,* **92**, A13: 14987.
145. Lindal, G.F., Lyons, I.R., Sweetnam, D.N., *et al.* (1990) *Geophys. Res. Lett.,* **17**, 10: 1733.
146. Lindal, G.F. (1992) *Astronom. J.,* **103**, 3: 967.
147. Marouf, E.A., Tyler, G.L., and Eshleman, R. (1982) *Icarus,* **49**, 2: 161.
148. Marouf, E.A., Tyler, G.L., Zebker, H.A., *et al.* (1983) *Icarus,* **54**, 2: 189.
149. Eshleman, R., Breakwell, I. Tyler, G.L., and Marouf, E.A. (1983) *Icarus,* **54**, 2: 212.
150. Zebker, H.A., Tyler, G.L., and Marouf, E.A. (1983) *Icarus,* **56**, 2: 209.
151. Tyler, G.L., Marouf, E.A., Simpson, R.A., *et al.* (1983) *Icarus,* **54**, 2: 160.
152. Zebker, H.A. and Tyler, G.L. (1984) *Science,* **223**, 4634: 396.
153. Simpson, R.A., Tyler, G.L., Marouf, E.A., *et al.* (1984) *IEEE Trans. on Geosci. and Remote Sensing,* **GE-22**, 6: 656.
154. Marouf, E.A., Tyler, G.L., and Rossen, P.A. (1986) *Icarus,* **68**, 1: 120.
155. Gresh, D.L., Marouf, E.A., Tyler, G.L., *et al.* (1989) *Icarus,* **78**, 1: 131.
156. Tyler, G.l. (1987) *Proc. IEEE,* **75**, 10: 1404.
157. Vasil'ev, M.B., Vinogradov, V.A., Vyshlov, A.S., *et al.* (1974) *Kosmicheskie Issledovaniya* (Cosmic Res.), **12**, 1: 115 (in Russian)*.
158. Savich, N.A. (1976) *Space Research,* **16**: 941.
159. Howard, H., Tyler, L., Esposito, P., and Anderson, I. (1974) *Science,* **185**, 4146: 179.
160. Fjeldbo, G., Kliore, A., Sweetnam, D., *et al.* (1976) *Icarus,* **29**, 4: 439.
161. Kliore, A.I., Fjeldbo, G., Seidel, B.L., et. al. (1975) *Icarus,* **24**, 4: 407.
162. Johnson, T. Matson, D.G., and Carlson, R.W. (1976) *Geophys. Res. Lett.,* **3**, 6: 293.
163. Andreev, V.E. and Gavrik, A.L. (1990) *Kosmicheskie Issledovaniya* (Cosmic Res.), **28**, 2: 293 (in Russian)*.
164. Andreev, V.E. and Gavrik, A.L. (1991) *Kosmicheskie Issledovaniya* (Cosmic Res.), **29**, 3: 458 (in Russian)*.
165. Ladreiter, H.P. and Kirchengast, G. (1996) *Radio Sci.,* **31**: 877.
166. Kursinski, E.R., Hajj, G.A., Hardy, K.R., *et al.* (1997) *J. Geophys. Res.,* **102**, D19: 23429.
167. Rocken, C., Anthes, R., Exner, M., *et al.* (1997) *J. Geophys. Res.,* **102**, D25: 29849.
168. Karayl, E.T. and Hinson, D.P. (1997) *Radio Sci.,* **32**, 2: 411.
169. Mortensen, M.D. and Hoeg, P. (1998) *J. Geophys. Res.,* **25**, 13: 2441.
170. Ahmad, B. and Tyler, G.L. (1998) *Radio Sci.,* **33**, 1: 129.
171. Hocke, K. (1997) *Annales Geophysical.,* **15**, 443.
172. Kucheryavenkov, A.I., Yakovlev, O.I., Kucheryavenkova, I.L., and Samoznaev, L.N. (1998) *Radiotekhnika Elektronika* (J. Communications Technology and Electronics), **43**, 8: 945 (in Russian)*.
173. Leroy, S. (1997) *J. Geophys. Res.,* **102**, 6971.
174. Pavelyev, A.G., Volkov, A.V., Zakharov, A.I., *et al.* (1996) *Acta Astronautica,* **39**, 9–12: 721.
175. Gorbunov, M.E., and Gurvich, A.S. (1998) *J. Geophys. Res.,* **103**, D12: 13819.
176. Syndergaard, S. (1998) *J. Atmosph. Solar-Terrestrial Phys.,* **60**, 2: 171.
177. Tsuda, T., Nishida, M., Rocken, C., and Ware, R. (2000) *J. Geophys. Res.,* **105**, 7257.
178. Schreiner, S., Sokolovsky, S., Rocken, C., and Hunt, D. (1999) *Radio Sci.,* **34**, 949.
179. Kursinski, E., Hajj, G., Leroy, S., and Henman, B. (2000) *Atmospheric Oceanic Sci.,* **11**, 53.
180. Zou, X., Vandenberger, F., Wang, B., *et al.* (1999) *J. Geophys. Res. Atmosphere,* **104**, D18: 22301.

181. Solheim, F., Vivekanandan, J., Ware, R., and Rocken, C. (1999) *J. Geophys. Res. Atmosphere*, **104**, D8: 9663.
182. Vorobev, V., Gurvich, A., Kan, V., *et al.* (1999) *Earth Observation and Remote Sensing*, **15**, 4: 609.
183. Hocke, K., Pavelyev, A., Yakovlev, O., *et al.* (1999) *J. Atmospheric and Solar Terrestrial Phys.*, **61**, 1169.
184. Sokolovskiy, S. (2000) *Radio Sci.*, **35**, 1: 97.
185. Steiner, A., Kirchengast, G., and Ladreiter, H. (1999) *Ann. Geophys.*, **17**, 122.
186. Anufriev, V., Matyugov, S., and Yakovlev, O. (2000) *J. Communications Technology and Electronics*, **45**, 1: 48.
187. Kuo, Y., Sokolovskiy, S., Anthes, R., and Vandenberger, F. (2000) *Terrestrial Atmospheric and Oceanic Sci.*, **11**, 1: 157.
188. Rocken, C., Kuo, Y., Schreiner, D., *et al.* (2000) *Terrestrial Atmospheric and Oceanic Sci.*, **11**, 1: 21.
189. Anthes, R., Rocken, C., and Kuo, Y. (2000) *Terrestrial Atmospheric and Oceanic Sci.*, **11**, 1: 115.
190. Hajj, G., Lee, L., Pi, X., *et al.* (2000) *Terrestrial Atmospheric and Oceanic Sci.*, **11**, 1: 235.
191. Gorbunov, M., Gurvich, A., and Kornbluch, L. (2000) *Radio Sci.*, **35**, 4: 1025.

Chapter 3

Radio sounding of the circumsolar and interplanetary plasma

3.1 Radio wave time delay and electron density distribution in plasma

Interplanetary and circumsolar plasma represents a continuous outflow of ionized gases from the solar photosphere. Plasma traveling at a high velocity outwards from the Sun is called solar wind. This is an immense phenomenon: the Sun emits about one million tons of matter per second into outer space. The solar wind plasma, which originates from the solar photosphere, accelerates to supersonic velocities, passes through the solar system, and diffuses in interstellar space. The origination and primary acceleration of the plasma in the solar photosphere and corona can be investigated by optical methods. Steady plasma in the vicinity of Venus, Earth, and Mars is studied by spacecraft-borne instruments. The region at distances from 5 to 60 solar radii from the Sun, where the solar wind is formed, is observable only with remote sensing facilities.

The remote sensing of the solar wind was first accomplished through the use of noiselike signals from natural radio sources. Launchings of interplanetary vehicles allowed the solar wind by frequency-stable monochromatic radio waves to be remotely sensed. The radio sounding of plasmas is accompanied by a number of effects, such as the time delay, rotation of the polarization plane of radio waves, ray refraction, amplitude and frequency fluctuations, recurrent variations in radio frequency, and the spectral line broadening. All these effects are used to study the solar wind.

In this Chapter, we shall consider the propagation of radio waves through the circumsolar and interplanetary plasma and give an account of the main results of the remote sensing of the solar wind.

The solar-wind plasma is very turbulent, so that fluctuations in the plasma density are comparable with its mean value. As a result, the dependence of the plasma density N_e on the distance from the Sun's center, r, and on the heliolatitude χ can be derived only by averaging experimental data over extended time intervals. In this section, we shall concentrate on the phenomenon of the radio wave time delay and experimental determination of the dependence $N_e(r, \chi)$.

The first data on the plasma distribution in circumsolar space were obtained by optical methods. The light emitted from the solar photosphere is scattered by free plasma electrons, giving rise to a visible solar corona. The corona brightness pattern may provide information on the plasma concentration. When solar activity is at a minimum, the solar corona has a regular shape, and the respective dependence $N_e(r, \chi)$ can be described by a simple empirical formula. A comparative analysis of relevant data performed in [1, 2] showed that the electron density distribution is not spherically symmetric: when the Sun is quiet, the solar corona is more bright at the Sun's poles than at its equator. In this case the

149

dependence of the electron density on distance r and latitude χ can be given by the empirical formula [1]

$$N_e = A\left(\frac{a}{r}\right)^6 (1 - 0.95\sin\chi) + B\left(\frac{a}{r}\right)^{2+\varepsilon} (1 - \sqrt{\sin\chi}), \qquad (3.1)$$

where $a = 6.97 \cdot 10^5$ km is the solar radius, $A = 1.58 \cdot 10^8$, $B = 2.51 \cdot 10^6$, and $\varepsilon = 0.5$. Setting $\chi = 0$, formula (3.1) gives the distribution $N_e(r)$ in the ecliptic plane. Formula (3.1) is not valid for polar regions, where $\chi = 90°$. It should be noted that optical measurements enable the determination of N_e for a range of $1.2a < r < 10a$. At the same time, expression (3.1) satisfactorily describes the dependence $N_e(r)$ for greater distances, since, in deriving this formula, it was taken into account that the mean density of plasma around the Earth's orbit ($r = 215a$) is equal to 6 cm^{-3}. According to optical measurements, during the periods of maximum solar activity, the electron density in the equatorial plane is twice as high as during the periods of minimum solar activity. Plasma is a quasineutral matter, i.e., the concentrations of ions and electrons in it are equal. Figure 3.1 summarizes the results of nine sets of the optical measurements of $N_e(r)$ in the solar corona [2].

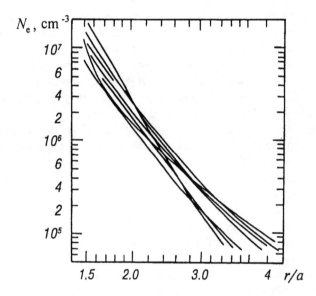

Fig. 3.1. Dependence of the plasma density on heliocentric distance as derived from the optical data [2].

The discovery of pulsars provided an opportunity for measuring the group delay ΔT of different radio frequencies propagating through the circumsolar plasma. The authors of

[3, 4] showed that the group delay is sufficiently high to allow its reliable measurement and, therefore, such measurements may yield information on the electron density. Let us briefly outline this method.

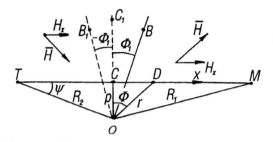

Fig. 3.2. Diagram illustrating the analysis of the radio wave propagation in the solar wind plasma.

In Fig. 3.2, points O, T, and M denote the centers of the Sun, the Earth, and a source of radio waves, respectively. The ray TCM is characterized by the impact parameter p. Let us introduce an (x, y, z) coordinate system whose origin is located at point T and whose x-axis is directed along the line TM. Using formulae (1.42) and (1.69) and taking into account that $\Delta T = \Delta L c^{-1}$, we arrive at the following formula for the group delay of radio waves in a plasma:

$$\Delta T = \frac{\gamma}{cf^2} \int_0^L N_e(r)dx = \frac{\gamma I}{cf^2},$$ (3.2)

where I is the integral electron density along the line TM, and parameter $\gamma = 40.4 \text{ m}^{-3}$ kHz^{-3}. As follows from (3.2), knowing ΔT for a few frequencies, one can determine the integral electron density as a function of the impact parameter, that is, $I(p)$. To derive the dependence of the electron density on heliocentric distance, one has to specify a distribution model $N_e(r)$, calculate the dependences $\Delta T(p)$, and compare them with experimental ones. Formula (3.2) suggests that the ray TM is straight. Actually, irregularities in the electron density bend the ray path; therefore, the apparent radio wave time delay must be greater (by a value of δT) than that predicted by formula (3.2). Analysis shows that $\delta T \sim \lambda^4$; therefore, one can determine and eliminate the component δT provided that a few radio frequencies are used in the experiment. Moreover, in centimeter and decimeter wavelength bands, the component δT can be neglected. Such a method was used to determine the dependence $N_e(r)$ by recording radio signals from pulsar NP0532 [5–7]. This pulsar is located very close to the ecliptic plane and is, therefore, appropriate to the radio occultation observations of the solar corona. The arrival times of radio pulses were measured at frequencies of 430, 196, 111, and 74 MHz; this enabled the component δT to be eliminated and the electron column density be determined as a function of the impact parameter. In this case, the following approximate distribution of electron density was used:

$$N_e = N_0 \left(\frac{r_0}{r}\right)^{2+\varepsilon} \left[1 - \frac{\xi_1}{2}(3\sin^2\chi - 1)\right], \tag{3.3}$$

where N_0 is the electron density at a distance of ten solar radii ($r_0 = 10a$). Parameter ξ_1 is a measure of the deviation of the distribution $N_e(r,\chi)$ from a spherically symmetric one. Indeed, at $\xi_1 = 0$, the distribution $N_e(r)$ is spherically symmetric; at $\xi_1 = 1$, the electron density at the poles (where $\chi = 90°$.) is equal to zero. By comparing the ΔT and I values calculated according to (3.2) and (3.3) with those obtained experimentally, the authors of [5–7] found N_0, ε, and ξ_1 (Table 3.1, the first series of experiments). Formula (3.3) and the parameters presented in Table 3.1 allow the dependence $N_e(r)$ to be found in an interval of $5a < r < 20a$.

New opportunities for estimating the electron density were offered by interplanetary vehicles, radio communications with which may involve the propagation of radio signals through the circumsolar plasma. The distance to a spacecraft can be estimated by measuring the round-trip time required for a radio wave to travel from the Earth to a spacecraft and back. The radial velocity of the spacecraft relative to the Earth can be measured from the Doppler effect. Such data enable a precise positioning of the spacecraft in terms of celestial mechanics. If, during the trajectory measurements, radio waves travel through circumsolar space, this leads to an apparent increase in the distance measured because of the radio wave time delay induced by the solar plasma. This phenomenon was used to determine ΔT and the integral electron density I as a function of the impact parameter p. In [8], the authors presented the results of the 6-month measurements of the time delay ΔT of radio waves travelling from the Earth to *Mariner 6* or *Mariner 7* and back. The Earth–spacecraft radio link operated in a 2120-MHz wavelength band, and the spacecraft–Earth radio link in a 2300-MHz band. Measurements were taken for a range of impact parameters of $3.6a < p < 14a$. Similar experiments were performed by means of the space vehicles *Helios 1* and *Helios 2* [9]. The dependence $N_e(r)$ was calculated using the approximate formula (3.1) with $\chi = 0$. Parameters B and ε were found by comparing the experimental values of the radio wave time delay ΔT with those calculated by formulae (3.1) and (3.2). Table 3.1 lists the relevant parameters determined in the second and third sets of experiments. It should be noted that the authors of [8, 9] determined parameter A from optical data, since they failed to find it from the time delay data. More precise estimations of the electron density from the group delay of radio waves were performed during the *Viking* mission in 1976–1977 [10]. Experiments were conducted at two radio frequencies, $f_1 = 2293$ MHz and $f_2 = 8340$ MHz. In these experiments, the delay ΔT was defined as the difference between the arrival times of the radio signals f_1 and f_2. According to (3.2), this difference can be given by the approximate formula

$$\Delta T = \frac{\gamma}{cf_1^2} \int_{R_1}^{R_2} \frac{N_e(r)r\,dr}{\sqrt{r^2 - p^2}} ,$$ (3.4)

where the length element of the ray, dx, is expressed through the heliocentric distance r and impact parameter p. For simplicity, when deriving formula (3.4), we neglected the time delay of the radio frequency f_2. Some difference in the analogous formula presented in [10] is understandable: the authors of that publication took into account the fact that direct and back radio waves had different frequencies.

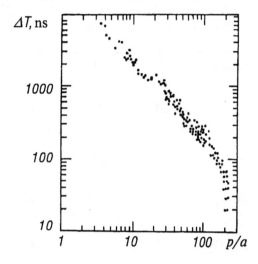

Fig. 3.3. Time delay of 2.2-GHz radio waves versus p/a [10].

Figure 3.3, taken from [10], presents ΔT values, expressed in nanoseconds, versus the impact parameter p. In the equatorial plane, the experimental dependence $N_e(r)$ can be given by the formula

$$N_e = D\left(\frac{a}{r}\right)^{2.7} + B\left(\frac{a}{r}\right)^{2.04} ,$$ (3.5)

which is valid for the interval $4a < r < 100a$. The parameters found in the fourth and fifth sets of experiments are also listed in Table 3.1. The experimental dependence $\Delta T(p)$ obtained with a high accuracy enabled the authors of [10] to solve the inverse problem of the determination of $N_e(r)$. The solution is based on expression (3.4) and the inverse Abel transform (2.70), from which it follows:

$$N_e(r) = \frac{cf^2}{\gamma\pi} \int_R^r \frac{d(\Delta T)}{dp} \cdot \frac{dp}{\sqrt{p^2 - r^2}}.$$ (3.6)

Formula (3.6), which is valid only for a spherically symmetric distribution of electron density, allows the determination of the function $N_e(r)$ from the experimental dependence $\Delta T(p)$.

Table 3.1. Determination of electron density from experimental data on the time delay

No.	Observation year	Region probed	$N_e \cdot 10^{-3}$ for $r = 5a$	$N_e \cdot 10^{-3}$ for $r = 10a$	Formula	Formula parameters; N_e in cm^{-3}
1	1969, 1970	$(5-20)a$	28	7.0	3.3	$N_0 = 7 \cdot 10^3$ $\varepsilon = 0.9$ $\xi_1 = 0.5$
2	1970	$(3.6-14)a$	28	5.3	3.1	$A = 1 \cdot 10^8$ $\xi_1 = 0.06$ $B = 0.6 \cdot 10^6$
3	1971, 1972	$(5-20)a$	30	5.3	3.3	$N_0 = 5.3 \cdot 10^3$ $\varepsilon_1 = 0$ $\xi_1 = 0.5$
4	1976	$(4-14)a$	31	6.3	3.1	$A = 0.3 \cdot 10^8$ $B = 1 \cdot 10^6$ $\varepsilon = 0.2$
5	1977	$(4-100)a$	26	4.7	3.5	$D = 1.32 \cdot 10^6$ $B = 2.3 \cdot 10^5$
6	1985	$(6-40)a$	64 and 30	9.6 and 6.1	3.7	$\alpha_1 = 2.6$ $B_1 = 3.8 \cdot 10^6$ $\alpha_2 = 1.94$ $B_2 = 0.53 \cdot 10^6$
7	1988	$(10-90)a$	100 and 180	24 and 35	3.7	$\alpha_1 = 2.08$ $B_1 = 2.93 \cdot 10^6$ $\alpha_2 = 2.28$ $B_2 = 7.08 \cdot 10^6$
8	1991	$(5-40)a$	67 and 53	10 and 8.9	3.7	$\alpha_1 = 2.54$ $B_1 = 3.64 \cdot 10^6$ $\alpha_2 = 2.41$ $B_2 = 2.26 \cdot 10^6$

The results of the sixth, seventh, and eighth sets of experiments devoted to the determination of electron density from experimental data on the time delay are described in [11–15] (the most detailed information is in [13]). The authors of these publications analyzed experimental data in terms of the empirical formula

$$N_e = A\left(\frac{a}{r}\right)^6 + B\left(\frac{a}{r}\right)^\alpha , \tag{3.7}$$

whose parameters B and α are also presented in Table 3.1.

Generally, Table 3.1 summarizes experimental data on electron density in the circumsolar plasma, $N_e(r)$, for $r = 5a$ and $r = 10a$. The table also gives observation years, the regions studied, and the parameters of approximating formulae. In the last column, the two values for the same parameter are for the rays approaching the Sun and receding from it. It should be noted that 1976 was the year of minimum solar activity and 1990 was the year of maximum solar activity (it is noteworthy that experiments in 1988 gave overestimated values for N_e). The authors of [25] compared the dependences $N(r)$ derived from optical and time delay data and found that $N_e(r)$ is best approximated by the formula

$$N_e = \left[5.79\left(\frac{a}{r}\right)^{16} + 1.6\left(\frac{a}{r}\right)^6 + 9.2 \cdot 10^{-3}\left(\frac{a}{r}\right)^2\right] \cdot 10^8 , \tag{3.8}$$

where N_e has the dimension of cm^{-3}. Figure 3.4 shows the results of the nine sets of experiments devoted to the determination of $N(r)$ from data on the radio wave time delay. The results of the experiments performed in 1988 are omitted, since they gave highly overestimated values for electron density. At $4a > r > 5a$, measurements were performed by both optical and radiometric methods. As is evident from Figs. 3.1 and 3.4, the data obtained by both methods show good agreement.

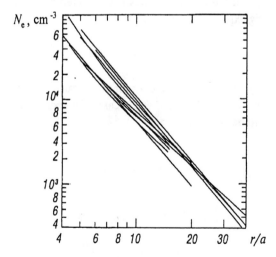

Fig. 3.4. Dependence of the electron density on the radial distance r/a as derived from the data on the radio wave time delay.

The above consideration was for the case of the radial distribution of electron density. When considering the solar wind density it should be taken into account that plasma is quasi-neutral, i.e., the concentrations of electrons and ionized hydrogen are equal. The experimental dependences $N_e(r)$ are usually approximated by power functions, whose parameters can be determined in one way or another. Such an approach enables the important characteristics of the radial distribution of electron density in the region of the maximum acceleration of the solar wind to be found. Using tabulated data on the integral electron density I presented in [13], we plotted curves shown in Fig. 3.5, in which the values for I are given in 10^{19} m^2. Recall that, as follows from (3.2), the electron column density I is proportional to the time delay ΔT. The experimental dependences $I(p/a)$ in this figure are shown for the rays approaching the Sun (dots) and receding from it (squares). One can see drastic changes in the integral electron density in the region $p = (8-14)a$ (the possible reasons for such changes were analyzed in [13]). Similar changes in the dependence $\Delta T(p)$ for $r = (10-14)a$ were observed by the authors of [10]. It should be noted in relation to this that, at $p = (8-14)a$, the solar wind becomes supersonic and, therefore, may give rise to shock waves, which are presumably responsible for the aforementioned changes in the dependence $N_e(r)$ in the region where $p = (8-14)a$. In circumsolar space, there may occur current layers and plasma flows that have very different electron densities. In general, the behavior of $N_e(r)$ is well understood except for the peculiarity in the region $p = (8-14)a$.

For $r > 30a$, the dependence of the electron density and radio wave time delay on heliocentric distance can be given by simple expressions.

The distribution $N_e(r)$ is closely related to the speed of the outflow of matter from the Sun as a function of radial distance, $V(r)$. If the flux of matter in a unit solid angle is constant, which implies that the matter that has been ejected from the Sun will never fall downward, the following expression is valid:

$$N_e(r)\, V(r)\, r^2 = \text{const},$$

where $V(r)$ is the solar wind speed as a function of the distance r. It is shown at the end of this chapter that, at $r > 30a$, the velocity $V(r)$ weakly depends on the distance r. Therefore, $V(r)$ can be taken to be constant; hence

$$N_e = B\left(\frac{a}{r}\right)^2. \tag{3.9}$$

If so, formula (3.4), which gives the group delay of radio waves, takes the form

$$\Delta T = \frac{\gamma B a^2}{c f^2} \int\limits_{p}^{R_2} \frac{dr}{r\sqrt{r^2 - p^2}} + \int\limits_{p}^{R_1} \frac{dr}{r\sqrt{r^2 - p^2}}. \tag{3.10}$$

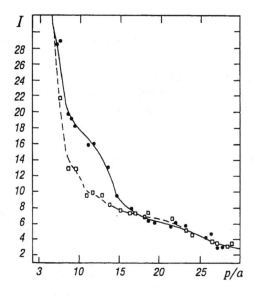

Fig. 3.5. Dependence of the electron column density on the parameter p/a [13].

Here p, R_1, and R_2 are the minimal distances from the Sun's center to the ray, the Earth, and the source of radio waves, respectively (Fig. 3.2). From (3.10), we can obtain the theoretical dependence of the time delay on the parameter p:

$$\Delta T = \frac{\gamma B a^2}{c p f^2}\left[\arccos\left(\frac{p}{R_2}\right) + \arccos\left(\frac{p}{R_1}\right)\right]. \tag{3.11}$$

The aforementioned average dependence $N(r)$ undergoes slow variations related to the solar cycle. The electron density measured at the same distance r from the Sun but in various years may differ by a factor two to three. The actual distribution of the electron density at an arbitrary moment in time may also differ considerably from the average one. The solar photosphere nearly always contains regions with enhanced activity, which are responsible for regular plasma emissions with an increased density and speed. Because of the Sun's rotation, plasma streams are bent, so that the circumsolar space looks as though it is divided by these streams into sectors that have very different electron densities. Such a sector-type pattern of the circumsolar plasma affects the propagation of radio waves [16]. During the measurements of the group delay in the circumsolar plasma, slow irregular variations in the integral electron density may reach a value of 100%. These variations lead to irregular variations in the phase of radio waves. This phenomenon is considered in Section 3.6.

3.2 Refraction and frequency shift in the circumsolar plasma

The radial decrease in the electron density of the circumsolar plasma must cause a refraction of radio waves, or, in other words, bending of their ray path.

In Fig. 3.6, the ray *TCM* passes through a source of radio waves located at point M and a receiving station located at point *T*; the Sun's center is at point *O*; point *C* corresponds to the closest approach of the ray to the Sun. Let us introduce a polar coordinate system (r, θ), where point *T* and the source of radio waves have coordinates $(R_2, 0)$ and (R_1, θ), respectively.

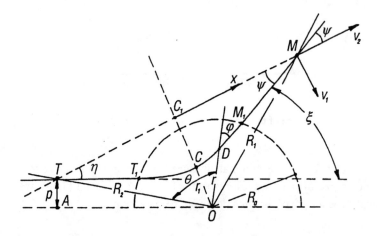

Fig. 3.6. Schematic representation of the geometry of the problem of the refraction of radio waves in the circumsolar plasma.

In Chapter 2, we derived formulae that are applicable for the analysis of ray paths in a spherically symmetric medium. Taking into account relations (1.42), (2.1), (2.2), (2.13), and (2.14), we may write the following expression for the refraction of radio waves in the circumsolar plasma:

$$p = n(r)\sin\varphi = n(r_1)r_1 \; , \tag{3.12}$$

$$\Theta = \Theta_0 - p \int\limits_{r_1}^{R_0} \frac{dr}{r\sqrt{r^2 n^2 - p^2}} \; , \tag{3.13}$$

$$\xi = -2p \int\limits_{r_1}^{R_0} \frac{1}{n}\frac{dn}{dr}\frac{dr}{\sqrt{r^2 n^2 - p^2}} \; , \tag{3.14}$$

$$n = 1 - \gamma f^{-2} N_e. \tag{3.15}$$

Here (r, θ) are the coordinates of an arbitrary point D on the ray TCM, $r_1 = CO$ is the minimum distance of the ray from the Sun's center, $n(r_1)$ is the refractive index at the minimum distance of the ray, φ is the angle made by the length element of the ray and the radius-vector.

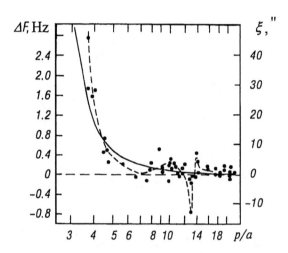

Fig. 3.7. Refraction angle and frequency shift of 32-cm radio waves versus p/a [21].

In Fig. 3.6, the dashed circle of radius R_0 bounds the space region that contributes the most to the refraction of radio waves. One may assume that $R_0 \approx 20a$. At $r > R_0$, refraction can be neglected; in this case the rays TT_1 and MM_1 can be assumed to be straight. The ray that is symmetric relative to the straight line OC is characterized by the impact parameter p defined as the distance between two parallel lines, one of which (line AO) passes through the solar center and the other (the so-called line of sight TT_1) coincides with the ray in the region where the effect of the plasma is negligible. At $p = 0$, the line of sight passes through the solar center. Given the minimum distance r_1 and the radio frequency f, the impact parameter p can be found by using the simple expression (3.12) and formula (3.15). The angle ξ between the tangents to the ray at points T and M expresses the refraction of radio waves; this angle can be estimated by formula (3.14) provided that $N_e(r)$ is known.

The solid line in Fig. 3.7 represents the refraction angle ξ calculated for 32-cm radio waves and various values of the ratio p/a. The refraction of 32-cm radio waves is small; thus, at p/a equal to 1.2, 1.8, and 3, the angle ξ is 1, 0.3, and 0.07°, respectively. Considerable bending of rays, such as that given in Fig. 3.6, is possible in the meter wavelength band, as follows from the relation $\xi \sim \lambda^2$. When the impact parameter p diminishes, the refraction angle ξ increases, and the distance r decreases.

In the radar investigations of the Sun, $p = 0$; therefore, as follows from (3.12), $n(r_1) = 0$. Allowing for (3.15), one can obtain the following simple expression relating the maximum depth of penetration of radio waves into the solar corona, r_m, and the radio frequency:

$$f^2 = \gamma N_e(r_m).$$ (3.16)

The penetration depth of radio waves into the solar plasma increases with their frequency. Thus, at $f = 27$ and 110 MHz, r_m is equal to 1.6 and 1.1 a, respectively. In the (r, θ) coordinate system, the ray path form is given by equation (3.13). The ray path depends on the wavelength λ and the impact parameter p; for instance, the ray path shown in Fig. 3.6 corresponds to $\lambda = 3$ m and $p = 0.8a$. More detailed analysis of ray paths in astronomy [17] showed that, apart from the refracted ray TCM, there is a straight ray represented by the dashed line TC_1M. The geometry presented in Fig. 3.6 yields the simple equation

$$\Theta_1 = \pi - \arcsin\frac{p}{R_1} - \arcsin\frac{p}{R_2} + \xi(p).$$ (3.17)

This equation has two solutions in the variable p. One solution corresponds to the straight ray TC_1M ($\xi = 0$ and $p_0 = CO_1$) and the other, with $p = AT$, corresponds to a ray bent by the angle ξ. Thus, there are two rays passing through points T and M.

The Doppler shifts for these two rays are different, which permits the experimental investigation of the refraction of radio wave by means of the frequency selection of radio signals. The remote sounding of the circumsolar plasma by signals from a spacecraft brings about a refractive frequency shift ΔF dependent on ξ. This shift is analogous to that occurring during the radio occultation observations of planetary ionospheres (see Chapter 2). The refraction of radio waves and the frequency shift ΔF during the radio occultation observations of the circumsolar plasma were analyzed in [18–21]. To establish a relationship between the refractive frequency shift ΔF and the refraction angle ξ, let us introduce the projection of the velocity of a spacecraft onto the straight line TM, V_2, and the velocity component V_1 normal to this line. The Doppler frequency shift along the ray TCM is given by

$$\Delta f = \lambda^{-1}(V_2 \cos\psi_1 - V_1 \sin\psi_1).$$ (3.18)

Here ψ_1 is the angle between straight lines TM and M_1M. If the effect of the plasma is absent, the frequency shift due to the motion of the spacecraft is given by the following formula:

$$\Delta f_0 = V_2\lambda^{-1}.$$ (3.19)

From formula (3.19) the frequency shift due to refraction, $\Delta F = \Delta f - \Delta f_0$, can easily be derived:

$$\Delta F = \lambda^{-1}\left[V_2(\cos\psi_1 - 1) - V_1\sin\psi_1\right].\tag{3.20}$$

For decimeter and centimeter radio waves, the angle ψ_1 is small; therefore, equation (3.20) can be simplified to

$$\Delta F = -V_1\psi_1\lambda^{-1}.\tag{3.21}$$

As is evident from Fig. 3.6, $\xi = \psi_1 + \eta$. Then, at $R_1 = R_2$, we have $\psi_1 = \eta$ and $\xi = 2\psi_1$. But if $R_2 \gg R_1$, then $\xi = \psi_1$. Equations (3.20) and (3.21) show that the experimental determination of the frequency shift ΔF is equivalent to the determination of the refraction angle ξ.

At high frequencies, the refractive bending of ray paths is small; this enables a simple formula relating the frequency shift ΔF and the radial distribution of electron density to be derived [20, 21]. The frequency shift ΔF in a spherically symmetric medium can be found by differentiating, with respect to time, the phase increment in the plasma along the path TC_1M:

$$\Delta F = \frac{2uV\gamma}{cf}\frac{d}{dp_0}\int_0^{R_0}N_e(r)dx.\tag{3.22}$$

In this formula, integration is performed along the C_1x-axis (Fig. 3.6), and the impact parameter changes at a rate of $u = dp_0/dt$. This approach is approximate, since the difference in the minimum distances OC_1 and OC is not taken into account and integration in (3.22) is performed along the straight line TM rather than along the bent ray. If the dependence $N(r)$ is approximated by an expression similar to (3.1) at $\chi = 0$, then from (3.22) we obtain

$$\Delta F = \frac{2u\gamma}{cf}\left[AA_1\left(\frac{a}{p}\right)^6 + BB_1\left(\frac{a}{p}\right)^{2+\varepsilon}\right],\tag{3.23}$$

where

$$A_1 = \frac{5\pi\,\Gamma(5)}{2^5\Gamma^2(3)},\quad B_1 = \frac{(1+\varepsilon)\pi\,\Gamma(1+\varepsilon)}{2^{1+\varepsilon}\,\Gamma^2\left(\dfrac{2+\varepsilon}{2}\right)}.$$

Here Γ is the gamma-function. The authors of [20, 21] found that, at $\varepsilon = 0.3$, $A = 2.21 \cdot 10^8$ cm^{-3}, and $B = 1.55 \cdot 10^6$ cm^{-3}, the experimentally and theoretically derived dependences $\Delta F(p)$ show good agreement. In this case formula (3.23) transforms into

$$\Delta F = \frac{\gamma u}{cf}\left[1.3 \cdot 10^9 \left(\frac{a}{p}\right)^6 + 5.3 \cdot 10^6 \left(\frac{a}{p}\right)^{2.3}\right]. \tag{3.24}$$

The comparison of this formula with (3.1) and (3.21) shows that the radial distribution of the electron density, $N(r)$, and the dependences $\Delta F(p)$ and $\psi_1(p)$ are given by similar power functions provided that the radio frequency is high. The frequency shift ΔF has opposite signs when the ray approaches the Sun ($\Delta F > 0$) or recedes from it ($\Delta F < 0$) at the beginning and end of occultation.

The authors of [19–21] presented experimental data on refractive changes in the frequency of radio signals, ΔF, and on the refraction angle ξ, which were obtained during the radio occultation observations of the circumsolar plasma performed from *Venera 10* in 1976 and from *Venera 15* and *Venera 16* in 1984. Observations were carried out using 32-cm radio waves; the velocity u was about 8 km s^{-1}. At $p > 15a$, no regular changes in the radio frequency were detected. However, at $p \leq 14a$, the authors recorded regular refractive variations in the frequency, ΔF, which drastically increased at $p < 6a$. Figure 3.7, which is taken from [21], shows the experimental dependences $\Delta F(p)$ (dots) and $\xi(p)$ (dashed line). In 1984, the refraction angle $\xi(p)$ comprised 14" at $p = 4a$ and 50" at $p = 3a$ [21]. At $p = 2.3a$, the refraction angle was 96" [20]. It should be noted that the authors of [20, 21] mistakenly considered that $\psi_1 = \xi$ (actually, ξ must be equal to $2\psi_1$ under the experimental conditions used); therefore, the refraction angle was underestimated by those authors twice. In an interval of $p = (10 - 14)a$, the function $\xi(p)$ changes its sign; this is an indication of changes in the gradient dN_e/dr in this region.

The quantities ΔF and ξ undergo rapid and strong variations due to the motion of the large-scale plasma inhomogeneities produced by the coronal prominences extending to $r \approx 10a$ or by transients in the form of loops or globules. The frequency variations induced by the large-scale plasma inhomogeneities can be estimated by formula (3.22) provided that the model distribution of the electron density and the speed of these inhomogeneities are known (such estimations were carried out in [20, 22]). Experimental and theoretical estimations show that, at $p > 10a$, variations in the radio frequency, ΔF, may considerably exceed their mean values at a given p; at $p < 6a$, however, slow variations in ΔF are comparable in magnitude with the mean value of the refractive frequency shift. The authors of [19] found that, at $p = 4a$, the frequency shift ΔF changes twofold over a period of 100 s. This was due to the coronal emission of the solar plasma flowing out at velocities of 600–1200 km s^{-1}.

Large variations in ΔF must strongly affect the ray paths. Because of the phase fluctuations induced by plasma irregularities, the equiphase surface becomes distorted. Some regions of the equiphase surface with linear sizes that are less than the phase-correlation radius can be approximated by planes. The normal to such regions on the equiphase surface represents a ray, which is characterized by the arrival angle of radio waves. A given point may have a variety of such planes at different moments in time; this variety represents the angular spectrum of radio waves with a bandwidth $\Delta\xi$. The bandwidth of the angular spectra was investigated in detail by the radio sounding of the circumsolar and interplanetary plasma using radio emissions from celestial point sources. Figure 3.8 shows the experimental dependence of the angular spectrum bandwidth $\Delta\xi$ of 3-m radio waves on the p/a ratio [23]. Line segments in this figure represent the dependence $\Delta\xi(p)$ obtained for different meter-band radio waves and converted to the wavelength $\lambda = 3$ m.

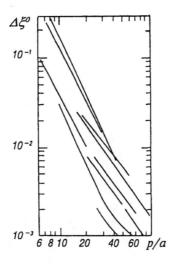

Fig. 3.8. Angular broadening bandwidth of meter radio waves versus the ratio p/a [23].

Let us compare the values of the regular refraction ξ and the angular broadening bandwidth $\Delta\xi$, taking into account that both of these parameters, ξ and $\Delta\xi$, are proportional to the square of the wavelength. The difficulty about this comparison is that reliable data on ξ and $\Delta\xi$ are available for different regions ($p < 6a$ and $p > 6a$, respectively). At $p = 6a$ and $\lambda = 3$ m, the angular broadening bandwidth $\Delta\xi$ is equal to 0.2–0.3°. At $p = 6a$ and $\lambda = 32$ cm, the refraction angle ξ is 0.002° and, consequently, can be estimated to be about 0.2° at $\lambda = 3$ m. In other words, the angular broadening bandwidth $\Delta\xi$ approximates the refraction angle ξ. It should be noted in relation to this that ray paths in the circumsolar plasma described in [17] or traced on the cover of a book on solar radio physics [24] are quite conventional. From the analysis of refraction and delay data it follows that the circumsolar plasma is a very irregular medium, for which the integral effects of refraction and time delay can be determined only by averaging data over sufficiently long time intervals. The propagation of radio waves in plasmas is governed by plasma irregularities.

3.3 Faraday effect

The spatial structure of the circumsolar plasma, which is strongly influenced by the solar magnetic field, is variable. This was first revealed during optical observations of the solar photosphere and corona. The photospheric magnetic field H is formed by a random superposition of its components possessing various scales and amplitudes. The strength of the photospheric magnetic field can reliably be determined by measuring the Zeeman splitting of spectral lines. The strength of the photospheric magnetic field averages 2 G, though may exceed this value by tens and even hundreds of times in active regions [24]. Unlike the photosphere, the solar corona exhibits a large-scale magnetic field that probably governs various large-scale events in the solar wind, such as circumsolar plasma streams and the sectorial structure of interplanetary plasma. The photographs of the solar corona usually display prominences, i.e., the regions of moving plasma whose density is higher than that of the surrounding medium. It can be suggested that it is the extension of the prominences to interplanetary space that determines the sectorial structure of the solar wind. Magnetometric measurements in interplanetary space from spacecraft showed that the magnetic field also has a sectorial structure. Vector H was found to be oriented radially, but in opposite directions in the neighboring sectors. At the boundaries of the plasma sectors, the radial component of the magnetic field, H_r, is zero.

   The magnetic field in the region of the maximum plasma acceleration can be investigated by the radio occultation technique. In an interval of $2a < r < 14a$, the magnetic field can be studied in terms of the Faraday effect, which lies in a rotation of the polarization plane of radio waves by angle Ω as a result of their propagation in the plasma occurring in the magnetic field. For high radio frequencies, the angle Ω depends on the strength of the magnetic field, electron density, and radio frequency. When applied to the problem of the radio sounding of the solar wind, equation (1.140) can easily be transformed into

$$\Omega = Kf^{-2} \int_{T}^{M} N_e(r) H_x(r) dx . \tag{3.25}$$

   Here $K = 1.35 \cdot 10^{-6}$ (if Ω, f, N_e, dx, and H are expressed in deg., MHz, cm^{-3}, cm, and G, respectively) and H_x is the projection of vector H onto the ray TM. When analyzing the Faraday effect, the refraction of radio waves can be neglected; therefore, integration in formula (3.25) can be performed along the straight line TM. Assume in calculations that vector H is directed radially (Fig. 3.2). If the distributions $N(r)$ and $H(r)$ are spherically symmetric, the angle Ω is equal to zero; this follows from the fact that the integrals along the lines CT and CM are equal in magnitude but have opposite signs. Thus, the

Faraday effect is not zero if at least one of the distributions, $N(r)$ or $H(r)$, is spherically asymmetric.

The first estimations of the Faraday effect in the circumsolar plasma performed in 1968 and 1970 from *Pioneer 6* and *Pioneer 9* gave unexpected results. In particular, it was found that at a radio frequency of 2.29 GHz and $p \approx 6a$, the angle Ω undergoes slow and large irregular variations, reaching 20–30° in about 2 h [25–27]. The effect was shown to be due to the sectorial structure of the solar corona [26]. This inference was confirmed by measurements of the polarized radiation from the Tau-A source in the Crab Nebula. Detailed investigations performed in 1971–1975 at f = 4.2 GHz showed that the angle Ω is variable even at this high frequency [28]. The annual probing of the circumsolar plasma conducted every year over a period from 12 to 19 June (the ray distance from the Sun was minimum ($p = 5a$) on June 15) showed that the angle Ω underwent slow irregular variations within $\pm 20°$. The comparison of these variations with the distribution of the magnetic field in the photosphere did not reveal any correlation between them. At $p = 7a$, the product $N_e H$ was estimated to be 400 G cm^{-3} [28]. Measurements of the Faraday effect performed in 1978 using the radiation with λ = 8 cm from the same Tau-A source in the Crab Nebula showed that variations in the angle Ω can reliably be recorded at $p \leq 12a$ [29, 30]. The angle Ω changed in opposite directions when the ray path was approaching and receding from the Sun. Apart from the irregular variations in the angle Ω, it was found to undergo regular variations $\Omega(p)$. In [29], attempts were made to estimate the magnetic field strength from the experimental values for angle Ω and the known dependence $N(r)$. To do this, the authors made quite arbitrary assumptions about the plasma sectors responsible for the Faraday effect. Evidence presented in [25–30] suggests that the large-scale solar magnetic field can be studied in terms of the Faraday effect at distances of $10a$ and less from the Sun. The most detailed investigations of the solar magnetic field by measuring the Faraday effect were carried out during the missions of *Helios 1* and *Helios 2* [31–34]. A remarkable result of these studies was the observation of strong regular variations in the dependence $\Omega(p)$. Figure 3.9 illustrates the experimental dependences $\Omega(p)$ derived at f = 2.29 GHz in 1975 (curve 2) and 1976 (curve 1). The magnitude of variations in the angle Ω may differ from year to year, but the main regularities of the plasma behaviour remain the same. The functions $\Omega(p)$ shown in Fig. 3.9 correspond to the ray path approaching the Sun. For the receding ray path, the dependences derived were analogous, but with opposite sign. It should be noted that in the latter case another region of the circumsolar plasma was investigated.

The above phenomenon can be explained as follows [31, 34]. Let two sectors, *TOB* and *BOM*, be separated by the boundary *OB* and have different directions of the magnetic field H (Fig. 3.2). For instance, vector H in the former sector is directed towards the Sun, and in the latter sector it points outwards from the Sun. The position of the boundary *OB* can be fixed by the angle Φ_1. Suppose for simplicity that the dependences $N_e(r)$ and $H(r)$ in these sectors do not differ and can be approximated by the following formulae

$$N_e = N_0 \left(\frac{a}{r}\right)^\beta \tag{3.26}$$

and

$$H = H_0 \left(\frac{a}{r}\right)^{\alpha_1}. \tag{3.27}$$

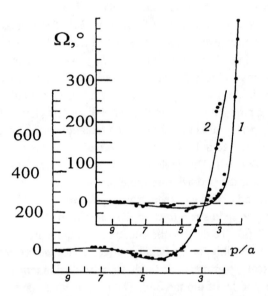

Fig. 3.9. Representative dependences of the angle of rotation of the polarization plane of radio waves on the impact parameter (dots represent experimental data, and curves are calculated) [31, 32].

From (3.25) it follows that the effect of the sector BOM on the angle Ω can be given by

$$\Omega_1 = Kf^{-2} \int_{x_1}^{\infty} N_e(r) H_r \sin \Phi \, dx, \tag{3.28}$$

where

$$\begin{aligned}
x_1 &= p \cdot \tan \Phi_1, \\
r &= p \cdot \cos^{-1} \Phi, \\
dx &= p \cdot \cos^{-2} \Phi \, d\Phi.
\end{aligned} \tag{3.29}$$

Allowing for equalities (3.29) and expressions (3.26) and (3.27), formula (3.28) transforms into

$$\Omega_1 = \frac{Kf^{-2}N_0H_0}{\left(\dfrac{p}{a}\right)^{\alpha_1+\beta-1}} \int\limits_{\cos\Phi_1}^{0} (\cos\Phi)^{\alpha_1+\beta-2}\, d(\cos\Phi)$$

and, after integration,

$$\Omega_1 = \frac{Kf^{-2}N_0H_0(\cos\Phi_1)^{\alpha_1+\beta-1}}{(\alpha_1+\beta-1)\left(\dfrac{p}{a}\right)^{\alpha_1+\beta-1}}. \tag{3.30}$$

Consider now the effect of the sector *TOB*. For this purpose, let us divide it into the subsector TOB_1 and subsector B_1OB with the vertex angle $2\Phi_1$ (Fig. 3.2). The effect of the subsector B_1OB on the angle Ω is zero, since the projections H_x in the regions BOC_1 and C_1OB_1 are directed oppositely. As a result, the contribution from the sector *TOB* (i.e., Ω_2), is determined only by the contribution from the subsector TOB_1; therefore, $\Omega_1 = \Omega_2$. The total effect of the two subsectors is twice as great as predicted by formula (3.30):

$$\Omega = \frac{2Kf^{-2}N_0H_0(\cos\Phi_1)^{\alpha_1+\beta-1}}{(\alpha_1+\beta-1)\left(\dfrac{p}{a}\right)^{\alpha_1+\beta-1}}. \tag{3.31}$$

The Faraday effect must be greatest at $\Phi_1 = 0$, when the sectorial boundary *OB* is perpendicular to the ray path *TM*. In this case,

$$\Omega_m = \frac{2Kf^{-2}N_0H_0}{(\alpha_1+\beta-1)\left(\dfrac{p}{a}\right)^{\alpha_1+\beta-1}}. \tag{3.32}$$

If the ray path traverses several boundaries between sectors, the angle Ω is determined by the algebraic sum of the terms Ω_i of the type (3.31) with allowance for the sign of the magnetic field in particular sectors and the angles Φ_i.

Such an approach implies a sufficiently rough assumption that dependences (3.26) and (3.27) are the same in all sectors and that the sectors differ only in their angles Φ_i. In the case of *m* sectors, we have

$$\Omega = \left[\frac{-2Kf^{-2}N_0H_0}{(\alpha_1 + \beta - 1)} \right] \sum_{i=1}^{m} (-1)^i \left(\frac{\cos\Phi_i}{\frac{p}{a}} \right)^{\alpha_1 + \beta - 1} . \tag{3.33}$$

From (3.31) or (3.33) it follows that the angular variation $\Omega(t)$ is due to two factors: changes in $p(t)$ induced by the motions of the spacecraft and the Earth and changes in the position of the sectorial boundaries $\Phi_i(t)$ because of the Sun's rotation. The first factor is decisive at $p \le 3a$, and the second factor becomes predominant at $p \ge 4a$. From (3.31) and (3.33), we can obtain the relation: $\Omega \sim p^{1-\alpha_1-\beta}$.

As was shown in [31], at $p \le 3a$, experimental and theoretical data agree well if $\alpha + \beta = 5.1$. For a wider range of the impact parameters p, good agreement between the experiment and theory was obtained after the number of the sectors considered had been raised to six. If $\alpha + \beta = 5.5$, the agreement is good at $p = (2 - 10)a$. In this case, the authors analyzed the photographs of solar prominences taken with a coronograph; this allowed the angles Φ_i to be determined more accurately. In Fig. 3.9, the solid lines show the theoretical dependences $\Omega(p)$. More detailed analysis of experimental data with the aim to determine the magnetic field was performed in [34]. To this end, the authors averaged the angle Ω over the observation time periods, which lasted from tens of minutes to one hour. The averaging made it possible to eliminate the fluctuations $\delta\Omega$ and to obtain a more reliable dependence $\Omega(p)$ for an interval of $p = (3 - 12)a$. Although the data thus obtained showed a wide scatter, the authors were able to derive the dependence of the maximum Ω_m values on the impact parameter p for the years 1975–1976, when the solar activity was minimum. The empirical dependence of Ω_m on p at $f = 2.29$ GHz is as follows

$$\Omega_m = 9.5 \cdot 10^4 \left(\frac{a}{p} \right)^{4.15} , \tag{3.34}$$

where Ω is expressed in degrees. The authors of [34] believed that the maximum experimental Ω_m values corresponded to the case of the sectorial boundary being normal to the ray path, i.e., when formula (3.32) is valid. A comparison of (3.34) and (3.32) yields the equality $\alpha + \beta = 5.15$. To find the dependence of the magnetic field (i.e., the parameters α and H_0 in formula (3.27)) on distance r, the authors used experimental data on the function $N_e(r)$ (see Section 3.1). Namely, setting $N_0 = 1.4 \cdot 10^6$ cm^{-3} and $\beta = 2.45$ for $3a < r < 10a$ yielded $\alpha = 2.7$ and $H_0 = 7.9$ G. As a result, the following empirical dependence of the magnetic field strength on distance r was derived:

$$H = 7.9 \left(\frac{a}{r} \right)^{2.7} , \tag{3.35}$$

where H is expressed in gauss. This formula is valid for $r = (2 - 12)a$. The circumsolar magnetic field is commonly represented as the sum of two components, one being proportional to r^{-2} and the other to r^{-3}. As is shown in [34], formula (3.35) can alternatively be represented as

$$H = 6\left(\frac{a}{r}\right)^3 + 1.18\left(\frac{a}{r}\right)^2, \tag{3.36}$$

where H is expressed in gauss.

Measurements of the Faraday effect resulted in the discovery of the so-called Alfvénic waves [35–38, 152–156], which are supposed to essentially contribute to the energy of the solar wind in the region of its maximum acceleration. In view of this, a brief account of this phenomenon should be given. Alfvénic waves in fact represent the magnetic field disturbances propagating in plasma. They are generated in the solar photosphere and propagate for distances of $(6 - 10)a$, almost without being absorbed. The authors of [35] revealed quasi-periodic changes in the angle Ω and assumed that they are due to the influence of Alfvénic waves whose magnetic field can bring about the rotation of the polarization plane of radio waves. The characteristic frequency of changes in Ω was estimated to be about 10^{-3} radian per second. The detailed analysis of relevant experimental data showed that the dispersion of the fluctuations $\delta\Omega$ decreases with p according to the power law [36]:

$$\langle\delta\Omega^2\rangle^{1/2} = 246\left(\frac{a}{p}\right)^{2.6}, \tag{3.37}$$

where $\delta\Omega$ is expressed in degrees. The empirical dependence (3.37) was derived from the data obtained in 1976 in a 13-cm wavelength band. From formula (3.25) it follows that fluctuations in the rotation angle of the polarization plane, $\delta\Omega$, may be due to the fluctuations in both electron density, δN_e, and magnetic field, δH. Analysis of variations in the integral electron density δI and fluctuations $\delta\Omega$, which were recorded concurrently, showed that the fluctuations $\delta\Omega$ cannot be related to variations in the integral electron density δI, but are presumably due to variations in the magnetic field, i.e., are produced by Alfvénic waves [36]. It should be noted that the authors of this work made some speculative assumptions in order to obtain a theoretical dependence consistent with formula (3.37).

More convincing evidence that Alfvénic waves are responsible for the fluctuations $\delta\Omega$ was presented in [37]. The comparison between the fluctuations $\delta\Omega$ recorded at two spaced ground-based stations showed that these fluctuations are similar with a correlation coefficient of about 0.9, but are separated in time by some interval τ. The spaced reception technique enables one to elucidate the velocity V_a as a factor that affects the angle Ω. If ray paths in the circumsolar space are at a distance Δp apart, the velocity V_a is defined as the ratio $\Delta p/\tau$. In the experiments described above, the spacing between the rays, Δp, was 2015 km and the time delay τ was 7.5 s at $p = 4a$; this gave $V_a = 270$ km s^{-1}. For the impact parameters $p = (3.5 - 6.3)a$, the velocity V_a varied from 50 to 1000 km s^{-1}. Such high values of V_a suggest that the fluctuations $\delta\Omega$ are due to the effect of Alfvénic waves on the radio signals [37]. According to [24], the dependence of the speed of Alfvénic waves on the magnetic field strength and the plasma density ρ is given by

$$V_a = H(4\pi\rho)^{-1/2}. \tag{3.38}$$

Substituting H in the form (3.35) and taking into account that the plasma density ρ is proportional to the electron density and can be found from (3.26), we arrive at the empirical formula

$$V_a = A\left(\frac{a}{r}\right)^{1.67}, \tag{3.39}$$

where $A \approx 2800$ km s^{-1}. This formula is valid for distances $2a \leq r \leq 14a$.

3.4 Autocorrelation functions of the amplitude and phase fluctuations of radio waves

Radio observations of radioastronomical sources through the circumsolar and interplanetary plasmas showed that plasma irregularities cause the broadening of the angular spectrum of radio waves and give rise to fluctuations in their amplitude. These observations provided valuable experimental information about the dependence of these characteristics of radio waves on the impact parameter p and wavelength λ. Figure 3.8 presents the dependence of the angular broadening bandwidth on the impact parameter p. The radioastronomical studies of plasma inhomogeneities, which are reviewed in [40–45], showed that the scale of these inhomogeneities ranges from tens to thousands of km and that the magnitude of fluctuations in the electron density rapidly decreases with increasing the heliocentric distance r.

Analysis of signals from space vehicles can provide information about variations in the integral electron density, δI. Analysis of these variations showed that the characteristic size of large-scale plasma formations can reach one million km [46–49]. Generally, the interplanetary and circumsolar plasmas can be considered as random inhomogeneous media with a wide range of inhomogeneity scales. The motion of plasma irregularities leads to the phase and amplitude fluctuations of radio waves and broadens their power spectra.

Like the turbulence of ionized gases [50], the spatial spectrum of the refractive index fluctuations can be given by

$$\Phi_n(\text{æ}) = 0.033 c_n^2 \left(\text{æ}^2 + \text{æ}_0^2\right)^{-\alpha/2} \exp\left(-\frac{\text{æ}^2}{\text{æ}_m^2}\right) . \tag{3.40}$$

Here $\text{æ} = 2\pi/\Lambda$ is the spatial wavenumber, $\text{æ}_0 = 2\pi/\Lambda_0$, and $\text{æ}_m = 2\pi/l$. The denominators in these formulas, Λ_0 and l, are the outer and inner scales of the refractive index irregularities, respectively. In the interval $\text{æ}_0 < \text{æ} < \text{æ}_m$, expression (3.40) takes the form

$$\Phi_n(\text{æ}) = 0.033 c_n^2 \text{æ}^{-\alpha} , \tag{3.41}$$

where α is the spatial spectrum index of plasma irregularities. The spectral parameters Φ_n for the interplanetary and circumsolar plasmas have the following approximate values: $\Lambda_0 \approx 10^6$ km, $l \approx 10$ km, and $\alpha \approx 11/3$. These approximations are rough, since the spectral parameters may differ considerably in space. The magnitude of refractive fluctuations is expressed by the structural coefficient c_n or variance σ_n. If $\alpha = 11/3$, then, according to [50],

$$c_n^2 = 1.9 \sigma_n^2 \Lambda_0^{-\frac{2}{3}}, \tag{3.42}$$

where σ_n^2 is the variance of the refractive index fluctuations of radio waves. For other α values, c_n is proportional to σ_n, but the coefficient of proportionality depends both on Λ_m and Λ_0. The refractive index of radio waves, n, is related to the electron density N_e as is given by formula (1.42). Therefore, the root-mean-square refractive fluctuations σ_n are related to the root-mean-square electron density fluctuations σ_N as follows

$$\sigma_n = \gamma f^{-2} \sigma_N,$$

$$c_n^2 = 1.9 \Lambda_0^{-\frac{2}{3}} \gamma^2 f^{-4} \sigma_N^2. \tag{3.43}$$

As was shown in Section 3.1, the electron density rapidly decreases with the distance from the Sun's center, r. Parameters c_n and σ_n also depend strongly on r. Assuming that the degree of the plasma heterogeneity, σ_n/N_e, depends insignificantly on r and taking into account that $N_e \sim r^{-2}$ at high heliocentric distances, the approximate dependence

$$c_n = c_0 a^2 r^{-2} \tag{3.44}$$

can be suggested to be valid within the interval $30a < r < 200a$. It should be noted that the spectrum $\Phi_n(\text{æ})$ has an uncertainty in the vicinity of $\text{æ} \leq \text{æ}_0$, since large-scale plasma inhomogeneities are anisotropic. For this reason, formula (3.40) poorly describes plasma irregularities in the aforementioned interval of heliocentric distances.

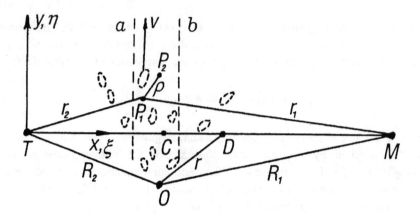

Fig. 3.10. Diagram illustrating the analysis of the amplitude and phase fluctuations of radio waves.

For the theoretical analysis of radio wave fluctuations to be possible, it is necessary to obtain general initial relationships describing the temporal autocorrelation functions of the amplitude and phase fluctuations of radio waves propagating along the path TM (Fig. 3.10). This can be done by invoking the wave equation

$$\nabla^2 E + k^2 n^2 E = 0 , \tag{3.45}$$

where E denotes the electric field strength of radio waves, n is the refractive index, and $k = 2\pi / \lambda$ is the wavenumber. We neglect here the minor effects related to changes in the polarization of radio waves; that is why equation (3.45) is scalar. Let us also neglect the refraction of radio waves, i.e., assume that, in the absence of the electron density irregularities ΔN_e, the refractive index $n = 1$. Now we can write

$$n = 1 - \Delta n = 1 - \gamma\, f^{-2} \Delta N_e . \tag{3.46}$$

Here Δn and ΔN_e are fluctuations in the refractive index and electron density, respectively. The solution to the problem of radio wave fluctuations in the interplanetary and circumsolar plasmas can be sought by the method of smooth perturbations [50]. With this method, the solution to equation (3.45) can be written in the form

$$E_1(\bar{r}_2,t) = E(\bar{r}_2,t)r_2^{-1} \exp\{i[\omega t - \varphi(r_2,t)]\} \;, \tag{3.47}$$

where r_2 is the distance from the transmitter to an arbitrary point in the plasma; t is time; E and φ are the fluctuating amplitude and phase of radio waves. The solution (3.47) can be rewritten in the form

$$E_1(r_2,t) = E_0 r_2^{-1} \exp\{i[\omega t - \psi(r_2,t)]\} \;, \tag{3.48}$$

where E_0 is the constant amplitude of a radio wave in a uniform plasma. From the comparison of (3.47) and (3.48) it follows:

$$\psi(\bar{r}_2,t) = \varphi(\bar{r}_2,t) + i \cdot \ln[E_0^{-1} E(\bar{r}_2,t)] \;, \tag{3.49}$$

where the real and imaginary parts of the function ψ refer to the amplitude and phase of radio waves, respectively. Substituting (3.48) into (3.45), we get the following equation for the function ψ :

$$(\nabla \psi)^2 + i\nabla^2 \psi - \frac{2i}{r_2}\frac{\partial \psi}{\partial r_2} = k^2 n^2(\bar{r}_2,t). \tag{3.50}$$

If the plasma is uniform, this equation takes the form

$$(\nabla \psi_0)^2 + i\nabla^2 \psi_0 - \frac{2i}{r_2}\frac{\partial \psi_0}{\partial r_2} = k^2 \;. \tag{3.51}$$

Subtracting expression (3.51) from (3.50) and introducing the differential function $\psi_1 = \psi - \psi_0$, we obtain the following equation for ψ_1 :

$$2(\nabla \psi_0 \nabla \psi_1 + i\nabla^2 \psi_1 - \frac{2i}{r_2}\frac{\partial \psi_1}{\partial r_2} = 2k^2 \Delta n + k^2(\Delta n)^2 + (\nabla \psi_1)^2. \tag{3.52}$$

Here Δn is a small deviation in the refractive index. Taking into account that $(\nabla \psi_1)^2$ is of the order of $(\Delta n)^2$ and neglecting all the terms of the order $(\Delta n)^2$ in expression (3.52), we have

$$2(\nabla \psi_0 \nabla \psi_1) + i\nabla^2 \psi_1 - \frac{2i}{r_2}\frac{\partial \psi_1}{\partial r_2} = 2k^2 \Delta n(\bar{r}_2,t) \;. \tag{3.53}$$

Assuming that the phase of the undisturbed wave in (3.53) is $\psi_0 = kr_2$, we obtain the following equation for the function ψ_1:

$$2k\frac{\partial \psi_1}{\partial r_2} + i\nabla^2 \psi_1 - \frac{2i}{r_2}\frac{\partial \psi_1}{\partial r_2} = 2k^2\Delta n(\bar{r}_2,t). \tag{3.54}$$

Let us introduce a new function W, such that

$$\psi_1 = r_2 W \exp(ikr_2) \ . \tag{3.55}$$

From (3.54) and (3.55), we have the following equation for the function W:

$$\nabla^2 W - k^2 W = 2ik^2 r_2^{-1}\Delta n(r_2,t)\exp(-ikr_2). \tag{3.56}$$

The solution to the nonuniform equation (3.56) has the form

$$W = \frac{ik^2}{2\pi}\int_V r_2^{-1}r_1^{-1}\exp[-ik(r_2 + r_1)]\Delta n(\xi,\eta,\zeta)dV \ , \tag{3.57}$$

where (ξ, η, and ζ) are the coordinates of the irregularity Δn, and integration is performed over the space parcel V between a transmitter and receiving station. Recall that r_2 is the distance from the transmitter positioned at point T to an arbitrary point P in the inhomogeneous medium and r_1 is the distance from point P to the receiving station positioned at point M. The coordinate system depicted in Figure 3.10 has the coordinate axis (x, ξ) directed along the ray path TM and the coordinate axis (y, η). The axis (z, ζ) is normal to the plane of the figure. From the geometry presented in Fig. 3.10 it follows:

$$r_2^2 = \xi^2 + \eta^2 + \zeta^2,$$
$$r_1^2 = (L - \xi)^2 + \eta^2 + \zeta^2, \tag{3.58}$$

where L is the length of the ray path TM. In Fig. 3.10, the arbitrary point P in the plasma has coordinates (ξ, η, ζ), the source of radio waves is positioned at the origin of the coordinate system (point T), the ground-based receiving station located at point M has coordinates (L, O, O), and the Sun is at point O. Making use of (3.55) and (3.57) and assuming that $r_2 = L$ in the former formula, we obtain

$$\psi_1 = \frac{ik^2 L}{2\pi}\int_V r_1^{-1}r_2^{-1}\exp[-ik(r_2 + r_1 - L)]\Delta n(\xi,\eta,\zeta)dV \ . \tag{3.59}$$

Taking into account that $\psi_1 = \psi - \psi_0$ and using (3.49) and (3.59) with separated real and imaginary parts, we obtain the following formulae for the phase and log-amplitude fluctuations of radio waves:

$$\Delta\varphi = \frac{k^2}{2\pi}\int_V r_1^{-1}r_2^{-1}\sin[k(r_2 + r_1 - L)]\Delta n\, dV, \tag{3.60}$$

$$\ln\left(\frac{E}{E_0}\right) = \frac{k^2}{2\pi}\int_V r_1^{-1}r_2^{-1}\cos[k(r_2 + r_1 - L)]\Delta n\, dV. \tag{3.61}$$

The characteristic size of plasma irregularities, Λ, is considerably larger than the wavelength of the radio waves used; this enables formulae (3.60) and (3.61) to be simplified. Indeed, fluctuations in the radio waves are produced primarily by the plasma inhomogeneities concentrated near the ray path *TM*. The diffraction of radio waves from plasma irregularities essentially affects the field pattern within a solid angle of about $\lambda\Lambda^{-1}$ in size. This allows the integrand functions in (3.60) and (3.61) to be approximated by the expressions that are valid within a small solid angle of about $\lambda\Lambda^{-1}$ in size with the vertex at point *M*:

$$r_1 = L - \xi + \frac{1}{2}(\eta^2 + \zeta^2)(L - \xi)^{-1},$$

$$r_2 = \xi + \frac{1}{2}(\eta^2 + \zeta^2)\xi^{-1}, \tag{3.62}$$

$$r_2^{-1} = \xi^{-1},\ r_1^{-1} = (L - \xi)^{-1}.$$

From expressions (3.60) and (3.61), allowing for the approximate relations (3.62), it follows:

$$\Delta\varphi = \frac{Lk^2}{2\pi}\int_V \xi^{-1}(L - \xi)^{-1}\sin\left[\frac{1}{2}Lk(\eta^2 + \zeta^2)\xi^{-1}(L - \xi)^{-1}\right]\Delta n\, dV,$$

$$\ln\left(\frac{E}{E_0}\right) = \frac{Lk^2}{2\pi}\int_V \xi^{-1}(L - \xi)^{-1}\cos\left[\frac{1}{2}Lk(\eta^2 + \zeta^2)\xi^{-1}(L - \xi)^{-1}\right]\Delta n\, dV. \tag{3.63}$$

Here the volume element $dV = d\xi\, d\eta\, d\zeta$, and Δn is a stochastic function of coordinates (ξ, η, ζ) and time t. Making use of (3.49) and (3.63), we can write the following expression for ψ_1:

$$\psi_1 = \frac{Lk^2}{2\pi} \int_V \xi^{-1}(L-\xi)^{-1} \exp\left[\frac{1}{2}ikL(\eta^2+\zeta^2)\xi^{-1}(L-\xi)^{-1}\right]\Delta n dV . \qquad (3.64)$$

The real part of function ψ_1 expresses log-amplitude fluctuations, and its imaginary part describes the phase fluctuations of radio waves at a moment t. Let us introduce two new functions:

$$\begin{aligned} B_1 &= \langle \psi_1(t)\psi_1^*(t+\tau)\rangle, \\ B_2 &= \langle \psi_1(t)\psi_1(t+\tau)\rangle. \end{aligned} \qquad (3.65)$$

Here t and $t+\tau$ are arbitrary moments in time, the asterisk symbol indicates the complex conjugate function, and symbols $\langle\ \rangle$ indicate averaging over time t. Making use of (3.49) and (3.65), the temporal autocorrelation functions of fluctuations in amplitude, $B_E(\tau)$, and phase, $B_\varphi(\tau)$, can be given in terms of B_1 and B_2 as:

$$B_E = \langle \ln(E(t)E_0^{-1})\ln(E(t+\tau)E_0^{-1})\rangle = \frac{1}{2}(\operatorname{Re}B_1 + \operatorname{Re}B_2),$$

$$\qquad (3.66)$$

$$B_\varphi = \langle \varphi(t)\varphi(t+\tau)\rangle = \frac{1}{2}(\operatorname{Re}B_1 - \operatorname{Re}B_2).$$

From (3.64) and (3.65), we have the following expression for B_1 and B_2:

$$\begin{aligned} B_b &= \frac{L^2k^4}{4\pi^2} \int_V\int_V \xi_1^{-1}\xi_2^{-1}(L-\xi_1)^{-1}(L-\xi_2)^{-1} \times \\ &\quad \times B_n(\xi_1-\xi_2, \eta_1-\eta_2, \zeta_1-\zeta_2) \times \\ &\quad \times \exp\left\{\frac{1}{2}ikL\left[\xi_1^{-1}(L-\xi_1)^{-1}(\eta_1^2-\zeta_1^2)+\right.\right. \\ &\quad \left.\left. +(-1)^b\xi_2^{-1}(L-\xi_2)^{-1}(\eta_2^2-\zeta_2^2)\right]\right\}dV_1 dV_2. \end{aligned} \qquad (3.67)$$

Here (ξ_1, η_1, ζ_1) and (ξ_2, η_2, ζ_2) are the coordinates of points P_1 and P_2, dV_1 and dV_2 are the volume elements near these points, and index b equals either 1 or 2. The expression

$$B_n = \langle \Delta n(\xi_1,\eta_1,\zeta_1,t)\,\Delta n(\xi_2,\eta_2,\zeta_2,t+\tau)\rangle$$

is the autocorrelation function of the refractive index fluctuations of radio waves, which depends on the time interval t and the distance between points P_1 and P_2.

Fluctuations in the amplitude and phase of radio waves can be calculated using (3.63) and introducing a concrete autocorrelation function for the refractive index fluctuations. Such an approach is, however, valid only for the Gaussian autocorrelation functions and, consequently, for the Gaussian spatial spectra of plasma inhomogeneities [51, 52], since this approach suggests that all plasma irregularities are uniform in size, whereas actually the irregularity scales are diverse.

For this reason, the spectral approach described in [50] is more appropriate. Let us simplify expression (3.67) by introducing the differential coordinate system (ξ, η, ζ) and the coordinates of the 'center' of plasma inhomogeneities, (x, y, z):

$$\xi = \xi_1 - \xi_2, \eta = \eta_1 - \eta_2, \zeta = \zeta_1 - \zeta_2,$$
$$x = \frac{1}{2}(\xi_1 + \xi_2), y = \frac{1}{2}(\eta_1 + \eta_2), z = \frac{1}{2}(\zeta_1 + \zeta_2). \tag{3.68}$$

Taking into account that

$$\eta_1^2 + \zeta_1^2 = \left(y + \frac{1}{2}\eta\right)^2 + \left(z + \frac{1}{2}\zeta\right)^2,$$
$$\eta_2^2 + \zeta_2^2 = \left(y - \frac{1}{2}\eta\right)^2 + \left(z - \frac{1}{2}\zeta\right)^2 \tag{3.69}$$

expression (3.67) can be rearranged to the form

$$B_b = \frac{L^2 k^4}{4\pi^2} \int_0^L \int_0^L \int_{-\infty}^{\infty} \int\int \left(x^2 - \frac{1}{4}\xi^2\right)^{-1} \left[(L-x)^2 - \frac{1}{4}\xi^2\right]^{-1} \times B_n(\xi,\eta,\zeta,\tau) \times \exp\left\{\frac{1}{2}ikL\right. \tag{3.70}$$

$$\left[\frac{\left(y + \frac{1}{2}\eta\right)^2 + \left(z + \frac{1}{2}\zeta\right)^2}{\left(x - \frac{1}{2}\xi\right)\left(L - x + \frac{1}{2}\xi\right)} + (-1)^b \frac{\left(y - \frac{1}{2}\eta\right)^2 + \left(z - \frac{1}{2}\zeta\right)^2}{\left(x + \frac{1}{2}\xi\right)\left(L - x - \frac{1}{2}\xi\right)}\right]\right\} dx\, dy\, dz\, d\xi\, d\eta\, d\zeta.$$

The calculation of the cumbersome integral (3.70) is facilitated by the fact that the autocorrelation function of the refractive index fluctuations, B_n, depends only on differential coordinates and vanishes when coordinates ξ, η, and ζ are of the order of ten characteristic scales of plasma inhomogeneities; therefore, integration with respect to these coordinates within variable limits can be replaced by integration from $-\infty$ to $+\infty$. Integration with respect to variable x is performed along the section of the ray path *TM* occupied by plasma inhomogeneities, that is, from 0 to L. The limits of integration with respect to co-ordinates y and z can also be taken between $-\infty$ and $+\infty$, since the characteristic scale of plasma irregularities is considerably larger than the wavelength, and the space region where the fluctuations of radio signals are essential lies close to the ray path TM (Fig. 3.10). Moreover, these circumstances have been taken into account when deriving the relations (3.62). Integrating expression (3.70) with respect to coordinates y and z, we obtain

$$B_1 = -\frac{iLk^3}{2\pi}\int_0^L\int\int_{-\infty}^{+\infty}\int F_1^{-1}B_n\exp\left[\frac{ikL}{2}(\eta^2+\zeta^2)F_1^{-1}\right]dx\,d\xi\,d\eta\,d\zeta , \qquad (3.71)$$

$$B_2 = \frac{iLk^3}{2\pi}\int_0^L\int\int_{-\infty}^{+\infty}\int F_2^{-1}B_n\exp\left[\frac{ikL}{2}(\eta^2+\zeta^2)F_2^{-1}\right]dx\,d\xi\,d\eta\,d\zeta , \qquad (3.72)$$

where the following designations are used:

$$F_1 = \xi(L-2x) ,$$
$$F_2 = 2x(L-x)-\frac{\xi^2}{2} . \qquad (3.73)$$

Let us go over from the autocorrelation function B_n to the spatial spectrum of the refractive index fluctuations, $\Phi(\ae,\tau)$, using the known statistical relation

$$B_n(\rho,\tau) = \int\int_{-\infty}^{+\infty}\int\Phi_n(\ae,\tau)\exp[i(\overline{\ae},\overline{\rho})]d\ae_x d\ae_y d\ae_z \qquad (3.74)$$

with the following designations:

$$\ae^2 = \ae_x^2+\ae_y^2+\ae_z^2 ,$$
$$(\overline{\ae},\overline{\rho}) = \ae_x\xi+\ae_y\eta+\ae_z\zeta .$$

Here parameter $\ae$ is a three-dimensional wavenumber, $\ae_x$, $\ae_y$, and $\ae_z$ are the projections of this wavenumber, and ρ is the distance between points P_1 and P_2 in a nonuniform plasma.

By integrating (3.71) and (3.72) with respect to η and ζ and taking into account (3.74), we have

$$B_1 = k^2 \int\limits_0^L \int\limits_{-\infty}^{+\infty}\int\int\int \Phi_n(\text{æ},\tau)\exp\left[\frac{iF_1}{2kL}\left(\text{æ}_y^2 + \text{æ}_z^2\right)+\right.$$

$$\left. + i\text{æ}_x\xi \right]dxd\xi d\text{æ}_x d\text{æ}_y d\text{æ}_z, \tag{3.75}$$

$$B_2 = k^2 \int\limits_0^L \int\limits_{-\infty}^{+\infty}\int\int\int \Phi_n(\text{æ},\tau)\exp\left[\frac{-iF_2}{2kL}\left(\text{æ}_y^2 + \text{æ}_z^2\right)+\right.$$

$$\left. + i\text{æ}_x\xi \right]dxd\xi d\text{æ}_x d\text{æ}_y d\text{æ}_z. \tag{3.76}$$

To integrate ξ and æ_ξ, one has to take into account the known relations for the delta function:

$$\frac{1}{2\pi}\int\limits_{-\infty}^{+\infty}\exp\left[i\left(\text{æ}_x - \frac{(2x-L)\left(\text{æ}_y^2 + \text{æ}_z^2\right)}{2kL}\right)\right] = \delta\left[\text{æ}_x - \frac{(2x-L)\left(\text{æ}_y^2 + \text{æ}_z^2\right)}{2kL}\right] \approx \delta(\text{æ}_x), \tag{3.77}$$

$$\int\limits_{-\infty}^{+\infty}\int\int \Phi_n(\text{æ},\tau)\delta(\text{æ}_x)d\text{æ}_x d\text{æ}_y d\text{æ}_z = \int\limits_{-\infty}^{+\infty}\int \Phi_n(\text{æ}_x = 0,\text{æ}_y,\text{æ}_z,\tau)d\text{æ}_y d\text{æ}_z. \tag{3.78}$$

Finally we have

$$B_1 = 2\pi k^2 \int\limits_0^L \int\limits_{-\infty}^{+\infty}\int \Phi_n(\text{æ}_x = 0,\text{æ}_y,\text{æ}_z,\tau)dxd\text{æ}_y d\text{æ}_z,$$

$$B_2 = -2\pi k^2 \int\limits_0^L \int\limits_{-\infty}^{+\infty}\int \Phi_n(\text{æ}_x = 0,\text{æ}_y,\text{æ}_z,\tau)\times \tag{3.79}$$

$$\times\exp\left[\frac{-ix}{kL}\left(\text{æ}_y^2 + \text{æ}_z^2\right)(L-x)\right]dxd\text{æ}_y d\text{æ}_z.$$

Fluctuations in the amplitude and phase of radio signals are due to the transfer of plasma irregularities across the ray path of the radio signals. If plasma inhomogeneities do not considerably change their structural characteristics while crossing the region that is essential for the propagation of radio waves (this implies that the so-called 'frozen-in' assumption is valid [50]), the relation (3.74) can be rearranged through the following sub-stitution

$$\overline{\rho} \rightarrow \overline{\rho} - \overline{V}\tau,$$

which gives

$$B_n(\rho,\tau) = \int\int\int_{-\infty}^{+\infty} \Phi_n(æ) \exp\left[i\left(æ \cdot \overline{\rho} - æ \cdot \overline{V}\tau\right)\right] dæ_x dæ_y dæ_z .\qquad(3.80)$$

Here V is the velocity of plasma irregularities carried away by the solar wind (plasma irregularities are shown in Fig. 3.10 by dashed lines). If plasma irregularities are 'frozen-in', formulae (3.79) transform into

$$B_1 = 2\pi k^2 \int_0^L\int\int_{-\infty}^{+\infty} \Phi_n(æ_y,æ_z) \cdot \exp\left[-i(æ_y V_y + æ_z V_z)\tau\right] dx\,dæ_y\,dæ_z,\qquad(3.81)$$

$$B_2 = -2\pi k^2 \int_0^L\int\int_{-\infty}^{+\infty} \Phi_n(æ_y,æ_z) \cdot \exp\left\{-i\left[\frac{x(L-x)(æ_y^2 + æ_z^2)}{kL} + \right.\right.$$
$$\left.\left. + (æ_y V_y + æ_z V_z)\tau \right]\right\} dx\,dæ_y\,dæ_z .\qquad(3.82)$$

Assuming that the spectrum $\Phi_n(æ)$ is isotropic, we can introduce new variables ρ_1 and φ, such that $æ_y = \rho_1 \cos\varphi$ and $æ_z = \rho_1 \sin\varphi$. From (3.74) it follows that $\rho_1 = æ$; then (3.81) and (3.82) yield

$$B_1 = 2\pi k^2 \int_0^L\int_{-\infty}^{+\infty} \Phi_n(\rho_1)\rho_1 \int_0^{2\pi} \exp\left[-i\rho_1\tau(V_y \sin\varphi + V_z \cos\varphi)\right] dx\,d\rho_1\,d\varphi,\qquad(3.83)$$

$$B_2 = -2\pi k^2 \int_0^L\int_{-\infty}^{+\infty} \Phi_n(\rho_1)\rho_1 \exp\left[\frac{-ix}{kL}(L-x)\rho_1^2\right] \times$$
$$\times \int_0^{2\pi} \exp\left[-i\rho_1\tau(V_y \sin\varphi + V_z \cos\varphi)\right] dx\,d\rho_1\,d\varphi.\qquad(3.84)$$

In terms of the zeroth-order Bessel function

$$J_0(\beta_1) = \frac{1}{2\pi} \int\limits_0^{2\pi} \exp(-i\beta_1 \cos\varphi) \, d\varphi, \tag{3.85}$$

expressions (3.83) and (3.84) transform into

$$B_1 = (2\pi k)^2 \int\limits_0^{L} \int\limits_{-\infty}^{+\infty} \Phi_n(\rho_1) \rho_1 J_0(\rho_1 V\tau) \, dx d\rho_1, \tag{3.86}$$

$$B_2 = -(2\pi k)^2 \int\limits_0^{L} \int\limits_{-\infty}^{+\infty} \Phi_n(\rho_1) \rho_1 J_0(\rho_1 V\tau) \exp\left[\frac{-ix}{kL}(L-x)\rho_1^2\right] dx d\rho_1. \tag{3.87}$$

Here $V = (V_y^2 + V_z^2)^{1/2}$ is the component of the plasma irregularity velocity that is perpendicular to the ray path TM. Using relations (3.86), (3.87), and (3.66) and taking into account that $\rho_1 = \text{æ}$, we obtain the following formula for the temporal autocorrelation functions of fluctuations in the amplitude and phase of radio signals (B_E and B_φ, respectively):

$$B_{\varphi,E}(\tau) = 2\pi^2 k^2 \int\limits_0^{L} \int\limits_{-\infty}^{+\infty} \Phi_n(\text{æ}) \text{æ} J_0(\text{æ} V\tau) \times \left[1 \pm \cos\left(\frac{x}{kL}(L-x)\text{æ}^2\right)\right] dx d\text{æ}. \tag{3.88}$$

The plus and minus signs correspond to B_φ and B_E, respectively. Formula (3.88) is basic for the analysis of the amplitude, frequency, phase fluctuations, and power spectra of the radio waves propagating through the circumsolar plasma.

3.5 Amplitude fluctuations of radio waves and plasma inhomogeneities

Let us derive theoretical formulas for the variance and temporal spectra of amplitude fluctuations in order to establish their relation to the statistical characteristics of plasma inhomogeneities. Consider first the variance of the log-amplitude fluctuations of radio waves. In this case the time interval in (3.88) is $\tau = 0$. Taking into account that $B(0) = \sigma^2$ and $J_0(0) = 1$, we obtain

$$\sigma_E^2 = (2\pi k)^2 \int\limits_0^{L} \int\limits_0^{\infty} \Phi_n(\text{æ}) \text{æ} \cdot \sin^2\left[\frac{x(L-x)\text{æ}^2}{2kL}\right] dx d\text{æ}. \tag{3.89}$$

Formula (3.89) shows that amplitude fluctuations are influenced primarily by the portion of the spectrum $\Phi_n(\mathbf{æ})$ with $\mathbf{æ} \approx 2\pi (L\lambda)^{-1/2}$. Therefore, plasma inhomogeneities with the scale close to the size of the first Fresnel zone make the greatest contribution to fluctuations in the amplitude of radio signals. This zone acts as an original spatial filter that selectively passes plasms inhomogeneities with a characteristic size $\Lambda \approx (L\lambda)^{-1/2}$. Large-scale plasma clumps affect the amplitude fluctuations of radio signals negligibly, which is indicated by the presence of the term $\sin^2(e\mathbf{æ})$ in formula (3.89). Let us apply formula (3.41) to spectrum $\Phi_n(\mathbf{æ})$ with $\alpha = 11/3$ and integrate (3.89) over $\mathbf{æ}$. Allowing for

$$\int_0^\infty y^{-11/6} \sin^2 y \, dy = 2.72 \cdot 10^{-2} \tag{3.90}$$

yields the variance of the log-amplitude fluctuations of radio signals in the following form

$$\sigma_E^2 = 0.56 k^{7/6} L^{-5/6} \int_0^L c_n^2(r)[x(L-x)]^{5/6} \, dx \ . \tag{3.91}$$

This formula is valid when fluctuations are small, i.e., at $\sigma_E < 0.3$. Let us take the dependence $c_n(r)$ in the form (3.44) and introduce the following designations based on the geometry represented in Fig. 3.10:

$$TC = L_2, \quad CM = L_1,$$

$$OC = p, \quad OD = r,$$

$$L = L_1 + L_2,$$

$$OM = R_1, \quad OT = R_2.$$

In Fig. 3.10, point C is the closest to the Sun, and p is the impact parameter of the ray path TM. From (3.44) it follows that $c_n \approx r^{-2}$ and, therefore,

$$c_n^2(r) = c_n^2(p)\left[1 + p^{-2}(L_2 - x)^2\right]^{-2} . \tag{3.92}$$

Formula (3.92) can be analyzed more easily in terms of an alternative analytical form for $c_n(r)$. As is shown in [53], the expression

$$c_n^2(r) = c_n^2(p)\exp\left[-\left(\frac{L_2 - x}{\beta p}\right)^2\right] \tag{3.93}$$

gives a good approximation for the function $c_n{}^2(r)$ taken in the form (3.92), if we assume that $\beta = 0.85$. From (3.91) and (3.92) it follows:

$$\sigma_E^2 = 0.563 k^{7/6} L^{-5/6} c_n^2(p) \int_0^L [x(L-x)]^{5/6} \exp\left[-\left(\frac{L_2-x}{\beta_1 p}\right)^2\right] dx . \tag{3.94}$$

Consider now how the components of the integrand in formula (3.94) depend on x. It can easily be seen that the power function changes slowly with increasing x, whereas the exponential function changes rapidly; therefore, the exponential function can be integrated within infinite limits. Taking $x(L - x) = L_1 L_2$, we obtain the following approximation for the integral in formula (3.94):

$$\int_0^L [x(L-x)]^{5/6} \exp\left[-\left(\frac{L_2-x}{\beta_1 p}\right)^2\right] dx = (L_1 L_2)^{5/6} p\beta_1 \sqrt{\pi} .$$

Now formula (3.94) for the variance of the log-amplitude fluctuations of radio signals takes the form

$$\sigma_E^2 = 0.563 k^{7/6} c_n^2(p) [\beta_1 \sqrt{\pi} p] \left(\frac{L_1 L_2}{L_1 + L_2}\right)^{5/6} , \tag{3.95}$$

where the bracketed term $L_e \approx 1.5p$ denotes the relative thickness of a layer of plasma irregularities that induce fluctuations of radio waves. The distribution of plasma inhomogeneities is close to being spherically symmetric. Nevertheless, formulas (3.94) and (3.95) possess the structure typical of the radio wave fluctuations in the relatively thin layer of plasma irregularities with thickness $L_e \approx 1.5p$. In Fig. 3.10, this layer is shown by the dashed straight lines a and b.

Let us treat the dependence of the amplitude fluctuations of radio signals on their wavelength theoretically. According to formulae (3.43), $c_n \sim \lambda^2$; then from expression (3.95) it follows that $\sigma_n \sim \lambda^{1.42}$. This dependence is valid provided that the spectrum $\Phi_n(æ)$ is of Kolmogorov-type, that is, if $\alpha = 11/3$. As is shown below, this condition is satisfied at $r/a \geq 25$. Near the Sun, at $r/a \leq 6$, the spectrum of plasma irregularities is characterized by $\alpha = 3$ and, therefore, the dependence of σ_E on the wavelength must be different.

The variance of amplitude fluctuations depends on the spectral index of the spatial spectrum of inhomogeneities, α. At $3 < \alpha < 4$, the following formula for the variance is valid [77]:

$$\sigma_E^2 = A c_n^2(p) L_e \lambda^{\frac{\alpha+2}{2}} \left(\frac{L_1 L_2}{L_1 + L_2}\right)^{\frac{\alpha-2}{2}} . \tag{3.96}$$

Here A is the following numerical coefficient, which weakly depends on α:

$$A = \frac{(2\pi)^{\frac{6-\alpha}{2}} r_e^2 \Gamma\left(\frac{6-\alpha}{2}\right)\sin\left(\frac{\alpha\pi}{4}\right)}{(\alpha - 2)(4 - \alpha)},$$

where r_e is the classical electron radius. In this case, which is more general than the above case, the dependence of σ_E on the wavelength is given by

$$\sigma_E \sim \lambda^{\frac{\alpha+2}{4}}.$$

It should be noted that the expressions for σ_E available in the literature slightly differ because of the different methods used for the approximate calculations of integrals. In radio astronomical observations, $L_2 \gg L_1$; this should be taken into account by substituting L_1 for the parenthesized term in formulas (3.95) and (3.96).

Let us now analyze the temporal spectrum of amplitude fluctuations, $\Phi_E(\omega)$. This spectrum is related to the autocorrelation function $B_E(\tau)$ as is given by the expression

$$\Phi_E(\omega) = 4\int_0^\infty B_E(\tau)\cos(\omega\tau)d\tau, \tag{3.97}$$

where $\omega = 2\pi F$ is the frequency of fluctuations. From (3.88) and (3.97) follows

$$\Phi_E(\omega) = 8\pi^2 k^2 \int_0^L\int_0^\infty \Phi_n(\ae)\ae\left[1-\cos\left(\frac{x(L-x)\ae^2}{kL}\right)\right]\times$$

$$\times \int_0^\infty J_0(\ae V\tau)\cdot\cos(\omega\tau)d\tau dx d\ae. \tag{3.98}$$

Taking into account that the integral with respect to τ in expression (3.98) is equal to $(\ae^2 V^2 - \omega^2)^{-1/2}$ at $\ae V > \omega$ and vanishes at $\ae V < \omega$, we obtain

$$\Phi_E(\omega) = 8\pi^2 k^2 \int_0^L\int_{\omega/V}^\infty \Phi_n(\ae)\cdot\ae(\ae^2 V^2 - \omega^2)^{-1/2}\sin^2\left(\frac{x(L-x)\ae^2}{2kL}\right)dx d\ae. \tag{3.99}$$

This expression can be simplified by replacing the integration over x with a statistically nonuniform layer of thickness L_e with a center at point C (Fig. 3.10). The value of Φ_n within this layer should be taken at point C, i.e., for the impact parameter p. According to [53, 54], the arbitrary thickness of the layer, which is equivalent to an integration with respect to x, equals $1.2p$ to $1.5p$; therefore, one can put $L_e \approx 1.3p$. The transformation of (3.99) allowing for the fact that the integrand should be taken at point C, where $r = p$ and $x = L_e$, gives

$$\Phi_E(\omega) = 8\pi^2 k^2 L_e \int\limits_{\omega/V}^{\infty} \Phi_n(\text{æ})\text{æ}(\text{æ}^2 V^2 - \omega^2)^{-1/2} \sin^2\left(\frac{L_1 L_2 \text{æ}^2}{2k(L_1 + L_2)}\right) d\text{æ} . \qquad (3.100)$$

Using formula (3.41) for spectrum $\Phi_n(\text{æ})$, we can rearrange expression (3.100) as follows

$$\Phi_E(\omega) = 0.264\pi^2 k^2 L_e c_n^2(p) J , \qquad (3.101)$$

where

$$J = \int\limits_{\omega/V}^{\infty} \text{æ}^{1-a}(\text{æ}^2 V^2 - \omega^2)^{-1/2} \sin^2\left(\frac{L_1 L_2 \text{æ}^2}{2k(L_1 + L_2)}\right) d\text{æ} .$$

By changing the variables in accordance with $t = \text{æ}^2 V^2 \omega^{-2} - 1$, integral J can be transformed into

$$J = \frac{1}{2V}\left(\frac{\omega}{V}\right)^{1-a} \int\limits_0^{\infty} (t+1)^{-a/2} t^{-1/2} \sin^2[\beta(t+1)] dt , \qquad (3.102)$$

where

$$\beta = \frac{L_1 L_2 \omega^2}{2kV^2(L_1 + L_2)} . \qquad (3.103)$$

From formulas (3.101) and (3.102) it follows that the frequency dependence of the amplitude fluctuation spectrum, $\Phi_E(\omega)$, is determined by the function $J(\omega)$ alone. Let us express this function through the confluent hypergeometric function $\psi(a, b, z)$ and gamma function Γ using the relation

$$\sin^2 z = \frac{1}{2}[1 - \text{Re} \exp(2iz)]$$

and dividing expression (3.102) into two integrals

$$J = \frac{1}{2V} \left(\frac{\omega}{V} \right)^{1-\alpha} (J_1 + J_2),$$

$$J_1 = \frac{1}{2} \int_0^\infty (t+1)^{-\alpha/2} t^{-1/2} dt = \frac{\Gamma\left(\frac{1}{2}\right) \Gamma\left(\frac{\alpha-1}{2}\right)}{2\Gamma\left(\frac{\alpha}{2}\right)}, \qquad (3.104)$$

$$J_2 = -\frac{1}{2} \operatorname{Re} \left[e^{2i\beta} \int_0^\infty (t+1)^{-\alpha/2} t^{-1/2} e^{2i\beta t} dt \right].$$

From (3.102) and (3.104) it follows:

$$J = \frac{\Gamma\left(\frac{1}{2}\right)}{4V} \left(\frac{\omega}{V} \right)^{1-\alpha} \left\{ \frac{\Gamma\left(\frac{\alpha-1}{2}\right)}{\Gamma\left(\frac{\alpha}{2}\right)} + J_2 \right\}. \qquad (3.105)$$

Thus, the dependence of the spectrum Φ_E on the fluctuation frequency ω is determined by formula (3.105). Let us analyze this formula in more detail, as in [55]. The dimensionless parameter β of the function J_2 depends on frequency ω. According to (3.103), we introduce a frequency ω_0, such that $2\beta = \omega_2^2 \omega_0^{-2}$, where ω_0 is given by

$$\omega_0 = 2\pi F_0 = V_1 \left[\frac{k(L_1 + L_2)}{L_1 L_2} \right]^{1/2}. \qquad (3.106)$$

Here V_1 is the velocity of plasma irregularities. From (3.101) and (3.105), we can derive the following final formula for the temporal spectrum of amplitude fluctuations:

$$\Phi_E(\omega) = 0.264 \pi^2 k^2 L_e c_n^2(p) \frac{\Gamma\left(\frac{1}{2}\right)}{4V_1} \left(\frac{\omega}{V_1} \right)^{1-\alpha} \left\{ \frac{\Gamma\left(\frac{\alpha-1}{2}\right)}{\Gamma\left(\frac{\alpha}{2}\right)} + J_2 \right\}. \qquad (3.107)$$

The dependence of the complex expression (3.107) on ω was analyzed in detail by the authors of [55], who expanded J_2 into a series, found its asymptotic terms, and derived dependences $\Phi_E(\omega)$. At $\omega\omega_0^{-1} > 1$, the second braced term in formula (3.107) is negligible; therefore, the dependence of spectrum Φ_E on ω is determined by the term $\omega^{1-\alpha}$ (spectral region *1* in Fig. 3.11). If this region is approximated by $\Phi_E \approx \omega^{-m}$, then

$$m = \alpha - 1.$$ (3.108)

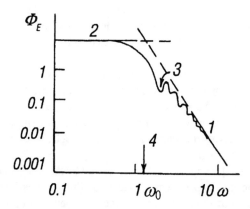

Fig. 3.11. Theoretical spectrum of amplitude fluctuations.

At $\omega\omega_0^{-1} < 1$ (spectral region 2 in Fig. 3.11), the spectrum $\Phi_E(\omega)$ does no longer depend on frequency. The asymptotes *1* and *2* intersect at $\omega = 1.43\,\omega_0$ (this frequency is indicated by arrow *4* in Fig. 3.11). The method described makes it possible to find the frequency ω_0 and, using formula (3.106), the velocity of plasma inhomogeneities, V_1, from experimental spectra $\Phi_E(\omega)$. In the interval $1 < \omega_0^{-1} < 2$, the spectrum $\Phi_E(\omega)$ has indistinct maxima and minima, whose occurrence is due to the term $\sin^2 y$ in the integrand of formula (3.101). The first minimum, laveled by numeral *3* in Fig. 3.11, corresponds to the condition $\sin^2 y = 0$; i.e.,

$$\frac{L_1 L_2 \mathfrak{x}^2}{2k(L_1 + L_2)} = \pi.$$ (3.109)

Note that the three-dimensional wavenumber $\mathfrak{x} = 2\pi\Lambda^{-1}$, and $\Lambda F_1 \approx V_1$; therefore, $\mathfrak{x} \approx \omega_l V_l^{-1}$. Then from (3.109), we have

$$\omega_1 \approx 2\pi V_1 \left(\frac{L_1 + L_2}{\lambda L_1 L_2} \right)^{1/2}.$$ (3.110)

This expression indicates that there is an alternative approach for approximating the velocity of plasma inhomogeneities, V_1, through the frequency of the first minimum, ω_1, in the spectrum $\Phi_E(\omega)$.

It should be noted that we analyzed the spectrum of amplitude fluctuations caused by the motion of plasma inhomogeneities across the path of radio waves by assuming that the spatial spectrum of fluctuations in the electron density is given by the power function (3.41) and that the magnitude of amplitude fluctuations is low. From (3.106) and (3.108) it follows that the spectrum of amplitude fluctuations can provide information about the spectral index of the spatial spectrum of plasma inhomogeneities, α, and their velocity V_1. More detailed analysis of radio wave fluctuations caused by the solar wind plasma can be found in [60–66].

Plasma irregularities are due to various processes, including wave phenomena. In particular, some irregularities in the circumsolar plasma are produced by magnetosonic waves with different wavelengths and vector directions. The authors of [56–59] analyzed the effect of magnetosonic waves occurring in plasma on the temporal spectrum of amplitude fluctuations and velocity V_1 and found that the electron density but not the shape of the spectrum $\Phi_E(\omega)$ can be modulated by these waves (Fig. 3.11). Like the spectral shape, the characteristic frequency ω_0 is also retained. On the other hand, the velocity V_1, which can be found from the experimental spectra of fluctuations by formula (3.106), is no longer the true velocity of the solar wind, V, since it depends on the velocity of magnetosonic waves, V_s. As is shown in [56, 57], the apparent velocity V_1 given by formula (3.106) is related to the velocities of the solar wind, V, and magnetosonic waves, V_s, as

$$V_1^2 = \left| V^2 - V_s^2 \right|. \tag{3.111}$$

if $\alpha = 3$.

From (3.111) it follows that, within the region of heliocentric distances r_s, where $V = V_s$, the apparent velocity V_1 vanishes. Since only part of the plasma inhomogeneities are produced by magnetosonic waves, the apparent velocity V_1 must have a minimum at $r = r_s$ rather than be equal to zero. From (3.111) it follows that, in the regions where $V > V_s$, the velocities V_1 and V are equal.

Let us consider the results of the experimental study of amplitude fluctuations. We should preliminarily note that researchers usually calculate the power scintillation index s_4, rather than the standard deviation of amplitude fluctuations, σ_E. At $\sigma_E < 0.5$, when the above theory is valid, $s_4 = 2\sigma_E$. Earlier, it was common practice in radio astronomy to study fluctuations in the amplitude of the radio waves propagating through the circumsolar plasma by observing small-angle discrete sources of radio waves [42, 45, 67–70]. As the Earth orbits around the Sun, the ray path of radio waves passes different distances from the Sun; this fact was used to derive experimental dependences of the scintillation index s_4 on the impact parameter p for various wavelengths. In these experiments, the noiselike sources of radio waves, which are not point sources, were at an infinite distance, that is, $L_2 \to \infty$. Figure 3.12 illustrates the dependences of s on the p/a obtained by the aforementioned method for three wavelengths, $\lambda = 3.7$, 21, and 154 cm (curves *1*, *2*, and *3*, respectively) [45, 71].

In the case of the signals emitted from a spacecraft moving in the ecliptic plane, their source can be considered as a point source located at a finite distance L_2 from the receiving station. Then the dependences $s(p/a)$ can be deduced. Figure 3.13 illustrates the dependences of the scintillation index s_4 on the ratio p/a obtained in different years using radio signals sent out from space vehicles. Dependence 1 was derived during the mission of *Helios*, which sent 13-cm radio signals [72]. Dependences 2 and 3 were derived from the data provided by *Venera 10* ($\lambda = 32$ cm) [73] and *Pioneer 9* ($\lambda = 71$ cm) [74]. As is evident from Figs. 3.12 and 3.13, the dependences $s(p/a)$ obtained by the two methods noticeably differ. Thus, radio astronomical observations indicate that, as p decreases, the scintillation index s reaches a maximum and then diminishes. At the same time, spacecraft data suggest that the index s tends to unity, that is, to saturation. According to the data presented in [72–74], the fluctuations of 13-, 32-, and 71-cm radio waves become saturated at p/a values equal to 12, 17, and 36, respectively. The maxima on the $s(p)$ dependences obtained by the radio astronomical method are due to the finite angular sizes of radio sources [64]. The results were found to be more correct when they were obtained based on the analysis of the spacecraft signals. At $s < 0.4$, however, both methods gave close results.

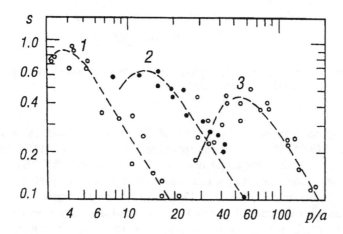

Fig. 3.12. Dependence of the scintillation index on p/a for three different wavelengths as derived from radio astronomical data [45, 71].

Let us compare the experimental dependences $s(p/a)$ for various λ (Fig. 3.13) and respective dependences calculated by formulas (3.95) and (3.96). If the experimental dependences in Fig. 3.13 are approximated by the power function $(a/p)^n$, the exponent n has a value of 1.6. The approximation of radio astronomical data gives similar results [45, 71]. The theoretical dependence $s(p/a)$ is defined by integral (3.91) and, at $r > 20a$, by formula (3.95). The slope of the theoretical curve $s(p/a)$ at $\sigma_E < 0.7$ is also well described by the power function with $n = 1.6$. As is shown in [78], the experimental dependences obtained in 1984 can be approximated by the function

$$s = A\,(a/p)^n.$$

If $\lambda = 32$ cm, then $A = 132$ and $n = 1.6$; but if $\lambda = 5$ cm, then $A = 15.7$ and $n = 1.7$. Amplitude fluctuations become small ($s \leq 0.3$) when $p \geq 8a$ (centimeter radio waves) or $p \geq 22a$ (decimeter radio waves). Centimeter and decimeter radio waves undergo extensive fluctuations at $p < 4a$ and $p < 18a$, respectively.

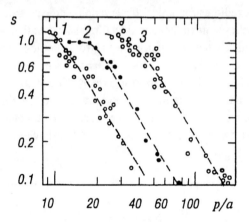

Fig. 3.13. Dependence of the scintillation index on p/a for three different wavelengths as derived by the analysis of radio signals sent from spacecraft [72–74].

Theoretically, the wavelength dependence of the scintillation index is defined by formulas (3.95) and (3.96). At $p \geq 25a$, the spatial spectrum index $\alpha \approx 11/3$; therefore, according to (3.95), σ_E and s are proportional to λ^{n_1} with $n_1 = 17/12$. The experimental value of the scintillation index is $n_1 \approx 1$, as was found from the comparison of these indices for different wavelengths (cf. Figs. 3.12 and 3.13). The difference between the theoretical and experimental values of n_1 may be due to the inaccuracy of experimental data, since they were obtained in years of different solar activities. The authors of [76–78] showed that dependences $s(\lambda)$ differ for different p. It should be noted that the experimental dependences presented in Fig. 3.13 were obtained in the years when the intensities of plasma fluctuations were different. The dependence $s(\lambda)$ can be determined more accurately if amplitude fluctuations are measured simultaneously for different wavelengths. Figure 3.14 shows the experimental dependences of the scintillation index s on the ratio p/a derived in 1984 from the results of the simultaneous measurements of radio signals sent out from *Venera 15* and *Venera 16* in two wavebands, $\lambda_1 = 32$ cm (data are shown by squares) and $\lambda_2 = 5$ cm (data are shown by circles) [78]. The authors of [78] compared theoretical and experimental data for the interval $p = (25 \div 35)a$, where the fluctuations of decimeter radio waves are not yet deep, while those of centimeter radio waves are sufficiently deep to be reliably recorded. For these values of the impact parameter p, the experimentally derived ratio $s(\lambda_1)/s(\lambda_2) = 11.6$. From (3.96) it follows that the theoretical dependence $\sigma(\lambda)$ has the form $\lambda^{(a+2)/4}$; therefore, the ratio $s(\lambda_1)/s(\lambda_2) = 11.6$ implies that the spatial spectrum index of plasma irregularities is $\alpha = 3.36$. A similar value, $\alpha = 3.4$, was obtained by the analysis of the amplitude fluctuation spectra in the indicated interval of p values. Therefore, the experimental data fit the theoretical dependence $\sigma \sim \lambda^{(a+2)/4}$ well.

The scintillation index s can be used as an approximate indicator of the intensity of plasma fluctuations. As is shown in [78, 79], the intensity of plasma fluctuations is increased in the range of distances between 9 and 15 solar radii. It should be noted that the standard deviation of fluctuations in electron density, σ_N, is a more adequate criterion of the intensity of plasma irregularities than s. However, the derivation of σ_N from the experimentally determined values for s is difficult for the following reasons. Theoretical expressions (3.95) and (3.96) involve the plasma irregularity parameter c_n but not σ_N. From formula (3.43), which relates these parameters, it follows that in order to go over from c_n to σ_N, one has to know not only the spatial spectral index of plasma inhomogeneities, α, but also the outer scale Λ_0. It should be noted that expressions (3.43) are valid only for $\alpha = 11/3$. For other values of α, the relationship between c_n and σ_N is described by other formulae. The estimation of σ_N from the s values obtained experimentally showed that $\sigma_N \approx (1 \div 2) \cdot 10^2$ cm^{-3} in the interval $p = (16 \div 25)a$ [78].

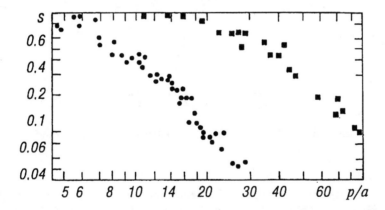

Fig. 3.14. Scintillation index versus p/a for the decimeter and centimeter radio signals recorded simultaneously [78].

Let us consider the results of the experimental investigation of the temporal spectra of amplitude fluctuations, Φ_E. From (3.108) it follows that the temporal spectrum Φ_E can yield information about the spatial spectrum of irregularities in the electron density of plasma, Φ_N. For this reason, the spectrum Φ_E has received considerable study [41, 44, 45, 62, 67–70, 73]. In earlier publications [41, 60, 75], experimental data were compared with the theoretical spectrum Φ_E found for the Gaussian form of the spectrum Φ_N. As was convincingly demonstrated in [62, 66–68], theoretical and experimental data show the best agreement if the power-law spectrum Φ_N has the form of (3.41). The author of [62] emphasized the importance of relation (3.108). In view of this, researchers concentrated their efforts on the determination of the index m for the temporal spectra of amplitude fluctuations. As is found in [67], at a frequency of 430 MHz, $m = 2.9\pm0.3$ in the interval $40a < p < 100a$. The analysis of the spectrum Φ_E for 408-MHz radio signals showed that $m = 2.4\pm0.2$ in the interval $40a < p < 86a$ and decreases when $p < 40a$ [66]. The spectra Φ_E were also analyzed in [54, 63, 80, 81].

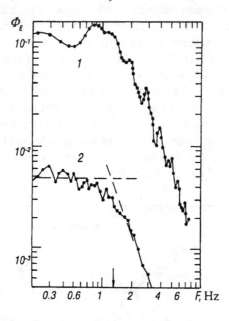

Fig. 3.15. Experimental spectra of amplitude fluctuations derived from the spacecraft radio signals [83].

The temporal spectra of amplitude fluctuations derived from the data provided by *Mars 2*, *Mars 7*, and *Venera 10* were studied in [82–85]. Figure 3.15 shows the two spectra Φ_E obtained in a 32-cm wavelength band: curve *1* represents a typical spectrum of intense fluctuations at $p = 18a$ and curve *2* shows the typical spectrum of small amplitude fluctuations at $p = 44a$. In terms of the theory stated above, the temporal spectra of small amplitude fluctuations contain two informative parameters, namely, the frequency F_0 and index m. The frequency F_0 can be found from the intersection of the asymptotes corresponding to the low- and high-frequency portions of the spectrum, which are shown by dashed straight lines in Fig. 3.15. The index m can be found from the slope of the high-frequency portion of the spectrum. Knowing F_0, one can determine the velocity V_1 of plasma irregularities by formula (3.106); and knowing the spectral index m, one can find the spectral index of plasma irregularities, α, by formula (3.108). In the region where $p > 14a$, the index m has a value of 2.6–2.8 [83–85]. Figure 3.16 presents the results of the determination of m from the data of *Venera 10* (dark circles) and radio astronomical data (open circles). The dashed straight line corresponds to $m = 2.67$. As is evident from this figure, in the interval $20a < p < 80a$, the index $m = 2.5 \pm 0.2$; this value fits the power-law spectrum Φ_N with $\alpha = 11/3$ well.

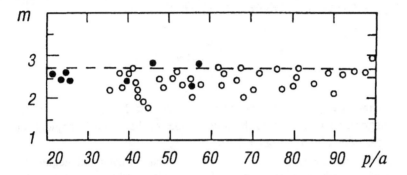

Fig. 3.16. Values of the index m derived from radio astronomical data and from the analysis of spacecraft radio signals [66, 85].

From Fig. 3.15 it follows that the experimental spectrum Φ_E occupies a narrow band of frequencies F from 0.2 to 10 Hz. For $V = 400$ km s^{-1}, these frequencies correspond to the plasma inhomogeneity scales ranging from 40 to 2000 km. Such inhomogeneities make the greatest contribution to amplitude fluctuations; therefore, the index α found from the amplitude fluctuation spectra characterizes only the small-scale portion of the spatial spectrum of fluctuations in the electron density. The spectrum $\Phi_E(F)$ can also yield information about the conventional minimum scale of plasma irregularities, l. The experimental spectra $\Phi_E(F)$ have a high-frequency limitation, so that the highest frequency of fluctuations, F, corresponds to the lowest scale of inhomogeneities, l. The authors of [86–90] determined α and V from the experimental spectra Φ_E and found that, at $p = (9 \div 14)a$, the velocity V_1 undergoes drastic changes, whereas changes in the spectral index α are low. These data are described in more detail in Section 3.8.

It should be noted that the theory presented here is applicable only for small fluctuations; therefore, information on plasma inhomogeneities can be gained only if $s < 0.3$. When distances p are small or wavelengths are large, the fluctuations are intense and saturated. Analysis of such fluctuations performed in [60, 80, 90, 91] showed that they can provide little information about plasma inhomogeneities.

3.6 Phase fluctuations of radio waves

In the same way as the amplitude of radio signals, their phase is susceptible to moving plasma irregularities. Let us derive, after the authors of [46–49, 53, 92–99], some analytical expressions for the variance and temporal spectra of phase fluctuations to elucidate their relationship with the characteristics of plasma inhomogeneities. To obtain the variance of phase fluctuations, $\sigma_\varphi^2 = B(\tau = 0)$, we put $\tau = 0$ in expression (3.88) and introduce the sum $B_\varphi(0) + B_E(0) = \sigma_\varphi^2 + \sigma_E^2$. It is known that fluctuations in the phase of radio signals are always more intense than in their amplitude; therefore, $\sigma_\varphi^2 >> \sigma_E^2$ and, consequently, $B_\varphi(0) + B_E(0) \approx \sigma_\varphi^2$. Then from (3.88) it follows:

$$\sigma_\varphi^2 = (2\pi k)^2 \int_0^L \int_0^{\ae_0} \Phi_n(\ae)\ae d\ae\, dx \,. \tag{3.112}$$

Here integration with respect to coordinate x from 0 to L is performed along the ray path, and integration with respect to the wavenumber of the plasma inhomogeneity spectrum, $\ae$, is performed from 0 to $\ae_0 = 2\pi\Lambda_0^{-1}$.

Small plasma inhomogeneities must affect the phase fluctuation negligibly; this is taken into account by the term $\cos^2(ex^2)$ in (3.88). In view of this we shall calculate σ_φ for the condition that the spectrum has the form (3.40) with $\alpha = 11/3$ and that the exponential coefficient accounting for the influence of small inhomogeneities is equal to unity. Allowing for (3.40), we obtain from (3.112) the following formula:

$$\sigma_\varphi^2 = 0.033(2\pi k)^2 \int_0^L \int_0^{\ae_0} \frac{c_n^2(r)\ae d\ae\, dx}{(\ae^2 + \ae_0^2)^{11/6}} \,, \tag{3.113}$$

where $\Lambda_0 = 2\pi/\ae_0$ is the outer turbulence scale. After integrating (3.113) with respect to $\ae$, we have

$$\sigma_\varphi^2 = 0.0365 k^2 \Lambda_0^{5/3} \int_0^L c_n^2(r)\, dx \,. \tag{3.114}$$

This formula is basic for the analysis of phase fluctuations. It should be noted that the authors of some publications (see, for instance, [50]) used the pre-exponential constant in the expressions analogous to (3.114) equal to 0.7 but not to 0.0365. This is due to the fact that we use the relation $\ae_0 = 2\pi/\Lambda_0$ between the outer turbulence scale Λ_0 and the respective wavenumber $\ae_0$, whereas the author of [50] used the relation $\ae_0 = 1/\Lambda_0$. It should be noted that formula (3.114) can be more easily derived in terms of the ray concept. In this case one can immediately write

$$B_\varphi(\xi,\tau) = k^2 \int_0^L \int_0^\infty B_n(\xi,\tau)d\xi dx \,. \tag{3.115}$$

Going over from the autocorrelation function of fluctuations in the refractive index, B_n, to the spectrum Φ_n and using (3.40), we arrive at formula (3.114).

Let us assume that the dependence $c_n(r)$ has the form (3.44). Integrating with respect to x we obtain the final formula

$$\sigma_\varphi^2 = \frac{0.0365 k^2 \Lambda_0^{5/3} c_0^2 a^4}{2p^3} \times \left[\frac{L_1 p}{p^2 + L_1^2} + \frac{L_2 p}{p^2 + L_2^2} + \arctan\left(\frac{L_1}{p}\right) + \arctan\left(\frac{L_2}{p}\right) \right]. \quad (3.116)$$

Phase fluctuations are maximum in the vicinity of point C (Fig. 3.10); therefore, we can put $L_1 = L_2 = L/2$. Now formula (3.116) is simplified to

$$\sigma_\varphi^2 = 0.0365 k^2 \Lambda_0^{5/3} c_0^2 a^4 p^{-3} \left[\frac{Lp}{2a^2} + \arctan\left(\frac{L}{2p}\right) \right]. \quad (3.117)$$

Furthermore, during the radio sounding of the circumsolar plasma, $L \gg p$. Hence

$$\sigma_\varphi^2 = 0.018 k^2 \Lambda_0^{5/3} c_0^2 a^2 p^{-2} L. \quad (3.118)$$

Parameter L_0 in formulae (3.114) and (3.117) reflects the predominant role of large-scale plasma inhomogeneities in inducing the phase fluctuations of radio signals. These formulae are valid if the signal recording sessions are durable, so that $\Delta T > \Lambda_0 V^{-1}$, where V is the velocity of the plasma transfer across the ray path. Typically, $\Lambda_0 = 3 \cdot 10^6$ km and $V = 400$ km s^{-1}; therefore, ΔT must be > 3 hours. If the phase fluctuations are recorded within shorter time periods, this is equivalent to a "filter" that cuts off low spatial frequencies in the spectrum Φ_n. In this case formulae (3.114) and (3.117) remain valid, but Λ_0 must be substituted by $\Lambda = V\Delta T$.

Formulae (3.116) and (3.117) were found for a particular distribution pattern of plasma irregularities under the assumptions that (3.44) is valid and that $\alpha = 11/3$. Phase fluctuations were analyzed in more detail in [53, 77, 95]. According to [77], at $\alpha > 3$, the variance of phase fluctuations, σ_φ, is related to the variance of the electron density fluctuations, σ_N, as

$$\sigma_\varphi^2 = \frac{\sqrt{\pi}(\alpha - 3)\Gamma\left(\dfrac{a}{2}\right)}{(\alpha - 2)\Gamma\left(\dfrac{\alpha - 1}{2}\right)} r_e^2 \lambda^2 p \Lambda_0 \sigma_N^2. \quad (3.119)$$

Here $r_e = 2.82 \cdot 10^{-13}$ cm is the classical electron radius and σ_N^2 is the variance of the electron density fluctuations at the closest approach of the ray path to the Sun (point C in Fig. 3.10).

Let us now consider the phase fluctuation spectrum Φ_φ. From (3.88) and (3.97), we have

$$\Phi_\varphi = 8\pi^2 k^2 \int_0^L \int_0^\infty \Phi_n(\text{æ})\text{æ}\left[1 + \cos\left(\frac{x(L-x)\text{æ}^2}{kL}\right)\right] \times$$

$$\times \int_0^\infty J_0(\text{æ}V\tau)\cdot\cos(\omega\tau)d\tau dx d\text{æ}.$$

(3.120)

Integrating with respect to τ, similar to that done in (3.98), and introducing the effective path length L_e, which is equivalent to the approximate integration with respect to x, we obtain

$$\Phi_\varphi = 8\pi^2 k^2 L_e \int_{\omega/V}^\infty \Phi_n(\text{æ})\text{æ}(\text{æ}^2 V^2 - \omega^2)^{-1/2} \cos^2\left(\frac{L_1 L_2 \text{æ}^2}{2k(L_1 + L_2)}\right)d\text{æ}.$$

(3.121)

Let us find an approximate expression for the spectrum Φ_φ. To this end, first introduce the sum $\Phi_\varphi + \Phi_E$ and take into account that $\Phi_\varphi \gg \Phi_E$. Then from (3.100) and (3.121) it follows that

$$\Phi_\varphi = \Phi_\varphi + \Phi_E = 8\pi^2 k^2 L_e \int_{\omega/V}^\infty \Phi_n(\text{æ})\text{æ}(\text{æ}^2 V^2 - \omega^2)^{-1/2} d\text{æ}.$$

(3.122)

Here the spectrum of fluctuations in the refractive index, Φ_n, is taken in the region of the closest ray approach to the Sun. Formula (3.122) implies that we go over to the ray concept because the wave concept does not contribute essentially to the analysis of phase fluctuations. Let us introduce the spectrum Φ_n in the form (3.41) into (3.122) and interchange the variables $\text{æ}^2 V^2 - \omega^2 = V^2 y^2$; then we obtain

$$\Phi_\varphi = 0.264\pi^2 k^2 L_e c_n^2(p)V^{-1}J_1,$$

(3.123)

$$J_1 = \int_0^\infty \left(y^2 + \frac{\omega^2}{V^2}\right)^{-\alpha/2} dy = \left(\frac{\omega}{V}\right)^{1-\alpha}\frac{\Gamma\left(\frac{1}{2}\right)\Gamma\left(\frac{\alpha-1}{2}\right)}{2\Gamma\left(\frac{1}{2}\right)}.$$

(3.124)

These formulae yield the following expression for the spectrum of phase fluctuations:

$$\Phi_\varphi = 0.264\pi^2 k^2 L_e c_n^2 (p) V^{\alpha-2} \omega^{1-\alpha} \frac{\Gamma\left(\frac{1}{2}\right)\Gamma\left(\frac{\alpha-1}{2}\right)}{2\Gamma\left(\frac{1}{2}\right)}.$$ (3.125)

This formula was found by the authors of the publication [96] under the assumption that the spectrum $\Phi_n(\mathit{æ})$ has the form of (3.41); therefore, the formula must be valid for frequencies $\omega > 2\pi V \Lambda_0^{-1}$, where Λ_0 is the outer scale of plasma irregularities. As is evident from (3.125), the spectrum $\Phi\varphi(\omega)$ is proportional to the power function ω^{-m} with the parameter m given by

$$m = \alpha - 1.$$ (3.126)

This relation shows how the index α of the spatial spectrum of plasma irregularities can be found from the index m of the power-law spectrum of phase fluctuations. The essential point is that formulae (3.125) and (3.126) are valid even when amplitude fluctuations become saturated.

Let us now discuss the results of the experimental investigation of phase fluctuations. Since the interplanetary and circumsolar plasma irregularities vary greatly in size, phase fluctuations also vary from being slow (with the characteristic time $\Delta T = \Lambda_0 V^{-1}$) to rapid. To avoid the effects of the phase instability of the transmitter and the motion of the spacecraft on the measurements of phase fluctuation, they were performed using the dual-frequency method. The first results of the spectral analysis of the phase fluctuations of radio signals were described in [47, 99]. The index m was found to be equal to 2.6 [47] or 2.68±0.3 [99]. Detailed information about the phase fluctuations of radio signals which was gathered during the missions of the *Viking*, *Helios*, and *Pioneer* spacecraft was published in [92, 98].

Figure 3.17 illustrates the phase fluctuation spectra of 13-cm radio signals from *Viking* [98]. Curves *1*, *2*, and *3* correspond to the impact parameters p/a equal to 2.17, 10.5, and 30.5, respectively. In the interval $20 < p/a < 200$, the variance of phase fluctuations changes with the impact parameter according to the law $\sigma_\varphi \sim p^{-2.45}$. The temporal spectral index of phase fluctuations, m, has a value of 2.65 ± 0.2 [98]. Analysis of the phase fluctuation spectra of 32-cm radio signals from *Venera 10* yielded a value of $m = 2.7 ± 0.2$ [83]. As is evident from Fig. 3.17, the spectral components of phase fluctuations cover the frequencies from 10 to 10^{-3} Hz; this corresponds to the size of plasma inhomogeneities ranging from 40 to $4 \cdot 10^5$ km.

Thus, at $p > 10a$, the experimental values for m fit the spectrum of the refractive index fluctuations with the spectral index $\alpha = 11/3$ well.

Slow variations in the phase of radio signals, $\Delta\varphi(t)$, are caused not only by random plasma inhomogeneities, but also by the variability of the large-scale distribution pattern of electron density. The emissions of plasma and the rearrangement of its sectorial structure lead to slow phase variations, which are produced by changes in the integral electron density J along the ray path.

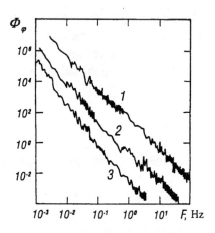

Fig. 3.17. Representative spectra of phase fluctuations [98].

Slow phase variations $\Delta\varphi$ and the related changes in frequency given by the expression

$$\Delta F = \frac{1}{2\pi}\frac{d\Delta\varphi}{dt} = \frac{\gamma}{cf}\frac{dI}{dt},$$

were investigated by the dual-frequency technique. The authors of [46] described the results of the measurements of variations in the integral electron density of interplanetary plasma conducted using the coherent 32-cm and 8-cm radio signals from *Venera 9* and *Venera 10*. These signals were analyzed with the ground-based dispersion interferometer. For the interval $45a < p < 140a$, the mean variation of the integral electron density, dI/dt, was found to change with the parameter p according to the power law $dI/dt \approx p^{-2.38}$. Similar results were obtained using two coherent signals sent from *Mariner 10* [47]. The long-term measurements of variations in the integral electron density of the circumsolar plasma were carried out at $\varphi = 12°$ (hereafter φ is the angle between the lines *TM* and *TO* in Fig. 3.2). Spectral analysis of fluctuations in the integral electron density showed that the spectrum of plasma irregularities whose size ranges from $\Lambda_0 = 3\cdot10^6$ km to $\Lambda = 6\cdot10^3$ km can be well described by a power function with an index close to 11/3. The authors of [48, 49] investigated variations in the integral electron density over time intervals of about one hour long using the 13-cm radio signals sent from *Mariner 6, 7,* and *9.* The electron column density was determined from the difference in the arrival times measured by the phase and group velocities of the radio waves propagating through the plasma [48, 49]. Analysis showed that variations in the electron column density represent a power-law spectrum with a spectral index equal to −2.9 and that the scale of the electron density irregularities changes with distance from the Sun as $\Delta N_e \approx p^{-2.2}$. Variations in the integral electron density were found to reach 100% in magnitude. The maximum scale of plasma inhomogeneities, Λ_0, which is responsible for slow and deep variations in the integral electron density, was found to be $(3-6)\cdot10^6$ km.

3.7 Frequency fluctuations and spectral line broadening

Moving plasma irregularities bring about fluctuations in the phase and amplitude of radio signals, which must lead to random variations in their frequency and induce the broadening of spectral lines. Let us, following the authors of [73, 82, 83, 85, 95–107, 121, 122], analyze fluctuations in radio frequency and, following the authors of [108–118], treat the problem of broadening the spectral lines of the radio signals propagating through the circumsolar plasma. It is known that the characteristic spectral frequencies of phase and amplitude fluctuations, $F = \omega/2\pi$, are considerably lower than the radio wave frequency f and that amplitude fluctuations are much lower than phase fluctuations. Therefore, amplitude fluctuations can be neglected, and we can consider the problem of the propagation of radio waves through irregular plasmas in terms of the influence of a random phase modulation on a monochromatic signal. With such an approach, the frequency fluctuations δf and the shape of the power spectrum $W(f - f_0)$ are determined by the temporal spectrum of phase fluctuations, $\Phi_\varphi(\omega)$. Fluctuations in the frequency and the power spectra of radio signals with a random phase modulation are the subjects of statistical radio physics [119, 120]. Further consideration follows this theory of random processes.

Let us consider frequency fluctuations in terms of the standard deviation σ_f and temporal spectrum $\Phi_f(\omega)$. From statistical radio physics it follows that

$$\Phi_f(\omega) = \omega^2 \Phi_\varphi(\omega),\qquad(3.127)$$

$$\sigma_f^2 = \int_0^\infty \omega^2 \Phi_\varphi(\omega)d\omega .\qquad(3.128)$$

Taking into account (3.127) and (3.125), we can derive the following formula for the frequency fluctuation spectrum:

$$\Phi_f = 0.264\pi^2 k^2 L_e c_n^2 (p) V^{\alpha-2} \omega^{3-\alpha} \frac{\Gamma\left(\frac{1}{2}\right)\Gamma\left(\frac{\alpha-1}{2}\right)}{\Gamma\left(\frac{\alpha}{2}\right)}. \tag{3.129}$$

It is evident from expression (3.129) that the fluctuation spectrum Φ_f depends on frequency as

$$\Phi_f \sim \omega^{-m_f},$$

$$m_f = \alpha - 3. \tag{3.130}$$

These formulae, which were derived in [96], suggest that the parameter α can be found through m_f, which, in turn, can be found from the experimental spectra of frequency fluctuations.

Taking into account (3.128) and (3.129), the variance of the frequency fluctuations can be given by

$$\sigma_f^2 = 0.264\pi^2 k^2 L_e c_n^2 (p) V^{\alpha-2} \frac{\Gamma\left(\frac{1}{2}\right)\Gamma\left(\frac{\alpha-1}{2}\right)}{\Gamma\left(\frac{\alpha}{2}\right)} \int\limits_{\omega_1}^{\omega_2} \omega^{3-\alpha}\, d\omega. \tag{3.131}$$

Let us discuss the integration limits in this formula. The formal expression (3.128) requires that all the spectral components of phase fluctuations, Φ_φ, be taken into account; experimentally, however, only fluctuations in frequencies limited by the minimum, ω_1, and maximum, ω_2, detectable frequencies can reliably be recorded. Actually, the signal frequency f is recorded at the regular time intervals Δt, for instance, every 1 s; therefore, the maximum frequency ω_2 of the spectrum Φ_f has an approximate value of $2\pi/\Delta t$. The lowest detectable frequency ω_1, which is determined by the duration of a signal recording session ΔT, can be expressed through the outer scale of plasma inhomogeneities, Λ_0, if we assume that $\Delta T \approx V^{-1}\Lambda_0$; then, $\omega_1 \approx 2\pi V/\Lambda_0$. Such a representation of ω_1 is valid if the duration of the frequency recording session $\Delta T \geq 3$ hours. For definiteness, assume that $\alpha = 11/3$; then from (3.131) we can derive the following expression for the variance of frequency fluctuations:

$$\sigma_f^2 = 0.264\pi^2 k^2 L_e c_n^2 (p) V^{5/3} \left(\omega_2^{1/3} - \omega_1^{1/3}\right) \times$$

$$\times \Gamma\left(\frac{1}{2}\right)\Gamma\left(\frac{4}{3}\right)\Gamma^{-1}\left(\frac{11}{6}\right). \tag{3.132}$$

If we put $\alpha = 3$, formula (3.132) transforms into

$$\sigma_f^2 = 0.264\pi^2 k^2 L_e c_n^2 (p) V(\omega_2 - \omega_1) \Gamma\left(\frac{1}{2}\right) \Gamma(1) \Gamma^{-1}\left(\frac{3}{2}\right). \tag{3.133}$$

It is evident from these formulae that the experimentally measured values of σ_f^2 can yield the product $c_n^2 \cdot V^{5/3}$ (if $\alpha = 11/3$) or $c_n^2 \cdot V$ (if $\alpha = 3$). Formulae (3.132) and (3.133) show that the spectral index α strongly affects the variance of frequency fluctuations (this problem is considered in more detail in [101]).

Now let us turn to the problem of the broadening of the power spectrum of radio signals. According to [119, 120], the shape of a spectral line, $W(\delta f)$, is entirely determined by the frequency fluctuation spectrum. The spectral line broadening can be expressed through the half-power bandwidth Δf_1 or the equivalent bandwidth Δf_2 given by

$$\Delta f_2 = \frac{1}{W_0} \int_{-\infty}^{\infty} W(\delta f) \, d(\delta f) \ , \tag{3.134}$$

where W_0 is the maximum spectral energy density and $\delta f = f - f_0$. In the case of a Gaussian spectrum, Δf_1 and Δf_2 are related as

$$\Delta f_2 = 1.065 \Delta f_1 \ . \tag{3.135}$$

If the frequency fluctuations follow the normal distribution, then

$$\Delta f_1 = \sqrt{2\pi} \cdot \sigma_f \ , \tag{3.136}$$

where σ_f is the r.m.s. frequency fluctuation. In this case the shape of the spectral line is given by

$$W(\delta f) = \frac{1}{\sigma_f \sqrt{2\pi}} \exp\left[-\frac{(\delta f)^2}{2\sigma_f^2} \right]. \tag{3.137}$$

If the frequency fluctuations follow the delta function, the shape of the spectral line is given by

$$W(\delta f) = \frac{\Delta f_1}{\pi\left[(\Delta f_1)^2 + (\delta f)^2 \right]}. \tag{3.138}$$

It should be noted that the spectra $W(\delta f)$ in (3.137) and (3.138) are normalized to unity:

$$\int_{-\infty}^{\infty} W(\delta f) d(\delta f) = 1. \tag{3.139}$$

Actual power spectra are intermediate between those described by (3.137) and (3.138). The shape of spectral wings depends on the parameter α and can be approximated by the power function [115]

$$W(\delta f) \sim (\delta f)^{1-\alpha}. \tag{3.140}$$

As is obvious from formula (3.136), the experimental determination of the bandwidth Δf yields the same information about plasma inhomogeneities as the variance of frequency fluctuations. The shape of the power spectrum wings enables the determination of the parameter α (according to formula (3.140)) and the inner scale of plasma inhomogeneities [115].

The r.m.s. frequency fluctuations and the power spectrum bandwidth depend on the velocity V of the plasma transfer across the ray path. In the radio sounding of plasmas, V is the sum of the solar wind velocity V_1 and the velocity of the ray path either approaching the Sun or receding from it, $u = dp/dT$; i.e., $V = V_1 \pm u$. Here the plus sign corresponds to the ray approaching the Sun, and the minus sign to the ray receding from it. As follows from (3.131), the ratio of the r.m.s. frequency fluctuations or spectral bandwidths depends only on the velocities V_1 and u, provided that the impact parameter p is the same:

$$\frac{\sigma_{f_1}^2}{\sigma_{f_2}^2} = \left(\frac{V_1 + u}{V_1 - u} \right)^{\alpha-2}. \tag{3.141}$$

As is shown in [113], formula (3.141) is convenient for the determination of the solar wind velocity through the measurements of frequency fluctuations or analysis of the spectral line broadening.

Consider now available experimental data on frequency fluctuations and the broadening of the spectral lines. These phenomena were studied with the 32-cm radio signals sent from the *Mars 2*, *Mars 7*, and *Venera 10* planetary probes [73, 82, 83, 96, 110, 116]. Figure 3.18 illustrates the temporal spectra of the frequency fluctuations of the radio signals received from *Venera 10*. Curves *1*, *2*, and *3* correspond to angles φ equal to 1.2, 5, and 12°, respectively. The dashed straight lines in this figure represent the approximating power-law spectra in the form (3.130) with $m_f = 0.8$. The experimental value of the spectral index was found to be $\alpha = 3.8$; this value is close to the theoretical value $\alpha = 11/3$. From Fig. 3.18 it follows that the spectral index α found from the frequency fluctuation spectrum corresponds to fluctuation frequencies F ranging from $2 \cdot 10^{-3}$ to 0.5 Hz or the plasma irregularity scale Λ ranging from $8 \cdot 10^{2}$ to $5 \cdot 10^{4}$ km.

The experimental values for the r.m.s frequency fluctuations σ_f times $(2\pi)^{1/2}$ are shown in Fig. 3.19 by dark circles. This figure illustrates a steep rise in the magnitude of frequency fluctuations when the ray path approaches the Sun.

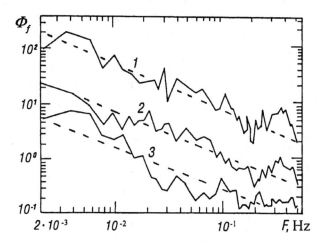

Fig. 3.18. Representative spectra of frequency fluctuations [85].

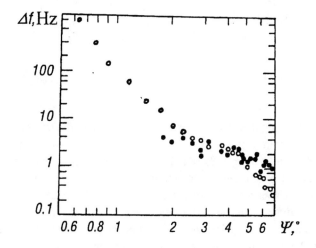

Fig. 3.19. Spectral line bandwidth and the r.m.s. frequency fluctuations of 32-cm radio signals versus the angle Ψ [73, 83].

The power spectrum also widens. Relevant experimental data obtained using the signals sent from three planetary probes are described in [50, 73, 82, 83, 110, 114, 116]. Figure 3.20 shows the appearance of the spectral lines obtained for various values of the angle ψ indicated by numerals along the curves. The figure only shows halves of the spectra, which are actually symmetric. The dependence of the equivalent bandwidth Δf on the angle ψ is shown in Fig. 3.19 by open circles. It can be seen that the spectral line broadening in the circumsolar plasma is great; for instance, as the angle ψ decreases from 6 to 0.6°, the equivalent spectral line bandwidth of 32-cm radio waves increases from 0.6 to 600 Hz. It is also evident from Fig. 3.19 that, in the angular interval $2 < \psi < 6°$, the r.m.s. frequency fluctuations σ_f times $(2\pi)^{1/2}$ (dark circles) and the equivalent bandwidth Δf (open circles) almost coincide, which implies the validity of formula (3.136) in the indicated interval of angles ψ.

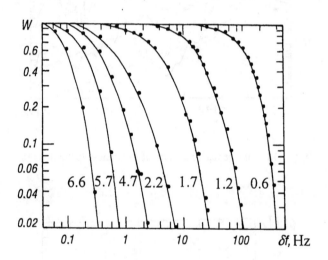

Fig. 3.20. Spectral broadening of 32-cm radio signals at various angles Ψ [110].

Detailed experimental data on frequency fluctuations and the broadening of the power spectrum of 13-cm radio signals were obtained during the missions of *Pioneers 10* and *11*, *Helios 1* and *2*, and *Viking* [98, 102, 111, 112]. Figure 3.21 summarizes data on the dependence of the r.m.s. frequency fluctuations, σ_f, on the impact parameter expressed in the solar radii, p/a. Experiments were performed during the years 1968–1976; the wide scatter of data is indicative of the variability of the plasma characteristics in this period. As is evident from Fig. 3.21, in an interval of impact parameters of $8 < p/a < 100$, the r.m.s. frequency fluctuations depend on the parameter p as $\sigma_f \sim p^{-1.3}$.

Important information on frequency fluctuations and the broadening of spectral lines was obtained from the analysis of the 32-cm and 5-cm radio signals transmitted from

Venera 15 and *Venera 16* [105–107, 118, 121, 122]. It was found that the r.m.s. frequency fluctuations change with the parameter p as $\sigma_f \sim p^{-1.85}$ [105, 106] or, in other years, as $\sigma_f \sim p^{-1.3}$. Such a difference in the variance of frequency fluctuations was likely to be due to the different activity of the Sun in these years. In full agreement with the theory, $\sigma_f \sim \lambda$. At the same p, σ_f was higher for the ray approaching the Sun than for the ray receding from it. Using these data and relation (3.141), the authors of [106] estimated the solar wind velocity. Analysis of the frequency fluctuation spectra showed that the index m_f in formula (3.130) averages 0.5; this corresponds to the spectral index of large plasma irregularities, $\alpha \approx 11/3$. Figure 3.22 presents the experimental dependence of the r.m.s. frequency fluctuations on the impact parameter (the squares correspond to $\lambda = 32$ cm and the circles correspond to $\lambda = 5$ cm) [106]. These dependences were derived from the data obtained in 1984 when solar activity was moderate (the mean number of sunspots was 40 to 80).

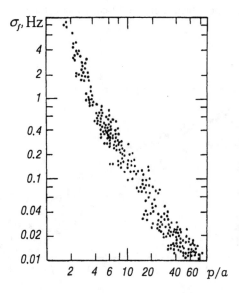

Fig. 3.21. The r.m.s. frequency fluctuations of 13-cm radio signals versus the parameter p/a [98, 102].

The authors of [121–123] analyzed the correlation function and the time delay of frequency fluctuations by the method using a spaced reception of radio signals and found that this method is very efficient to determine the radial dependence of the solar wind velocity.

Apart from random fluctuations, there exist regular variations in the radio frequency. To reveal the regular variations, the authors of [107] analyzed the autocorrelation func-

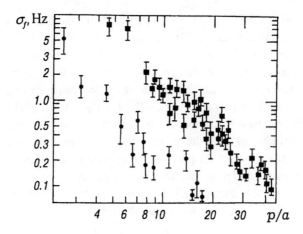

Fig. 3.22. The r.m.s. frequency fluctuations of 32-cm (squares) and 5-cm (circles) radio signals versus *p/a* [106].

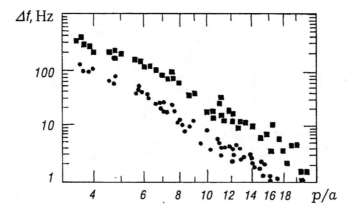

Fig. 3.23. The spectral line bandwidth of 32-cm radio signals transmitted in two modes as a function of *p/a* [118].

tions of frequency fluctuations for p varying from 3.5 a to 41 a and found the quasi-periodic component of frequency fluctuations and the arbitrary period of frequency variations. In the course of particular communication sessions lasting 20–60 min, the authors observed a nonstationarity, which was manifested as changes in the period and intensity of the periodic component of frequency fluctuations. The period T of frequency fluctuations was found to depend on the parameter p; for instance, at $p = 10$, 20, and $30a$, T was found to be 18±7, 60±30, and 120±60 s, respectively (these are the mean values of the period T averaged over several communication sessions). The authors of [107] inferred that the magnetosonic waves generated in plasma may affect the frequency fluctuations of propagating radio waves

The authors of [118] obtained the dependence of the spectral line bandwidth Δf on p/a for the 32-cm radio waves transmitted in two modes (Fig. 3.23). In the first, one-way, mode, radio signals propagated along the spacecraft–Earth path (data are shown in Fig. 3.23 by circles); and in the second mode of a coherent reply, radio waves made the round trip from the Earth-based transmitter to the spacecraft and back to the Earth-based receiver (data are shown by squares).

The spectral line bandwidth of decimeter radio waves can be approximated by the power function

$$\Delta f = B(a/p)^b ,$$

where $B = (6.1 \pm 0.6) \cdot 10^3$ Hz and $b = 3.0 \pm 0.2$.

These values refer to the one-way mode of propagation of radio signals. In the two-way propagation mode, spectral lines broaden greater, so that the ratio $\Delta f_2/\Delta f_1$ is equal to 3.1 ± 0.6 for p ranging from $6a$ to $22a$. For decimeter radio waves, $B = 1.9 \cdot 10^4$ Hz.

The power spectrum broadening of centimeter radio waves becomes noticeable at $p < 7\,a$. Again, the dependence $\Delta f(p)$ can be approximated by a power function with $b = 3.0$ and the value of parameter B that is 50 to 80 times less than in the case of decimeter radio waves.

Frequency fluctuations and the spectral broadening of radio signals can also be used to study the so-called plasma transients [103, 104].

3.8 Investigation of the solar wind by radio sounding methods

Let us briefly characterize the remote sensing methods that are used to study the solar wind plasma (Table 3.2).

Plasma density can be efficiently measured by determining the time delay of radio waves, ΔT. There are two modifications of this method, one based on the recording of the meter-band radio pulses of pulsars and the other based on the recording of the decimeter-band signals transmitted from space vehicles. Plasma density can be determined with a greater accuracy if ΔT is measured by means of the simultaneous recording of spacecraft signals in decimeter and centimeter wavelength bands. With such an approach, the time delay of radio waves is determined as a function of the impact parameter, $\Delta T(p)$, and then, assuming that the distribution of plasma density is spherically symmetric, the dependence of plasma density on heliocentric distance, $N_e(r)$, can be deduced (see Figs. 3.1 and 3.4).

Since plasma is a quasi-neutral matter, the concentrations of electrons and protons are equal.

Table 3.2. Radio sounding methods used to study the solar wind plasma

Parameter	Method	Ref.
Plasma density	Estimation of time delay in plasma	[3–14, 125, 143]
Solar wind velocity	Correlation analysis of amplitude fluctuations by the spaced reception technique; estimation of radio wave time delay	[43, 56, 73, 84–87, 89, 96, 113, 121–124, 130–140, 151]
	Correlation analysis of frequency fluctuations by the spaced reception technique; estimation of radio wave time delay	
	Analysis of the temporal spectra of amplitude fluctuations recorded at single ground stations; determination of characteristic frequency	
Magnetic field	Measurement of the angle of rotation of the polarization plane of radio waves	[25–39]
Random plasma irregularities; spectra and size scales	Analysis of the temporal spectra of fluctuations in amplitude, frequency, and phase of radio waves; evaluation of spectral indices	[43, 44, 46–49, 65, 73, 79, 81, 84–86, 92, 96, 98, 102, 115, 127, 128, 137–140, 142]
Magnetosonic and Alfvénic waves	Analysis of periodic frequency variations, the temporal spectra of amplitude fluctuations, and Faraday effect by the spaced reception technique	[36–38, 56, 58, 59, 107, 152–156]
Solar activity, disturbances, transients, shock waves	Changes in amplitude and frequency fluctuations; spectral line broadening of radio waves	[15, 16, 19, 27, 88, 103, 104, 125–127, 141]

A very important characteristic of the solar wind is the dependence of its velocity on heliocentric distance, $V(r)$; this characteristic shows where the solar wind plasma has the highest acceleration (it is the region where the Sun imparts its energy to the plasma) and where it moves at a constant speed. It should be noted that the effects governed by the motion of plasma irregularities do not always correspond to the actual mass velocity of the solar wind. Several methods were developed to determine the dependence $V(r)$. The first method uses the correlation analysis of the amplitude fluctuations of radio waves received at two ground-based stations located 100–200 km apart. The correlation analysis yields the delay time τ_1; then the velocity of the solar wind can be determined by the formula $V = b/\tau_1$, where b is the distance between two rays in the circumsolar plasma aimed at the ground stations (the region of point C in Fig. 3.2). With this method, the authors of

[125–127] estimated the velocities of the solar wind and shock waves at radial distances $r = (100–200)a$ and found that the velocity V in this region does not depend on heliocentric distance, averaging 370–400 km s^{-1}, whereas the velocity of shock waves in the solar wind plasma diminishes with increasing radial distance. The second method for determining the velocity V is based on the treatment of the temporal spectra of the amplitude fluctuations of radio signals. The data derived are then used to determine the characteristic frequency F_0 and, using formula (3.106), the velocity of plasma irregularities, V_1 [89]. The third method is based on the analysis of frequency fluctuations in the radio signals received at two ground-based stations located a few hundred km apart [121, 123, 151]. The velocity in this case is given by the formula $V = b/\tau_2$, where τ_2 is the delay time of frequency fluctuations. The fourth method is based on the analysis of the autocorrelation function of the frequency fluctuations of radio waves when radio communication occurs in the coherent reply mode [122]. With this method, radio signals are received at a single ground-based station, but, due to the round trip of radio signals from an Earth-based transmitter to a spacecraft and back to a ground receiver, this approach, which utilizes the orbital motion of the Earth, is equivalent to the spaced reception technique. The accuracy of the determination of $V(r)$ can be improved by using several methods at a time. Such an approach was first applied to the analysis of the radio signals sent out from *Venera 15* and *Venera 16* [56, 89, 121, 122].

Figure 3.24 shows the dependence of the velocity of plasma irregularities, V_1, on heliocentric distance derived by the second method, i.e., through the analysis of the temporal spectra of amplitude fluctuations. Figure 3.25 presents the dependence of the mass velocity of the solar wind on heliocentric distance derived by analyzing frequency fluctuations in terms of the third and fourth methods. As is evident from the comparison of Figs. 3.24 and 3.25, all three methods give close values for V at $r > 25a$. For radial distances $r = (25–35)\,a$, the velocity V slightly rises with heliocentric distance to become independent of it at $r > 35\,a$. The third and fourth methods give similar results at $r < 20a$. The solar wind velocity rises about tenfold as r increases from $5a$ to $20a$. At $r < 20a$, the second method gives a different dependence of V_1 on r (Fig. 3.24): at $r \approx 12a$, there is a minimum in the apparent velocity of plasma irregularities, V_1, and, at $r = (6–10)a$, the velocity V_1 is about twice as great as follows from Fig. 3.25. Analysis showed that the velocity V_1 determined from the spectra of amplitude fluctuations depends on both the mass velocity of the solar wind, V, and the velocity of magnetosonic waves, V_s [56–59]. As follows from these publications, the apparent velocity of plasma irregularities, V_1, must have a minimum at $V = V_s$ (this minimum is really present in Fig. 3.24). For $V_s > V$ (this takes place at $r < 6a$), the velocity V_1 is primarily determined by magnetosonic waves, i.e., $V_1 \approx V_s$. For $r < 20\,a$, the actual mass velocity of the solar wind can be reliably estimated from the frequency data. As for the amplitude data, they may provide information about the motion of plasma irregularities caused both by electron density waves and by the mass velocity of the solar wind.

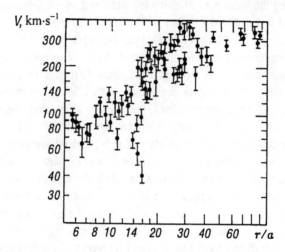

Fig. 3.24. Dependence of the velocity of plasma irregularities on heliocentric distance retrieved from the amplitude data [56, 89].

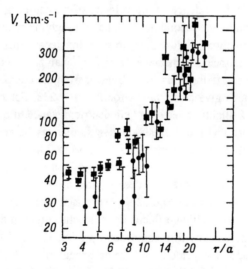

Fig. 3.25. Dependence of the solar wind velocity on heliocentric distance retrieved from the frequency data [121, 122].

The magnetic field pattern in the circumsolar plasma can be found from the rotation of the polarization plane of radio waves, i.e., from the Faraday effect. The interpretation of relevant data requires additional data on the sectorial structure of the circumsolar plasma, such as those deduced from the Sun's images taken with a coronograph.

Random plasma irregularities are characterized by the spatial spectral index α, minimum scale l, r.m.s. electron density σ_N, and the outer scale Λ_0. The spatial spectral index of plasma inhomogeneities, α, can be determined from the spectra of the amplitude, frequency, or phase fluctuations in radio signals using formulae (3.108), (3.130), and (3.126), respectively. These formulae were found to allow the index α to be reliably determined.

Figure 3.26 gives the spectral indices of large-scale plasma irregularities estimated from the spectra of frequency fluctuations [106, 139, 140]. The index α for large-scale irregularities is independent of the radial distance r, averaging 3.9. Conversely, the spatial spectral index α of small-scale plasma irregularities does depend on the heliocentric distance r (Fig. 3.27). For $r = (4-6)a$, $\alpha = 3$, and for $r > 25a$, $\alpha = 3.7$. For $r > 25a$, the spatial spectral indices of small-scale and large-scale irregularities become equal, approaching the spectral index of Kolmogorov, $\alpha = 11/3$. The minimum scale of plasma irregularities, l, determined from the spectra of amplitude fluctuations [137, 140, 142] was found to be $l = 4$ km at $r > 6a$. This scale increases with the heliocentric distance, so that $l = 24$ km at $r = 20\ a$. It was also found that l is of the order of the gyroradius of protons. The r.m.s. fluctuations of electron density, σ_N, are more difficult to determine. The ratio of the r.m.s. electron density fluctuations to the mean electron density, $\sigma_N/N = 0.1-0.2$, slightly depending on the heliocentric distance.

The outer turbulence scale of plasmas, Λ_0, can be reliably determined from the phase fluctuations of radio signals provided that signal recording sessions last for as long as 5-6 hours. As is shown in [157], the outer turbulence scale Λ_0 of the solar wind plasma for the heliocentric distances $r = 8a$ and $r = 70a$ is equal to $0.7 \cdot 10^6$ and $7 \cdot 10^6$ km, respectively.

Of great importance for studying the solar wind are plasma waves, since wave phenomena play a significant role in the energetics of plasma flows. Radio observations have provided evidence for the occurrence of intense Alfvénic [36-38, 152] and magnetosonic [56, 58, 59, 107] waves in the solar wind plasma. Magnetosonic waves, which are generated by intense Alfvénic waves in the region of the maximum acceleration of the solar wind plasma, are damped out in circumsolar space.

The solar wind plasma is subject to heavy alterations. Thus, plasma emissions, transients, and shock waves bring about fluctuations in the amplitude and frequency of radio waves and lead to the spectral broadening of radio signals. On the other hand, these phenomena allow the characteristics of plasma emissions, transients, and shock waves to be determined.

The solar wind velocity and plasma density retrieved through the analysis of radio signals enable the dependence of the solar wind power on the heliocentric distance to be deduced [144]. Estimations showed that the power of the solar wind is two orders of magnitude higher at $r = 20a$ than at $r = 6a$.

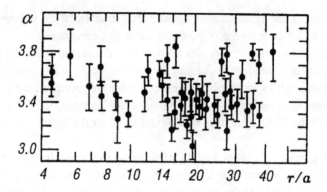

Fig. 3.26. Spatial spectral index of large-scale plasma irregularities versus heliocentric distance [106, 140].

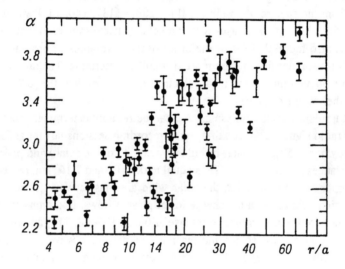

Fig. 3.27. Spatial spectral index of small-scale plasma irregularities versus heliocentric distance [139, 140].

Extensive radiophysical investigations made it possible to gain insight into the primary mechanisms of the formation of the solar wind. The maximum increase in the velocity and power of the solar wind, as well as the greatest changes in its turbulence, were found to occur at distances as great as $(8–20)a$ from the Sun. This phenomenon was found to be due to the long-range transfer of solar energy by intense Alfvénic waves. These waves give rise to magnetosonic waves, which undergo dissipation and thus can induce the transfer of energy to the solar wind plasma. It was found that the distance r_s at which the velocity of the solar wind becomes equal to the speed of sound is about 10 solar radii. In the region $r < r_s$, the velocity of the plasma flow is subsonic; whereas it is supersonic at $r > r_s$. At $r > 30a$, the motion of the solar wind plasma becomes inertial (with V = const), since here plasma is no longer acted upon by the accelerating force of the Sun.

The first radio observations were aimed at deriving evidence for the existence of plasma far away from the Sun. Later, efficient methods for studying the solar wind were developed, which provided substantial information about this immense phenomenon [12, 40–42, 44, 45, 145–150].

References

1. Saito, K., Poland, A., and Munro, R. (1977) *Solar Phys.*, **55**, 1: 121.
2. Dollfus, A. and Mouradian, Z. (1981) *Solar Phys.*, **70**, 1: 3.
3. Hollweg, J.V. (1968) *Nature*, **220**, 23: 771.
4. Goldstein, S.I. and Meisel, D.D. (1969) *Nature*, **224**, 5923: 349.
5. Counselman, C.C. and Rankin, I.M. (1972) *Astrophys. J.*, **175**, 3: 843.
6. Counselman, C. and Rankin, J. (1973) *Astrophys. J.*, **185**, 1: 357.
7. Weisberg, J.M., Rankin, J., Payne, R.R., and Counselman, C.C. (1976) *Astrophys. J.*, **209**, 1: 252.
8. Muhleman, D.O., Esposito, P.B., and Anderson, J.D. (1977) *Astrophys. J.*, **211**, 3: 943.
9. Edenhofer, P., Exposito, P., Hansen, R., et. al. (1980) *J. Geoph. Res.*, **85**: 3414.
10. Muhleman, D.O., and Anderson, J.D. (1981) *Astrophys. J.*, **247**, 3: 1093.
11. Anderson, L.D., Krisher, T.P., Borutzki, S.E., *et al.* (1987) *Astrophys. J.*, **323**, 2: 141.
12. Bird, M. and Edenhofer, P. (1990) *Physics of the Inner Heliosphere*, Schwenn R. (Ed.), Berlin: Springer-Verlag, p. 13.
13. Bird, M.K., Volland, H., Patzold, M., *et al.* (1994) *Astrophys. J.*, **426**, 1: 373.
14. Krisher, T.R., Anderson, J.D., Morabito, D.D., *et al.* (1991) *Astrophys. J.*, **375**, 1: 57.
15. Bougeret, L.G., King, J.H., and Schwenn, R. (1984) *Solar Phys.*, **90**, 2: 401.
16. Efimov, A.I. (1969) *Radiotekhnika i Elektronika*, **14**, 12: 2091 (in Russian).
17. Zheleznyakov, V.V. (1964) *Radio Emission from Sun and Planets*, Moscow: Nauka (in Russian).
18. Kucheryavenkov, A.I., Pavelyev, A.G., Rubtsov, S.N., and Yakovlev, O.I. (1985) *Radiofizika* (Radiophysics and Quantum Electronics), **28**, 7: 807 (in Russian)*.
19. Pavelyev, A.G., Yakovlev, O.I., and Kucheryavenkov, A.I. (1988) *Radiofizika* (Radiophysics and Quantum Electronics), **31**, 2: 127 (in Russian)*.
20. Dolbejev, G.G., Efimov, A.I., Tichonov, V.F., and Yakovlev, O.I. (1986) *Radiotekhnika i Elektronika* (Soviet J. Communication Technology and Electronics), **31**, 2: 354 (in Russian)*.
21. Efimov, A.I., Yakovlev, O.I., Vyshlov, A.S., *et al.* (1990) *Radiofizika* (Radiophysics and Quantum Electronics), **33**, 9: 729 (in Russian)*.
22. Chastel, A.A. and Heyverts, J. (1976) *Astron. and Astrophys.*, **51**, 2: 171.
23. Hollweg, J.V. (1970) *J. Geoph. Res.*, **75**, 19: 3715.
24. Kruger, A. (1979) *Introduction to Solar Radio Astronomy and Radio Physics*, D. Reidel Publishing Co.
25. Levy, G.S., Sato, T., Seidel, B.L., and Steilzrid, C.T. (1969) *Science*, **166**, 3905: 596.
26. Stelzrid, C.T., Levy, G.S., Sato, T., *et al.* (1970) *Solar Phys.*, **14**, 2: 440.

27. Bird, M.K. (1976) *Space Res.,* **16:** 711.
28. Sofue, J.,Kawabata, K.,Takahashi, F., and Kawajiri, N. (1976) *Solar Phys.,* **50,** 2: 465.
29. Parijsky, Yu.N., Soboleva, N.A., and Timofeeva, G.M. (1980) *Izv. Spek. Astrof. Observ., Astrofiz. Issled.,* **12:** 56, (in Russian).
30. Soboleva, N.A. and Timofeeva, G.M. (1983) *Pisma v Astronomicheskii Zhurnal,* **9,** 7: 409 (in Russian).
31. Volland, H., Bird, M.K., Levy, G.S., *et al.* (1977) *J. Geophysics,* **42,** 6: 659.
32. Bird, M., and Volland, H. (1982) *Das Faraday-rotations-experiment der Helios-Mission.* BMFT-FB-W82-014. (Radioastronomische Institut Universitat Bonn).
33. Bird, M.K., Volland, H., Howard, R.A., *et al.* (1985) *Solar Phys.,* **98,** 2: 341.
34. Patzold, M., Bird, M.K., Volland, H., *et al.* (1987) *Solar Phys.,* **109,** 1: 91.
35. Avdyushin, S.N., Krylov, V.A., and Poperechenko, B.A. (1982) *Doklady Akademii Nauk SSSR,* **266,** 1: 49 (in Russian).
36. Hollweg, J.V., Bird, M.K., Volland, H., *et al.* (1982) *J. Geoph. Res.,* **87,** A1: 1.
37. Bird, M.K., Volland, H., Efimov, A.I., *et al.* (1992) *Solar Wind,* Marsch, E. and Schwenn, R. (Eds.), Pergamon Press, p. 147.
38. Efimov, A.I., Chashei, I.B., and Bird, M.K. (1993) *Pisma v Astronomicheskii Zhurnal,* **19,** 2: 143 (in Russian).
39. Bekker, V.I., and Tokarev, M.V. (1990) *Geomagnetizm i Aeronomiya,* **30,** 1: 163 (in Russian).
40. Lotova, N.A. (1968) *Uspekhi Fizicheskikh Nauk,* **95,** 2: 292 (in Russian).
41. Hewish, A. and Symonds, M. (1969) *Planet Space Sci.,* **17,** 3: 603 (in Russian).
42. Buckley, R. (1971) *Planet Space Sci.,* **19,** 5: 421.
43. Vitkevich, V.V., and Vlasov, V.I. (1972) *Astronomicheskii Zhurnal* (Soviet Astronomy J.), **46,** 3: 595 (in Russian).
44. Lotova, N.A. (1975) *Uspekhi Fizicheskikh Nauk,* **115,** 4: 603 (in Russian).
45. Coles, W. (1978) *Space Sci. Rev.,* **21,** 4: 411.
46. Vasil'ev M.B., Vyshlov, A.S., Kolosov, M.A., *et al.* (1981) *Kosmicheskie Issledovaniya* (Cosmic Res.), **19,** 1: 83 (in Russian)*.
47. Woo, R., Yang, F., Yip, K., and Kendall, W. (1976) *Astrophys. J.,* **210,** 2: 568.
48. Callahan, P.S. (1974) *Astrophys. J.,* **187,** 1: 185.
49. Callahan, P.S. (1975) *Astrophys. J.,* **199,** 1: 227.
50. Ishimaru, A. (1978) *Wave Propagation and Scattering in Random Media,* Academic Press.
51. Rytov, S.M. (1971) *Radiofizika,* **14,** 5: 645 (in Russian).
52. Efimov, A.I. and Yakovlev, O.I. (1971) *Radiotekhnika i Elektronika,* **16,** 9: 1554 (in Russian).
53. Woo, R. (1975) *Astrophys. J.,* **201,** 1: 238.
54. Cronyn, W.M. (1972) *Astrophys. J.,* **171,** 1: 101.
55. Mitchell, D.G., and Roelof, E.C. (1976) *J. Geoph. Res.,* **81,** 28: 5071.
56. Efimov, A.I., Chashei, I.V., Shishov, V.I., and Yakovlev, O.I. (1990) *Kosmicheskie Issledovaniya* (Cosmic Res.), **28,** 4: 498 (in Russian)*.
57. Chashei, I.V. (1993) *Geomagnetizm i Aeronomiya,* **33,** 4: 145 (in Russian).
58. Efimov, A.I., Rudach, W.K., and Chashei, I.V. (1993) *Astronomicheskii Zhurnal* (Astronomical Reports), **37,** 5: 542 (in Russian)*.
59. Efimov, A.I., Chashej, I.V., Shishov, V.I., and Yakovlev, O.I. (1994) *Adv. Space Res.,* **14,** 4: 493.
60. Cohen, M.H., Gunderman, E.I., Hardebeck, H.E., and Sharp, L.E. (1967) *Astrophys. J.,* **147,** 2: 449.
61. Jokipii, J.R. (1970) *Astrophys. J.,* **161,** 9: 1147.

62. Gronyn, W.M. (1970) *Astrophys. J.*, **161**, 2: 755.
63. Readhead, A.C. (1971) *Mont. Notic. Roy. Astron. Soc.*,**155**, 2: 185.
64. Shishov, V.I. (1972) *Radiofizika*, **15**, 9: 1277 (in Russian).
65. Rickett, B.I. (1973) *J. Geoph. Res.*, **78**, 10: 1543.
66. Milne, R.G. (1976) *Austral. J. Phys.*, **29**, 3: 201.
67. Lovelace, R.V., Salpeter, E.E., and Sharp, L.E. (1970) *Astrophys. J.*, **159**, 3: 1047.
68. Jokipi, I.R. and Holweg, I.V. (1970) *Astrophys. J.*, **160**, 2: 745.
69. Young, A.T. (1971) *Astrophys. J.*, **168**, 3: 543.
70. Matheson, D.H. and Little, L.T. (1971) *Planet. Space Sci.*, **19**, 12: 1615.
71. Cohen, M.H. and Gunderman, E.I. (1969) *Astrophys. J.*, **155**, 2: 645.
72. Woo, R. and Armstrong, G. (1980) *Solar Wind Measurements Close to the Sun*, Preprint, Smolenize, Czechoslovakia.
73. Kolosov, M.A., Yakovlev, O.I., Efimov A.I., *et al.* (1982) *Radio Sci.*, **17**, 3: 664.
74. Chang, H. (1976) *Analysis of Dual-Frequency Observations of Interplanetary Scintillations Taken by the Pioneer 9 Spacecraft*, Techn. Rep. 13552-1, Stanford Univ. Press.
75. Rao P., Bhandari, S., and Ananthakrishnan, S. (1974) *Austral. J. Phys.*, **27**, 2: 105.
76. Readhead, A.C., Kemp, M.C., and Hewish, A. (1978) *Mont. Notic. Roy. Astron. Soc.*, **185**, 1: 207.
77. Armand, N.A., Efimov, A.I., and Yakovlev, O.I. (1987) The Radio Physical Investigations of the Turbulent Structure of Interplanetary and Solar Wind Plasma, in: *Problems of Current Radioengineering and Electronics*, Moscow: Nauka, p. 86 (in Russian).
78. Yakovlev, O.I., Efimov, A.I., Molotov, E.P., *et al.* (1988) *Radiofizika* (Radiophysics and Quantum Electronics), **31**, 1: 3 (in Russian)*.
79. Lotova, N.A., Blums, D.F., and Vladimirskii, K.V. (1985) *Astron. and Astrophys.*, **150**, 2: 266.
80. Armstrong, J.M., Coles, W.A., and Rickett, R.I. (1972) *J. Geoph. Res.*, **77**, 16: 2739.
81. Coles, W.A. and Harmon, I.K. (1978) *J. Geoph. Res.*, **A83**, 4: 1413.
82. Yakovlev, O.I., Molotov, E.P., Efimov, A.I., *et al.* (1977) *Radiotekhnika i Elektronika* (Radio Engineering and Electronic Physics), **22**, 2: 29 (in Russian)*.
83. Kolosov, M.A., Yakovlev, O.I., Efimov, A.I., *et al.* (1978) *Radiotekhnika i Elektronika* (Radio Engineering and Electronic Physics), **23**, 9: 1829 (in Russian)*.
84. Yakovlev, O.I., Efimov, A.I., Razmanov, V.M., and Shtyrkov, V.K. (1980) *Acta Astronautica*, **7**: 235.
85. Kolosov, M.A., Yakovlev, O.I., Rogal'sky, V.I., *et al.* (1978) *Doklady Akademii Nauk*, **241**, 3: 555 (in Russian).
86. Scott, S.L., Coles, W.A., and Bourgois, G. (1983) *Astron. and Astrophys.*, **123**, 1: 207.
87. Scott, S.L., Rickett, B.I., and Armstrong, I.W. (1983) *Astron. and Astrophys.*, **123**, 1: 191.
88. Coles, W.A. and Filice, I.P. (1985) *J. Geophys. Res.*, **A90**, 6: 5082.
89. Efimov, A.I., Yakovlev, O.I., Rubzov, S.N., *et al.* (1987) *Radiotekhnika i Elektronika* (Radio Engineering and Electronic Physics), **32**, 10: 2025 (in Russian)*.
90. Rubzov, S.N., Yakovlev, O.I., and Efimov, A.I. (1990) *Radiofizika* (Radiophysics and Quantum Electronics), **33**, 2: 135 (in Russian)*.
91. Shishov, V.I. (1971) *Radiofizika*, **14**, 1: 85 (in Russian).
92. Tyler, G.L., Vesecky, I.F., Plume, M.A., *et al.* (1981) *Astrophys.J.*, **249**, 1: 318.
93. Rees, W.C. and Duffett-Smith, P.I. (1985) *Mont. Notic. Roy. Astron. Soc.*, **212**, 3: 463.
94. Efimov, A.I., Korsak, O.M., Yakovlev, O.I., and Tserenin, I.D. (1988) *Radiotekhnika i Elektronika* (Soviet J. Communication Technology and Electronics), **33**, 12: 2640 (in Russian)*.

95.), **32**Efimov, A.I. and Yakovlev, O.I. (1986) Features of the Radio Wave Propagation in the Near-Sun and Interplanetary Space, in: *Electromagnetic Waves in the Atmosphere and Space*, Moscow: Nauka, p. 171 (in Russian).

96. Efimov, A.I., Yakovlev, O.I., Razmanov, V.M., *et al.* (1978) *Kosmicheskie Issledovaniya* (Cosmic Res.), **16**, 3: 337 (in Russian)*.

97. Croft, T.A. (1971) *Radio Sci.*, **6**, 1: 55.

98. Woo, R. and Armstrong, I.W. (1979) *J. Geoph. Res.*, **84**, A12: 7288.

99. Kolosov, M.A., Savich, N.A., Vasyl'ev, M.B., *et al.* (1977) *Uspekhi Fizicheskikh Nauk*, **123**, 4: 700 (in Russian).

100. Shishov, V.I. (1976) *Radiofizika* (Radiophysics and Quantum Electronics), **19**, 10: 1507 (in Russian)*.

101. Armand, N.A. (1982) *Radiotekhnika i Elektronika* (Radio Engineering and Electronic Physics), **27**, 9: 1683 (in Russian)*.

102. Woo, R. (1978) *Astrophys. J.*, **219**, 2: 727.

103. Woo, R., Armstrong, I.W., Sheeley, N.R., *et al.* (1985) *J. Geoph. Res.*, **90**, A1: 154.

104. Woo, R. (1988) *J. Geoph. Res.*, **93**, A5: 3919.

105. Savich, N.A., Azarch, S.L., Vyshlov, A.S., *et al.* (1987) *Kosmicheskie Issledovaniya* (Cosmic Res.), **25**, 2: 243 (in Russian)*.

106. Armand, N.A., Efimov, A.I., Yakovlev. O.I., *et al.* (1988) *Radiotekhnika i Elektronika* (Soviet J. Communications Technology and Electronics), **33**, 12: 54 (in Russian)*.

107. Yakubov, V.P., Yakovlev. O.I., Efimov, A.I., *et al.* (1989) *Kosmicheskie Issledovaniya* (Cosmic Res.), **27**, 5: 772 (in Russian)*.

108. Kolosov, M.A., Yakovlev, O.I., and Efimov, A.I. (1965) On the Radio Waves Propagation in the Interplanetary Medium and Near the Sun, in: *Space Research*, Moscow: Nauka, p. 227 (in Russian).

109. Hollwed, I.V. and Harington, I.V. (1968) *J. Geoph. Res.*, **73**, 23: 7221.

110. Yakovlev, O.I., Trusov, B.P., Vinogradov, V.A., *et al.* (1974) *Kosmicheskie Issledovaniya* (Cosmic Res.), **12**, 4: 600 (in Russian)*.

111. Woo, R., Yang, F., and Ishimaru, A. (1976) *Astrophys. J.*, **210**, 2: 593.

112. Tyler, G.L., Brenkle, J.P., Komarek, T.A., and Zygielbaum, A.T. (1977) *J. Geoph. Res.*, **82**, 28: 4335.

113. Efimov, A.I., Yakovlev, O.I., Razmanov, V.M. (1977) *Pisma v Astronomicheskii Zhurnal*, **3**, 7: 322 (in Russian).

114. Kolosov, M.A., Yakovlev, O.I., Efimov, A.I., *et al.* (1977) *Uspekhi Fizicheskikh Nauk*, **123**, 4: 698 (in Russian).

115. Razmanov, V.M., Efimov, A.I., and Yakovlev, O.I. (1979) *Radiofizika* (Radiophysics and Quantum Electronics), **22**, 9: 1051 (in Russian)*.

116. Aleksandrov, Yu.N., Vasil'ev, M.B., Vyshlov, A.S., *et al.* (1979) *Radiotekhnika i Elektronika* (Radio Engineering and Electronic Physics), **24**, 5: 1 (in Russian)*.

117. Laptev, N.V. (1981) *Radiotekhnika i Elektronika* (Radio Engineering and Electronic Physics), **26**, 11: 2241 (in Russian)*.

118. Efimov, A.I., Yakovlev, O.I., Vyshlov, A.S., *et al.* (1989) *Radiotekhnika i Elektronika* (Radio Engineering and Electronic Physics), **34**, 15: 31 (in Russian)*.

119. Middleton, D. (1960) *Communication Theory*, N.Y.

120. Rytov, S.M., Kravtsov, Yu.A., and Tatarskii, V.I. (1987) *Principles of Statistical Radiophysics*, **2**: Correlation Theory of Random Processes, Berlin: Springer Verlag.

121. Yakovlev, O.I., Efimov, A.I., Yakubov, V.P., *et al.* (1989) *Radiofizika* (Radiophysics and Quantum Electronics, **5**: 531 (in Russian)*.

122. Yakubov, V.P., Yakovlev, O.I., and Efimov, A.I. (1991) *Radiofizika* (Radiophysics and Quantum Electronics), **34**, 6: 615 (in Russian)*.

123. Efimov, A.I., Yakovlev, O.I., Shtrykov, V.K., *et al.* (1981) *Radiotekhnika i Elektronika* (Radio Engineering and Electronic Physics), **26**, 2: 311 (in Russian)*.

124. Efimov, A.I. (1981) *Geomagnetizm i Aeronomiya*, **21**, 5: 769 (in Russian).

125. Vlasov, V.I. (1979) *Astronomicheskii Zhurnal* (Soviet Astronomy), **56**, 1: 96 (in Russian)*.

126. Vlasov, V.I., Shishov, V.I., and Shishova, T.D. (1983) *Geomagnetizm i Aeronomiya*, **23**, 6: 888 (in Russian).

127. Vlasov, V.I. (1988) *Geomagnetizm i Aeronomiya*, **28**, 1: 1 (in Russian).

128. Lotova, N.A., Pisarenko, YA.V., and Korelov, O.A. (1993) *Geomagnetizm i Aeronomiya*, **33**, 6: 10 (in Russian).

129. Lotova, N.A., Vladimirskii, K.V., Yurovskii, I.V., *et al.* (1995) *Astronomicheskii Zhurnal* (Astronomy Reports), **72**, 1: 103 (in Russian)*.

130. Ekers, R.D. and Little, L.T. (1971) *Astron. and Astrophys.*, **10**, 2: 310.

131. Armstrong, I.W. and Goles, W.A. (1972) *J. Geoph. Res.*, **77**, 25: 4602.

132. Coles, W.A. and Maagol, S. (1972) *J. Geoph. Res.*, **77**, 28: 5622.

133. Coles, W.A. and Kaufman, I.I. (1978) *Radio Sci.*, **17**, 3: 591.

134. Kojuma, M., and Kakinuma, T. (1987) *J. Geoph. Res.*, **A92**, 7: 7269.

135. Tokumaru, M., Mori, H., Tanaka, T., *et al.* (1991) *J. Geomagn. Geoelectr.*, **43**: 619.

136. Goles, W.A., Esser, R., Lovhaug, U., and Markkanen, I. (1991) *J. Geoph. Res.*, **96**, A8: 13849.

137. Yakovlev, O.I., Efimov, A.I., Razmanov, V.M., and Shtrykov, V.K. (1980) *Astronomicheskii Zhurnal* (Soviet Astronomy), **57**, 4: 790 (in Russian)*.

138. Armand, N.A., Efimov, A.I., and Yakovlev, O.I. (1987) *Astron. and Astrophys.*, **183**: 135.

139. Yakovlev, O.I., Efimov, A.I., and Rubtsov, S.N. (1987) *Kosmicheskie Issledovaniya* (Cosmic Res.), **25**, 2: 201 (in Russian)*.

140. Yakovlev, O.I., Efimov, A.I., and Rubtsov, S.N. (1988) *Astronomicheskii Zhurnal* (Soviet Astronomy), **65**, 6: 1290 (in Russian)*.

141. Efimov, A.I., Yakovlev, O.I., Shtrykov, V.K., and Rogalskii, V.I. (1984) *Radiotekhnika i Elektronika* (Radio Engineering and Electronic Physics), **29**, 7: 1274 (in Russian)*.

142. Mullan, D.I. (1990) *Astron. and Astrophys.*, **232**, 2: 520.

143. Rubtsov, S.N., Yakovlev, O.I., and Efimov, A.I. (1987) *Kosmicheskie Issledovaniya* (Cosmic Res.), **25**, 4: 475 (in Russian)*.

144. Yakovlev, O.I., Shishov, V.I., and Chashej, I.V. (1990) *Pisma v Astronomicheskii Zhurnal*, **16**, 2: 163 (in Russian).

145. Lotova, N.A. and Chashej, I.V. (1977) *Trudy Fizicheskogo Instituta Akademii Nauk SSSR*, **93**: 78 (in Russian).

146. Vlasov, V.I., Chashej, I.V., Shishov, V.I., and Shishova, T.D. (1979) *Geomagnetizm i Aeronomiya*, **19**, 3: 401 (in Russian).

147. Bird, M.K. (1982) *Space Sci. Rev.*, **33**, 1-2: 99.

148. Lotova, N.A. (1988) *Itogi Nauki i Tekhniki, Astronomiya*, **33**: 121 (in Russian).

149. Shishov, V.I. (1990) *USSR Radiotelescopes and Radioastronomy of the Sun*, Moscow: Nauka, p. 170 (in Russian).

150. Mullan, D.J. and Yakovlev, O.I. (1995) *Irish Astr. J.*, **22**, 2: 119 and **23**, 1: 7.

151. Wohlmuth, R., Plettemeier, D., Edenhofer, P., *et al.* (1997) *Radio Sci.*, **32**, 2: 617.

152. Efimov, A.I., Andreev, V.E., and Samoznaev, L.N. (1997) *Advances in Space Res.*, **20**, 1: 65.

153. Efimov, A.I., Andreev, V.E., Samoznaev, L.N., *et al.* (1999), *Astronomy Reports*, **43**, 4: 267.

154. Chashei, I.V., Bird, M.K., Efimov, A.I., *et al.* (1999), *Solar Physics*, **189**, 399.
155. Chashei, I.V., Efimov, A.I., Samoznaev, L.N., *et al.* (2000), *Adv. Space Res.*, **25**, 9: 1973.
156. Samoznaev, L.N., Efimov, A.I., Andreev, V.E., *et al.* (2000), *Physics and Chemistry of the Earth*, part C, **25**, 1-2: 107.
157. Bird, M. K., Chashei, I. V., Efimov, A. I., *et al.* (2000), *Outer Scale of Coronal Turbulence Near the Sun*, Warsaw, the 33rd COSPAR Session.

Chapter 4

Radar observations of planets and asteroids

4.1 General relations

Depending on the positions of transmitting and receiving antennae relative the planet's surface under study, radar observations can be performed in three different modes. In Fig. 4.1, solid lines represent the rays of incident waves, and dashed lines represent the rays of reflected or scattered waves (echoes). The first mode of the remote sensing of a planet (Fig. 4.1, mode *1*) uses a ground-based radar whose distance from the planet's surface is much greater than the characteristic size of the planet; therefore, it can be assumed that the planet is irradiated with plane radio waves. The radar receives the waves that are backscattered from a large surface region with its center at point D. If the directional antenna of a spacecraft-borne radar is in a direction normal to the planetary surface, the radar receives the waves that are reflected from a small surface region with its center at point D_1.

In the second mode, the radio signals emitted from a spacecraft flying near the planet to be studied are reflected from a surface region located in the vicinity of the specular reflection point D and then received at a ground-based station (Fig. 4.1, mode *2*). This is the so-called bistatic radar observation mode, which is used for investigating planets and the Moon.

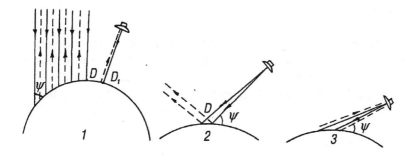

Fig. 4.1. Three modes of the radar observations of planets.

The third mode of remote sensing employs a side-looking spacecraft-borne radar which irradiates the planet's surface at some angle and receives the waves that are backscattered toward the spacecraft (Fig. 4.1, mode *3*).

Depending on the surface roughness and the wavelength of radar waves, they may either be reflected specularly or scattered diffusely from the surface. If the surface under study is relatively smooth and the radar operates at meter radio waves, then the second mode of radar observations can be described in terms of the reflection of radio waves from a smooth sphere. If the radar operates at ·centimeter or millimeter radio waves, the first and third modes of radar observations can be described in terms of the diffuse scattering of radio waves from the surface with randomly distributed irregularities. The specular reflection and diffuse scattering are undoubtedly conventional idealizations. Actually, a surface relief has both small-scale features such as stones and boulders and large-scale features such as hills and mountain slopes. The small-scale relief features whose size is comparable with the radar wavelength are responsible for the diffuse scattering of radar signals with a wide scattering indicatrix $F_d(\psi)$, where ψ is the glancing angle. Large-scale relief features are characterized by gently sloping quasi-flat regions whose sizes and curvature radii are much greater than the radar wavelength. These regions reflect radar waves specularly, so that the angle of the incidence of radar signals is equal to the angle of their reflection; in this case the scattering indicatrix $F(\psi)$ is narrow and has a maximum in the direction of the specular reflection of radar signals from the mean surface. The net scattering indicatrix of radar signals is the sum of $F_d(\psi)$ and $F(\psi)$.

In practice, one has to find the power of the returning signal, W_2, if the power of transmitting antenna, W_1, is known. To find the relationship between the power W_1 and W_2, let us introduce the effective scattering cross section of the target, σ. Then the power flux density of incident waves is given by

$$P_1 = \frac{W_1 G}{4\pi L_1^2},$$ (4.1)

where G is the power gain of the transmitting antenna, and L_1 is the distance from the antenna to the surface region under study. The irradiated surface may be considered conventionally as a source of reflected or scattered radio waves that has a power equal to $P_1 \sigma$. At a distance L_2, the power flux density of reflected radio waves is given by the formula

$$P_2 = \frac{P_1 \sigma}{4\pi L_2^2} = \frac{W_1 G \sigma}{16\pi^2 L_1^2 L_2^2}.$$ (4.2)

If the receiving antenna has an effective cross section A, the received power W_2 is equal to $P_2 A$; then, according to (4.2), we have

$$W_2 = \frac{W_1 G A \sigma}{16\pi^2 L_1^2 L_2^2},$$ (4.3)

$$\sigma = \frac{16\pi^2 L_1^2 L_2^2 W_2}{GAW_1}.$$ (4.4)

Formula (4.3) gives the relationship between the powers W_1 and W_2. Formula (4.4) can be considered as a formal definition of the effective cross section of the irradiated surface region, σ. If a radar system has one combined transmitting-and-receiving antenna, then

$$G = 4\pi A\lambda^{-2},$$
$$L_1 = L_2 = L.$$ (4.5)

From (4.3), (4.4), and (4.5) it follows that

$$W_2 = \frac{W_1 G^2 \lambda^2 \sigma}{64\pi^3 L^4},$$ (4.6)

$$\sigma = \frac{64\pi^3 L^4 W_2}{G^2 \lambda^2 W_1}.$$ (4.7)

These formulae describe monostatic radar observations schematically shown in Fig. 4.1 (modes *1* and *3*).

Let us compare the power flux density of reflected radio waves, P_2, with the value of this density that would be observed if the irradiated surface were an ideal reflecting surface tangent to the specular reflection point D. This point is characterized by the equal angles of the incidence and reflection of radar signals, which are made with the radar ray passing through the transmitting and receiving antennae. Such a comparison allows one to introduce a new parameter, the reflectance of radio waves, for the second radar observation mode (Fig. 4.1). Knowing that the power flux density of the radio waves reflected from an ideal surface is given by

$$P_2^* = \frac{W_1 G}{4\pi(L_1 + L_2)^2},$$ (4.8)

we may define the coefficient of the wave power reflection (or reflectance) η^2 as the ratio P_2/P_2^*. Then, allowing for (4.2) and (4.8), we have

$$\eta^2 = \frac{(L_1 + L_2)^2 \sigma}{4\pi L_1^2 L_2^2}.$$ (4.9)

If the transmitting and receiving antennae are placed at one point, then

$$\eta^2 = \frac{\sigma}{\pi L^2} \cdot \tag{4.10}$$

Formulae (4.9) and (4.10) describe the relationship between η^2 and σ.

Parameters η and σ depend on the size and relief of the surface under study, the relative position of the transmitting and receiving antennae, as well as on the dielectric permittivity ε and conductivity of the surface rock. The rock's conductivity affects primarily the depth of the radio wave penetration. If the conductivity is given by the tangent of the angle of loss, $\tan\Delta$, the effective depth of the radio wave penetration, or the skin-layer thickness L^*, is proportional to the wavelength [1]:

$$L^* = \frac{\lambda}{2\pi\sqrt{\varepsilon} \cdot \tan\Delta} \cdot \tag{4.11}$$

This formula makes it possible to estimate the thickness of the rock layer essential for the reflection and scattering of radio waves. For lunar soil, $L^* \approx 10\lambda$; at the same time, L^* is less than λ for wet Earth soils. The dielectric permittivity ε used in radar sensing is averaged over the surface layer of thickness L^*. The permittivity of dry soils depends only on their density ρ_1. For this reason, as is shown in [2.3], the following empirical formulae are valid:

$$\rho_1 = 2\left(\sqrt{\varepsilon} - 1\right) \tag{4.12}$$

or

$$\rho_1 = 1.52 \cdot \ln\varepsilon, \tag{4.13}$$

where ρ_1 has dimensions g cm^{-3}.

4.2 Reflection of radio waves from a sphere

The reflection of radio waves from a smooth spherical surface differs greatly from their reflection from real planetary reliefs. However, if the Rayleigh condition is satisfied (as is assumed in this section),

$$\langle \Delta z^2 \rangle^{\frac{1}{2}} < \frac{\lambda}{16\sin\psi}, \tag{4.14}$$

a rough surface reflects radio signals as an ideal plane or sphere [1]. In formula (4.14), the term $\langle \Delta z \rangle^{1/2}$ is the standard deviation of the relief heights.

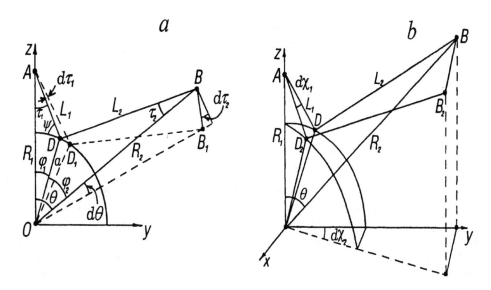

Fig. 4.2. Graphical representation of the reflection of radio waves from an ideal sphere.

Since the planetary radius a is considerably greater than the wavelength of the radio signals used, we may treat the problem in terms of the ray conception. Let a ray tube with angular sizes $d\tau_1$ in the plane of the figure and $d\chi_1$ in a direction perpendicular to this plane corresponds to the region of a spherical surface with its center at point D (Fig. 4.2). When reflected from the sphere, the radio beam has larger angular sizes and, consequently, has a lower power flux density than the beam reflected from the plane tangent to the sphere at point D. Denote the cross-sectional areas of the ray tube corresponding to reflection from the plane and sphere as ds and dS, respectively. The coefficient of the wave power reflection, η^2, is equal to the ratio of these areas times the Fresnel reflection coefficient M at point D:

$$\eta^2 = M_{1,2}^2 \frac{ds}{dS}. \qquad (4.15)$$

A rigorous consideration of this formula describing the reflection of the radio wave power can be found in Chapter 8 of the book [4].

The Fresnel reflection coefficients M_1 and M_2 for, respectively, vertically and horizontally polarized radio waves are given by the known formulae

$$M_1 = \frac{\sqrt{\varepsilon - \cos^2 \psi} - \sin \psi}{\sqrt{\varepsilon - \cos^2 \psi} + \sin \psi} \,,$$

$$M_2 = \frac{\varepsilon \cdot \sin \psi - \sqrt{\varepsilon - \cos^2 \psi}}{\varepsilon \cdot \sin \psi + \sqrt{\varepsilon - \cos^2 \psi}} \,,$$

(4.16)

where ψ is the glancing angle at point D. For vertically incident radio waves, $\psi = 90°$; then from (4.16) follows

$$M = \frac{\sqrt{\varepsilon} - 1}{\sqrt{\varepsilon} + 1} \,.$$

(4.17)

Formulae (4.16) and (4.17) correspond to homogeneous soils, for which ε does not depend on depth. Typically, however, the density and, consequently, the permittivity of the surface soil increases with depth. Such soils can be simulated by layered media. In this case formula (4.15) must have the respective Fresnel coefficient M. The theory of the wave reflection from layered media is given in the monograph [6]; the results of the calculation of the reflectance of radio waves in terms of a layered lunar soil model are described in [7–9].

To describe the reflectance of radio wave from a sphere in terms of formula (4.15), let us find cross-sectional areas ds and dS. When radio waves are reflected from a tangent plane, ds is given by

$$ds = (L_1 + L_2)^2 d\tau_1 \, d\chi_1 \,.$$

(4.18)

To find cross-sectional area dS, refer to Fig. 4.2, where the transmitter is located at point A; receiver is at point B; the center of the sphere is at the coordinate origin O; $AO = R_1$; $BO = R_2$; D is the point of specular reflection, at which the angle of incidence is equal to the angle of reflection; $AD = L_1$; $BD = L_2$; the angle $AOB = \Theta$; angle $AOD = \varphi_1$; angle $BOD = \varphi_2$; angle $OAD = \tau_1$; angle $OBD = \tau_2$; angle $d\tau_1$ lies in the zy-plane (Fig. 4.2, a); and angle $d\chi_1$ lies in the perpendicular plane (Fig. 4.2, b). Let us consider a ray tube confined within small angles $d\tau_1$ and $d\chi_1$. When reflected from the sphere, the incident rays AD, AD_1, and AD_2 follow the directions DB, D_1B_1, and D_2B_2. The ray tube corresponding to reflection from the sphere has greater angular sizes and, consequently, a lower power flux density at point B than the ray tube corresponding to reflection from the tangent plane. The cross-sectional area of the ray tube at point B should be sought for an equiphase surface, i.e., under the condition that $L_1 + L_2 = \text{const}$. The linear size of an element of this surface in the zy-plane (Fig. 4.2, a) is given by

$$BB_1 = R_2 \cos \tau_2 \, d\Theta$$

(4.19)

and in the perpendicular plane (Fig. 4.2, b) by

$$BB_2 = R_2 \sin \Theta \, d\chi_2 . \tag{4.20}$$

The area of the equiphase surface element, dS, is the product of BB_1 and BB_2; i.e.,

$$dS = R_2^2 \sin \Theta \cdot \cos \tau_2 \, d\Theta \, d\chi_2 . \tag{4.21}$$

Let us find the relationship between the angles $d\chi_1$ and $d\chi_2$ (Fig. 4.2, b). Since

$$DD_2 = L_1 \, d\chi_1 = a \cdot \sin \varphi_1 \, d\chi_2 ,$$

we have

$$d\chi_2 = \frac{L_1 \, d\chi_1}{a \cdot \sin \varphi_1} . \tag{4.22}$$

From (4.21) and (4.22) we have

$$dS = \frac{R_2^2 \sin \Theta \cdot \cos \tau_2 \, L_1 \, d\Theta \, d\chi_1}{a \cdot \sin \varphi_1} . \tag{4.23}$$

From (4.18) and (4.23) follows the ratio of the cross-sectional areas:

$$\frac{ds}{dS} = \frac{(L_1 + L_2)^2 \, a \cdot \sin \varphi_1}{L_1 R_2^2 \cos \tau_2 \sin \Theta \left(\dfrac{d\Theta}{d\tau_1} \right)} . \tag{4.24}$$

Taking into account that $a \sin \varphi_1 = L_1 \sin \tau_1$, equation (4.24) transforms into

$$\frac{ds}{dS} = \frac{(L_1 + L_2)^2 \sin \tau_1}{R_2^2 \cos \tau_2 \sin \Theta \left(\dfrac{d\Theta}{d\tau_1} \right)} . \tag{4.25}$$

Now calculate the derivative $d\theta / d\tau_1$. Since $\theta = \varphi_1 + \varphi_2$, then

$$\frac{d\Theta}{d\tau_1} = \frac{d\varphi_1}{d\tau_1} + \frac{d\varphi_2}{d\tau_1} . \tag{4.26}$$

Taking into account that

$$L_1^2 = R_1^2 + a^2 - 2aR_1 \cos\varphi_1,$$
$$L_1 \sin\tau_1 = a \cdot \sin\varphi_1,$$

we obtain

$$\frac{d\varphi_1}{d\tau_1} = \frac{\dfrac{dL_1}{d\tau_1}}{R_1 \sin\tau_1}. \tag{4.27}$$

Similarly, we have

$$\frac{d\varphi_2}{d\tau_1} = \frac{\dfrac{dL_2}{d\tau_2}}{R_2 \sin\tau_2}. \tag{4.28}$$

To find the derivatives $dL_1/d\tau_1$ and $dL_2/d\tau_1$ we can use the following geometrical relations:

$$L_1 = R_1 \cos\tau_1 - a \cdot \sin\psi,$$
$$L_2 = R_2 \cos\tau_2 - a \cdot \sin\psi,$$
$$L_1 \sin\tau_1 = a \cdot \sin\varphi_1, \tag{4.29}$$
$$L_2 \sin\tau_2 = a \cdot \sin\varphi_2,$$
$$a \cdot \cos\psi = R_2 \sin\tau_2 = R_1 \sin\tau_1.$$

From (4.29) it follows that

$$\frac{dL_1}{d\tau_1} = L_1 \cot\psi,$$
$$\frac{dL_2}{d\tau_2} = \frac{R_1 L_2 \cot\psi \cdot \cos\tau_1}{R_2 \cos\tau_2}. \tag{4.30}$$

Taking into account (4.26) – (4.28) and (4.30), we have

$$\frac{d\theta}{d\tau_1} = \frac{2L_1 L_2 + a(L_1 + L_2)\sin\psi}{a \cdot \sin\psi(L_2 + a \cdot \sin\psi)}. \tag{4.31}$$

Substituting (4.31) into (4.25) we obtain, after appropriate transformations, the desired expression for the ratio of the cross-sectional areas:

$$\eta_0^2 = \frac{ds}{dS} = \frac{a^2(L_1 + L_2)^2 \sin\psi \cdot \cos\psi}{[2L_1 L_2 + a(L_1 + L_2)\sin\psi]R_1 R_2 \sin\theta} . \tag{4.32}$$

According to (4.15), formula (4.32) gives the reflectance of radio waves for the Fresnel coefficient $M = 1$. Allowing for (4.32), formula (4.15) takes the form

$$\eta^2 = \frac{M_{1,2}^2 \sin\psi}{\left[\sin\psi + \dfrac{2L_1 L_2}{a(L_1 + L_2)}\right]\left[1 + \dfrac{2L_1 L_2 \sin\psi}{a(L_1 + L_2)}\right]} . \tag{4.33}$$

Formulae (4.32) and (4.33) were derived in [5] and [4], respectively.

Using (4.13) and (4.33), we can find the effective cross section of a planet:

$$\sigma = \frac{4\pi M_{1,2}^2 L_1^2 L_2^2 \sin\psi}{(L_1 + L_2)^2\left[\sin\psi + \dfrac{2L_1 L_2}{a(L_1 + L_2)}\right]\left[1 + \dfrac{2L_1 L_2 \sin\psi}{a(L_1 + L_2)}\right]} . \tag{4.34}$$

Formulae (4.33) and (4.34) allow the parameters σ and η to be calculated at an arbitrary relative position of the transmitting and receiving antennae.

To illustrate this, let us consider a particular case when a radar is located on the Earth surface (Fig. 4.1, mode *1*). Putting $L_1 = L_2 \gg a$ and $\psi = 90°$ in equations (4.33) and (4.34), we obtain

$$\sigma = \pi a^2 M^2 ,$$
$$\eta^2 = \frac{a^2 M^2}{L^2} . \tag{4.35}$$

Here M, given by formula (4.17), only depends on the soil permittivity. If the radar is installed aboard a lander at height H, then, putting

$$L_1 = L_2 = H, \quad \psi = 90° ,$$

and taking into account that $H \ll a$, we have

$$\sigma = \pi H^2 M^2 ,$$
$$\eta^2 = M^2 , \tag{4.36}$$

where the coefficient M, again, is given by formula (4.17).

In bistatic radar observations (Fig. 4.1, mode *2*), when one radar is located on the Earth surface and the other is installed aboard a spacecraft, $L_2 \gg L_1$ and from (4.33) and (4.34) we obtain

$$\sigma = \frac{4\pi M_{1,2}^2 L_1^2 \sin\psi}{\left(\sin\psi + \dfrac{2L_1}{a}\right)\left(1 + \dfrac{2L_1 \sin\psi}{a}\right)}, \tag{4.37}$$

$$\eta^2 = \frac{M_{1,2}^2 \sin\psi}{\left(\sin\psi + \dfrac{2L_1}{a}\right)\left(1 + \dfrac{2L_1 \sin\psi}{a}\right)}. \tag{4.38}$$

Here L_1 is the distance from the spacecraft to the point D of the radio wave reflection. The Fresnel reflection coefficients $M_{1,2}$ are given by formulae (4.16).

The above consideration is valid provided that the condition described by formula (4.4) is satisfied and the planetary surface can be considered to be smooth. The second condition is almost always satisfied for meter radio waves, while for decimeter radio waves it is satisfied only for relatively smooth surface regions. Actually, the planetary relief is quite rough, so that radar waves undergo random scattering from rough surfaces rather than specular reflection from smooth surfaces.

4.3 Scattering of radio waves from a rough surface

Let us consider the scattering of radio waves from a small region of a rough planetary surface. In this case the planet sphericity can be neglected, and this small surface region can be treated as flat. Assume that the linear sizes of this region are considerably greater than the radar wavelength and considerably less than the distance to the transmitter or receiver (Fig. 4.3, a). In the case of circular-polarized radio waves, the effective area of radio wave scattering, σ, can be presented as the sum

$$\sigma = s[\sigma_d F_d(\psi) + \sigma_0 F(\psi)], \tag{4.39}$$

where s is the surface area, σ_d and $F_d(\psi)$ are, respectively, the differential cross section and the indicatrix of the radio waves scattered from small-scale relief features such as stones or boulders, σ_0 and $F(\psi)$ are, respectively, the differential cross section and the indicatrix of scattering from large-scale topographic features such as hills or mountain slopes).

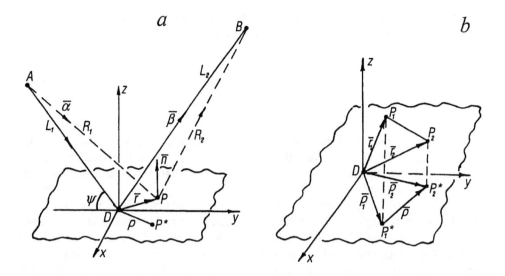

Fig. 4.3. Graphical representation of the radio wave scattering from a rough surface.

Formula (4.39) describes the reception of radio signals with a matched circular polarization. For the reception of radio signals with an inverse circular polarization, the second component in (4.39) is negligibly small; therefore,

$$\sigma^* = s\sigma_d^* F_d^*(\psi), \tag{4.40}$$

where ψ is the glancing angle, σ_d^* is the differential cross section of diffuse scattering for the case of the reception of radio signals with inverse polarization, $F_d^*(\psi)$ is the respective scattering indicatrix. For vertically incident radio waves, when $F(\psi) = 1$, the ratio σ_d^*/σ is an arbitrary characteristic of the small-scale surface roughness given by

$$\mu = \frac{\sigma_d^*}{\sigma_d + \sigma_0}. \tag{4.41}$$

Because of the absence of the satisfactory theory for the diffuse scattering of radio waves, σ_d and $F_d(\psi)$ must be determined experimentally. When surfaces are severely rough, the empirical Lambert law is satisfied:

$$F_d(\psi) = \sin^2\psi. \tag{4.42}$$

Experiments employing centimeter and meter radio waves showed that the following relations hold for the diffuse scattering of radio waves:

$$F_d(\psi) = \sin^m \psi$$

or   (4.43)

$$F_d(\psi) = (90° - \psi)^{-m},$$

where m ranges from 2 to 4. For the matched and inverse circular polarization of radio waves, the parameter m may have different values. If there are no large-scale relief features, the diffuse backscattering of radio waves is characterized by three experimental parameters, σ_d, m, and μ.

The diffuse scattering of decimeter and especially meter radio waves from some surface regions can be neglected, since the main contribution to the wave field comes from the radio waves that are reflected from the quasi-plane regions of large-scale reliefs. If the variance of the relief slopes, γ^2, of such regions is low, the half-power angular spectrum width of reflected radio waves, $\Delta\psi^2$, can be obviously given by the approximate formula

$$\Delta\psi^2 = 2\gamma^2.$$ (4.44)

In this case the maximum of the function $F(\psi)$ is directed in accordance with the law of specular reflection, which states that the angle of the incidence of radio waves is equal to the angle of their reflection. This situation is typical of the radar sensing of planets and their satellites, since the characteristic size of their relief roughness is much greater than the wavelength, and the relief slopes are not steep. In this case the backscattering of radio waves is characterized by the function $F(\psi)$ and the parameter σ_0.

Let us analyze the scattering of electromagnetic waves from a finite statistically rough surface after the authors of [10]. The scattered electromagnetic field can be determined in terms of the Kirchhoff method from the value of this field on the reflecting surface presented as the sum of the incident and reflected waves. The reflected wave can be calculated in terms of geometrical optics applied to a set of tangent planes to the rough surface. The Kirchhoff method gives a correct result if the wavelength is small when compared to the curvature radii of surface irregularities and if the line of sight is close to the direction of specular reflection from an arbitrary surface region. The second condition is satisfied only for some surface regions; nevertheless, as is shown in [11], the Kirchhoff method is applicable for the whole surface, since the main contribution to the scattered field comes from the regions that specularly reflect radio waves to the point of observation. In our treatment, the effect of polarization is neglected, i.e., the wave field is considered as being scalar.

Let us treat, following [10, 11], the problem of the radio wave scattering from a finite statistically rough (but on average flat) surface. The profile of the rough surface is given by a stochastic function $z(x,y)$, such that $\langle z \rangle = 0$. We assume that the heights of surface irregularities, z, are distributed by the normal law and that the respective autocorrelation function is known. Suppose also that the size of the scattering region is much less that the distance from the surface to points A and B (Fig. 4.3, a). Let a radio wave be incident from point A to the rough area s and the scattered scalar field E be found at point B. This can be done from the surface value of this field, E_s, using the Green function

$$E = -\frac{1}{4\pi} \int_s \left[E_s \frac{\partial}{\partial n} \left(\frac{e^{ikR_2}}{R_2} \right) - \frac{e^{ikR_2}}{R_2} \frac{\partial E_s}{\partial n} \right] ds , \tag{4.45}$$

where $\partial/\partial n$ is the derivative along the normal to the surface, R_2 is the distance from an arbitrary surface point P to point B, and the constant $k = 2\pi/\lambda$. The Green formula implies integration over a closed surface involving points A and B. Under the assumption that the field is generated by the reflecting surface area s, integration in (4.45) can be performed only over this area. The field on the area s can be considered as the result of the reflection of radio waves from quasi-flat areas with the respective Fresnel coefficient. If a radio wave with unit amplitude is reflected from a plane surface, the following relations are valid:

$$E_s = M(\psi) \frac{e^{ikR_1}}{R_1},$$

$$\frac{\partial E_s}{\partial n} = -M(\psi) \frac{\partial}{\partial n} \left(\frac{e^{ikR_1}}{R_1} \right), \tag{4.46}$$

where $M(\psi)$ is the Fresnel reflection coefficient, R_1 is the distance from the source of radio waves located at point A to the arbitrary point P in the area s. From (4.45) and (4.46) it follows that

$$E = \frac{1}{4\pi} \int_s M(\psi) \frac{\partial}{\partial n} \left[\frac{e^{ik(R_1+R_2)}}{R_1 R_2} \right] ds . \tag{4.47}$$

Let us introduce a coordinate system whose origin is located at point D of the specular reflection from the plane $z = 0$, the arbitrary point P on a rough surface defined either by the vector $\bar{r}$ or coordinates (x, y, z), the unit vectors of incident and reflected waves, $\bar{\alpha}$ and $\bar{\beta}$, and the respective glancing angle ψ. Since region s is far from the source of radio waves located at point A and the receiver located at point B, $\bar{\alpha}$ and $\bar{\beta}$ are constant for a rough surface; therefore,

$$R_1 = L_1 + (\bar{\alpha}\bar{r}),$$

$$R_2 = L_2 - (\bar{\beta}\bar{r}), \tag{4.48}$$

where $L_1 = AD$ and $L_2 = BD$. The differential of the integrand in (4.47) along the normal is as follows

$$\frac{\partial}{\partial n}\left[\frac{e^{ik(R_1+R_2)}}{R_1 R_2}\right] = \frac{ik(\bar{n}\bar{q})}{R_1 R_2} \cdot e^{ik(R_1+R_2)},\tag{4.49}$$

where $\operatorname{grad}(R_1 + R_2) = \bar{\alpha} - \bar{\beta} = \bar{q}$ represents the scattering vector. For gently sloping topographic features and the area s occurring in the remote zone, the glancing angle ψ can be taken as being approximately the same for all points P, so that $R_1 R_2 = L_1 L_2$. Allowing for expressions (4.48) and (4.49), we can rewrite equation (4.47) in the following form

$$E = \frac{ikM(\psi)e^{ik(L_1+L_2)}}{4\pi L_1 L_2}\int_s(\bar{n}\,\bar{q})\cdot e^{ik(\bar{q}\bar{r})}\,ds,\tag{4.50}$$

where the slowly changing quantities are removed from the integrand. The integral in formula (4.50), whose integrand represents an oscillatory function, can be calculated by the stationary phase method, which implies that the surface areas corresponding to the extrema of the oscillatory function make the greatest contribution to the integral. For such areas,

$$\operatorname{grad}(\bar{q}\bar{r}) = 0.\tag{4.51}$$

Taking into account that

$$(\bar{q}\bar{r}) = q_x x + q_y y + q_z z,$$

$$\frac{\partial}{\partial x}(\bar{q}\bar{r}) = q_x + q_z\frac{\partial z}{\partial x},$$

$$\frac{\partial}{\partial y}(\bar{q}\bar{r}) = q_y + q_z\frac{\partial z}{\partial y},$$

the following expressions can be written for the surface areas that satisfy the condition (4.51):

$$\frac{\partial z}{\partial x} = -\frac{q_x}{q_z}, \quad \frac{\partial z}{\partial y} = -\frac{q_y}{q_z}.\tag{4.52}$$

The projections of the normal n are given by

$$n_x = \frac{-\dfrac{\partial z}{\partial x}}{\sqrt{1 + \left(\dfrac{\partial z}{\partial x}\right)^2 + \left(\dfrac{\partial z}{\partial y}\right)^2}} \;,$$

$$n_y = \frac{-\dfrac{\partial z}{\partial y}}{\sqrt{1 + \left(\dfrac{\partial z}{\partial x}\right)^2 + \left(\dfrac{\partial z}{\partial y}\right)^2}} \;,$$

$$(4.53)$$

$$n_z = \frac{1}{\sqrt{1 + \left(\dfrac{\partial z}{\partial x}\right)^2 + \left(\dfrac{\partial z}{\partial y}\right)^2}} \;.$$

From (4.52) and (4.53) it follows that the stationary phase points are given by

$$\bar{n} = \frac{\bar{q}}{q} \;. \tag{4.54}$$

At these points, the scattering vector $\bar{q}$ is normal to the surface, which implies that the reflection of radio waves from small areas around these points is specular. From now on, q denotes the scattering vector magnitude:

$$q = \sqrt{2\left[1 - \left(\bar{\alpha} \cdot \bar{\beta}\right)\right]} \;. \tag{4.55}$$

At the stationary phase points, $(\bar{n}\bar{q}) = q$. Removing this quantity from the integrand and taking into account (4.52) and (4.53), we obtain from (4.50):

$$E = \frac{ikqMe^{ik(L_1 + L_2)}}{4\pi L_1 L_2} \int_s e^{ik(\bar{q}\bar{r})} \, ds \;. \tag{4.56}$$

The area element of the rough surface is given by $ds = dxdy/n_z$. According to (4.52) and (4.53), $n_z = q_z/q$ in the vicinity of the stationary phase points. Hence,

$$ds = \frac{dxdy}{n_z} = \frac{qdxdy}{q_z} \;. \tag{4.57}$$

Here $dxdy$ is the area element of the mean surface s, i.e., it is the projection of ds onto the plane $z = 0$. Substituting (4.57) into (4.56), we get

$$E = \frac{ikq^2 \, M e^{ik(L_1+L_2)}}{4\pi q_z L_1 L_2} \int_s e^{ik(\bar{q}\bar{r})} dxdy .$$ (4.58)

Thus, integration over a rough surface is replaced by integration over the plane $z = 0$. Taking into account (4.55), we can find the quantity $\langle EE^* \rangle$, which is proportional to the mean power flux density:

$$\langle EE^* \rangle = \frac{k^2 M^2 \left(1 - \bar{\alpha}\,\bar{\beta}\right)^2 I}{4\pi^2 L_1^2 L_2^2 q_z^2} ,$$

$$I = \left\langle \int_s \int_s \exp[ik(\bar{q}_1\bar{r}_1 - \bar{q}_2\bar{r}_2)]dx_1 dx_2 dy_1 dy_2 \right\rangle .$$ (4.59)

Here symbols $\langle \; \rangle$ denote averaging over the ensemble of rough surface realizations and indices 1 and 2 correspond to arbitrary points P_1 and P_2. Since we are not interested in the mean phase of the scattered radio wave, the term $\exp[ik(L_1 + L_2)]$ can be taken to be unity. We treat the scalar problem without accounting for the depolarization of radio waves brought about by surface irregularities. Such an approach is valid, because large-scale, gently sloping relief features do not lead to an appreciable depolarization of radio waves. In formula (4.59), the effect of the polarization of radio waves is taken into account by the Fresnel coefficient $M(\psi)$.

Since the linear size of the scattering area is assumed to be much less than the distance to the transmitter and receiver of radio waves, we can put in (4.59) $\bar{q}_1 = \bar{q}_2 = \bar{q}$:

$$I = \left\langle \int_s \int_s \exp[ik(\bar{q}_\rho \bar{\rho} + q_z(z_1 - z_2))]dx_1 dx_2 dy_1 dy_2 \right\rangle .$$ (4.60)

Here $\bar{\rho}$ and $\bar{q}_\rho$ are the projections of the vectors $\bar{r}$ and $\bar{q}$ onto the plane $z = 0$. In terms of these projections,

$$\bar{q} \cdot (\bar{r}_1 - \bar{r}_2) = \bar{q}_\rho \, \bar{\rho} + q_z (z_1 - z_2) .$$ (4.61)

Here z_1 and z_2 are the heights of the arbitrary points P_1 and P_2, $\bar{\rho}$ is the vector of the distance between two points, $P_1{}^*$ and $P_2{}^*$, in the plane $z = 0$ (Fig. 4.3, b). To average (4.60), one has to find the probability $U_1(z_1, z_2, \rho)$ that the height of point P_2 a distance ρ from point P_1 with height z_1 is z_2. If the heights are normally distributed, the desired probability is given by

$$U_1 = \frac{\exp\left[-\dfrac{z_1^2 - 2z_1 z_2 B(\rho) + z_2^2}{2z_0^2[1 - B^2(\rho)]}\right]}{2\pi z_0^2 \sqrt{1 - B^2(\rho)}}, \tag{4.62}$$

where $B(\rho)$ is the autocorrelation function of surface irregularities and z_0^2 is the variance of surface irregularity heights [10, 11]. From (4.60) and (4.62) it follows that

$$I = \int\limits_s \int\limits_s \int\limits_{-\infty}^{\infty} \int\limits_{-\infty}^{\infty} \exp\{ik[\bar{q}_p\,\bar{\rho} + q_z(z_1 - z_2)]\} U_1 dx_1 dx_2 dy_1 dy_2 dz_1 dz_2 \qquad . \tag{4.63}$$

Formulae (4.58) and (4.63) define the power flux density of the scattered radio waves provided that the distribution function $U_1(z_1, z_2, \rho)$ is known. Integrating (4.63) with respect to z_1 and z_2 and taking into account (4.62), we obtain

$$I = \int\limits_s \int\limits_s \exp\{ik\bar{q}_p\bar{\rho} - k^2 q_z^2 z_0^2 [1 - B(\rho)]\} dx_1 dx_2 dy_1 dy_2 . \tag{4.64}$$

Let us introduce into (4.64) the following coordinates

$$\rho_x = x_1 - x_2, \quad \rho_y = y_1 - y_2, \quad \xi = \frac{x_1 + x_2}{2}, \quad \zeta = \frac{y_1 + y_2}{2} .$$

Integration with respect to ξ and ζ yields the scattering surface area s. Taking into account that $dxdy = \rho\,d\rho\,d\varphi$, we have

$$I = s\int\limits_0^{\infty} \int\limits_0^{2\pi} \exp\{ik\bar{q}_p\bar{\rho} - k^2 q_z^2 z_0^2 [1 - B(\rho)]\}\rho d\rho d\varphi , \tag{4.65}$$

where the limits of integration with respect to ρ are taken to be infinite because of an insignificancy in the correlation radii of surface irregularities as compared to the linear size of the area s. The integration of (4.65) with respect to φ yields

$$I = 2\pi s \int\limits_0^{\infty} J_0\{k|q_p|\rho\}\rho \cdot \exp[-k^2 q_z^2 z_0^2 [1 - B(\rho)]]d\rho , \tag{4.66}$$

where J_0 is the zeroth-order Bessel function. Formulae (4.58) and (4.66) allow the power flux density of the scattered radio waves to be found at arbitrary relative positions of the scattering surface, source of radio waves, and observation point, provided that the variance of the surface irregularity heights, z_0^2, and the autocorrelation function $B(\rho)$ are known.

For further analysis of (4.66), one has to specify the function $B(\rho)$ in such a way that it could describe the statistical properties of real surfaces as accurately as possible. Typically, two types of autocorrelation functions are used

$$B(\rho) = \exp\left(-\frac{|\rho|}{l}\right) \tag{4.67}$$

and

$$B(\rho) = \exp\left(-\frac{\rho^2}{l^2}\right), \tag{4.68}$$

where l is the characteristic scale of surface irregularities.

The scattering of radio waves in terms of these two functions $B(\rho)$ was theoretically analyzed in [10, 11]. The authors of [12] treated experimental data on the scattering of radio waves from the lunar surface and found the best agreement between the theory and experiment for the function

$$B(\rho) = \exp\left[-\frac{\rho}{l_1} + \frac{\rho}{l_1}\exp\left(-\frac{\rho}{l_2}\right)\right]. \tag{4.69}$$

Here l_1 and l_2 are the characteristic scales of surface irregularities. At $\rho \ll l_2$, the autocorrelation function (4.69) behaves as (4.68), while as (4.67) at $\rho \gg l_1$. Let us perform integration with respect to ρ in formula (4.66). Due to the specific behavior of the function $J_0\{k|q_\rho|\rho\}$, the region essential for integration involves only small values of ρ. Then the expansion of (4.69) into a series gives

$$B(\rho) \approx 1 - \rho^2 l_1^{-1} l_2^{-1}.$$

Hence,

$$I = 2\pi s \int_0^\infty J_0\{k|q_\rho|\rho\}\rho \cdot \exp\{-k^2 q_z^2 z_0^2 l_1^{-1} l_2^{-1}\rho^2\}d\rho. \tag{4.70}$$

Here the term $\gamma^2 = 2z_0^2 l_1^{-1} l_2^{-1}$ can be interpreted as the mean square of surface slopes, $q_\rho^2 = q_x^2 + q_y^2$, where q_x and q_y are the projections of the scattering vector onto the x- and y-axis, respectively. If surface irregularities are described by the Gaussian autocorrelation function (4.68), then expression (4.70) is valid as well. In this case, however,

$\gamma^2 = 2z_0^2 l^{-2}$. Analysis of (4.70) shows that the major portion of the scattered energy collected by the receiver comes from the vicinity of the specular reflection point, since there $q_x^2 + q_y^2 = 0$.

Consider now the reflection coefficient η and the effective scattering cross section σ (see the section 4.1 of this book for their definitions). The planet surface can be represented as a set of the mean-flat areas tangent to the sphere. The radio waves scattered from these areas arrive at a receiving antenna with random phases; therefore, the power flux density at the receiving antenna can be found by integrating the scattered power over all of the surface areas that are visible from both the transmitting and receiving antennae. With such an approach, the role of the small area s is played by the area ds of the sphere, whereas L_1 and L_2 correspond to the distances between this region and the transmitter and receiver, respectively. According to (4.59), the power flux density at the receiver is given by

$$P = \frac{W_1 G k^2}{32\pi^3} \int_s \frac{UI\,ds}{\left(L_1^* L_2^*\right)^2}. \tag{4.71}$$

Here S is the portion of the spherical surface irradiated by the transmitting antenna and covered by the reception indicatrix of the directional receiving antenna; W_1 and G are the power of the transmitter and the gain of the transmitting antenna, respectively; L_1^* and L_2^* are the distances from points A and B to an arbitrary point on the sphere; and I is given by expression (4.70). Taking into account expression (4.59), we can write for parameter U:

$$U = \frac{M^2\left(1 - \overline{\alpha}\,\overline{\beta}\right)^2}{q_z^2}. \tag{4.72}$$

According to the definition given in Section 4.1, in order to determine the coefficient of the wave power scattering, expression (4.71) should be divided by expression (4.8). Then we have

$$\eta^2 = \frac{k^2 (L_1 + L_2)^2}{8\pi^2} \int_s \frac{UI\,ds}{\left(L_1^* L_2^*\right)^2}. \tag{4.73}$$

The comparison of (4.73) with formula (4.9) yields the effective cross section of the radio wave scattering:

$$\sigma = 2\pi L_1^2 L_2^2 \lambda^{-2} \int_s \frac{UI\,ds}{\left(L_1^* L_2^*\right)^2}. \tag{4.74}$$

In (4.73) and (4.74), U is given by formula (4.72), and I is given by (4.66) or (4.70). In many cases, we may take with good accuracy that $L_1' = L_1$, $L_2' = L_2$ and remove these quantities from the integrand:

$$\eta^2 = \frac{k^2(L_1 + L_2)^2}{8\pi^2 L_1^2 L_2^2} \int_S UI\,ds \,, \tag{4.75}$$

$$\sigma = \frac{2\pi}{\lambda^2} \int_S UI\,ds \,. \tag{4.76}$$

Here S is the region of the planetary surface that is irradiated by the transmitting antenna. The area and position of the region S depend on the directivity and relative orientation of the transmitting and receiving antennae.

4.4 Backscattering of radio waves from planets and asteroids

Let us consider the problem of the backscattering of radio waves for the case when a radar system, which is aboard a spacecraft or located on the Earth, uses the same antenna for emitting and receiving radio signals, In general, the scattering of radio waves is expressed by formula (4.39). If the radar operates at decimeter or meter radio waves, the diffuse component of scattering can be neglected; then, according to (4.39) and (4.76), the effective cross section of backscattering is given by

$$\sigma = \sigma_0 F(\psi) = \frac{2\pi}{S\lambda^2} \int_S UI\,ds \,, \tag{4.77}$$

where σ_0 is the effective unit cross section of backscattering for $\psi = 90°$; $F(\psi)$ is the dependence of the power flux density of scattered radio waves on the angle ψ, normalized to unity (that is, $F = 1$ at $\psi = 90°$).

Let us find U and I in formula (4.77). From the diagram in Fig. 4.4, in which a radar located at point A irradiates a surface at the glancing angle ψ, it follows that

$$\begin{aligned}
\bar{\alpha} &= -\bar{\beta}, \\
\bar{q} &= 2\bar{\alpha}, \\
q_z &= -2\sin\psi, \\
q_\rho &= 2\cos\psi.
\end{aligned} \tag{4.78}$$

Then from (4.66) follows

$$I = 2\pi S \int_0^\infty J_0(2k\rho\cos\psi)\rho \cdot \exp[-4k^2 z_0^2 \sin^2\psi(1 - B(\rho))]d\rho \ . \quad (4.79)$$

Formula (4.72) can be simplified by taking into account the fact that only the surface regions for which the vector $\bar{\beta}$ and local normal $\bar{n}$ essentially contribute to backscattering. Then the reflection coefficient M can be given by the simple formula (4.17), and expression (4.72) transforms into

$$U = \frac{M^2}{\sin^2\psi} \ . \quad (4.80)$$

Substituting expressions (4.79) and (4.80) into expression (4.77) for the effective cross section of backscattering, we have

$$\sigma_0 F(\psi) = \frac{4\pi^2 M^2 I_1}{\lambda^2 \sin^2\psi} \ , \quad (4.81)$$

where

$$I_1 = \int_0^\infty J_0(2k\rho\cos\psi)\rho \cdot \exp[-4k^2 z_0^2 \sin^2\psi(1 - B(\rho))]d\rho \ . \quad (4.82)$$

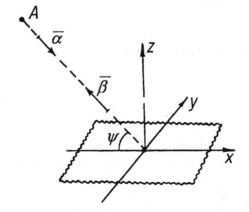

Formulae (4.81) and (4.82) allow the backscattering of radio waves to be analyzed provided that the autocorrelation function $B(\rho)$ is specified. As was mentioned in Section 4.3, integral I_1 depends on the behavior of the integrand at small ρ, when the difference $1 - B(\rho)$ is also small.

If $B(\rho)$ is given by formula (4.68), we have

$$1 - B(\rho) = \left(\frac{\rho}{l}\right)^2 , \qquad (4.83)$$

and relation (4.82) transforms to the tabular integral

$$I_1 = l^2 \int_0^\infty J_0\left[2k\left(\frac{\rho}{l}\right)l\cos\psi\right] \cdot \exp\left[-4k^2\sin^2\psi \cdot z_0^2\left(\frac{\rho}{l}\right)^2\right]\left(\frac{\rho}{l}\right)d\left(\frac{\rho}{l}\right) =$$
$$= \frac{l^2}{8k^2 z_0^2 \sin^2\psi} \exp\left(-\frac{l^2 \cot^2\psi}{4z_0^2}\right). \qquad (4.84)$$

From (4.81) and (4.84) follows

$$\sigma_0 F(\psi) = \frac{M^2}{2\gamma^2 \sin^4\psi} \exp\left(-\frac{\cot^2\psi}{2\gamma^2}\right), \qquad (4.85)$$

where $\gamma^2 = 4z_0^2 l^{-2}$ is the mean square of rough surface slopes. Formula (4.85) allows the backscattering indicatrix to be found,

$$F(\psi) = \sin^{-4}\psi \cdot \exp\left(-\frac{\cot^2\psi}{2\gamma^2}\right), \qquad (4.86)$$

and the differential backscattering cross section,

$$\sigma_0 = \frac{M^2}{2\gamma^2} . \qquad (4.87)$$

From (4.86) and (4.87) it follows that the backscattering indicatrix is determined only by the mean square of rough surface slopes γ^2, and that σ_0 depends on both γ and the soil permittivity. It should be emphasized that $F(\psi)$ for $B(\rho)$ in the form (4.68) or (4.69) does not depend on the wavelength of radio signals.

Let us now consider the exponential autocorrelation function (4.67), for which the following relation is valid:

$$1 - B(\rho) = \frac{\rho}{l} . \qquad (4.88)$$

Calculating the tabular integral (4.82) and taking into account (4.88) and (4.81), we have

$$\sigma_0 F(\psi) = \frac{C_0 M^2}{2} \left(\sin^4 \psi + C_0 \cos^2 \psi\right)^{-3/2}, \qquad (4.89)$$

$$\sigma_0 = \frac{C_0 M^2}{2} , \qquad (4.90)$$

$$F(\psi) = \left(\sin^4 \psi + C_0 \cos^2 \psi\right)^{-3/2} , \qquad (4.91)$$

where

$$C_0 = (4\pi \frac{z_0}{\lambda} \gamma)^2 ,$$

and the mean square of rough surface slopes is given by $\gamma^2 = z_0^2 \Gamma^2$.

From (4.90) and (4.91) it follows that the characteristics of backscattering for the exponential autocorrelation function $B(\rho)$ depend not only on the parameters of surface irregularities but also on the wavelength of radio waves.

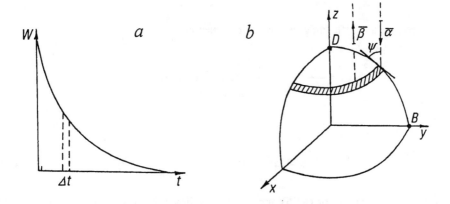

Fig. 4.5. Graphical representation of the backscattering of radio signals from a planetary surface.

The above theory is valid only if the curvature radius of surface irregularities is considerably greater than the wavelength, which largely limits the applicability of the theory. The two variants of this theory corresponding to two different autocorrelation functions involve two different parameters, γ and C_0. Analysis of experimental data shows that these parameters are related as $\gamma^2 \approx C_0^{-1}$. With allowance for this approximate relation, formulae (4.87) and (4.90) become identical. Parameters γ and C_0 depend on the wavelength; however, the theory does not give the correct form of this dependence. It is accepted that the relief irregularities of a particular scale scatter particular wavelengths; therefore, γ and C_0 must differ for different wavelengths.

Let us consider now the scattering of radio waves for the case of the monostatic radar observations of a planet from the Earth (the graphical representation of this problem is shown in Fig. 4.5). The vector of incident waves, $\bar{\alpha}$, is antiparallel to the z-axis and points toward the planet under study, whereas the vector $\bar{\beta}$ of the scattered radio waves received on the Earth is parallel to the z-axis. The effective scattering cross section area can be found by integrating (4.77) over the hemisphere:

$$\sigma = \sigma_0 \int_S F(\psi)ds. \tag{4.92}$$

For the hatched region in Fig. 4.5, the angles ψ are equal and its area is given by

$$ds = 2\pi a^2 \cos\psi \, d\psi \; ; \tag{4.93}$$

then from (4.92) it follows that

$$\sigma = 2\pi a^2 \sigma_0 \int_{\pi/2}^{0} F(\psi)\cos\psi \, d\psi. \tag{4.94}$$

Let us introduce a coefficient given by the expression

$$g_2 = \frac{\sigma}{\pi a^2 M^2} = \frac{2\sigma_0}{M^2} \int_{\pi/2}^{0} F(\psi)\cos\psi \, d\psi \; ; \tag{4.95}$$

then

$$\sigma = \pi a^2 M^2 g_2. \tag{4.96}$$

From the comparison of (4.96) and (4.35) it follows that the coefficient g_2 indicates how many times the parameter σ for a rough planetary surface differs from the parameter

σ for a smooth sphere. In the case of large-scale surface irregularities, when formulae (4.85) and (4.89) are valid, it can be easily shown that:

$$g_2 = 1 + \gamma^2. \tag{4.97}$$

Since $\gamma^2 \ll 1$, then g_2 does not significantly differ from unity. In this case formula (4.35) holds for meter and decimeter radio waves. This unexpected result is correct if the Kirchhoff approximation is applicable. As g_2 is close to unity, then from (4.96) and (4.17) it follows that the dielectric permittivity ε of the surface soil layers can be determined from the radar measurements of s in the meter and decimeter wavelength bands.

Unlike meter and decimeter radio signals, centimeter and millimeter radio signals are considerably scattered from the relief irregularities with small curvature radii and steep slopes. At present, there is no satisfactory theory of scattering for this case. It is clear that the parameter g_2 for centi- and millimeter radio waves must be much more than unity; however, the theoretical estimations of this parameter for these waves are not reliable. The empirical formula (4.43), which holds for very rough surfaces, suggests that $g_2 = 8/3$ at $m = 2$. It should be noted that the permittivity of soils can hardly be determined from radar data in centi- and millimeter wavelength bands, since in this case materials with different ε values may have nearly the same backscattering characteristics.

In the case of pulse radars, the trailing edge of their pulses is extended, which may be due to the scattering of radar signals from different regions of the planet investigated. This allows the determination of the backscattering indicatrix $F(\psi)$ in a way that is discussed below.

The approximate form of a reflected radio pulse is shown in Fig. 4.5a, whereas Fig. 4.5b illustrates the annular region of the surface that is responsible for the energy of the reflected pulse at time t. The major portion of the energy of the reflected pulse falls within a small initial time interval δt corresponding to the reflection of radio waves from the vicinity of point D. The time delay of the radio waves scattered from the hatched area relative to those scattered from the vicinity of point D is as follows

$$t = 2ac^{-1}(1 - \sin\psi). \tag{4.98}$$

The hatched annular scattering zone corresponds to the time interval Δt, which is related to the angle $d\psi$ as

$$\Delta t = 2ac^{-1}\cos\psi\, d\psi. \tag{4.99}$$

The area of this hatched annular zone is given by

$$\Delta s = \pi ac\Delta t. \tag{4.100}$$

This relation suggests that equal time intervals Δt correspond to equal areas Δs. Therefore, the time dependence of the wave power, $W(t)$, gives the unit-area angular dependence of the power flux density of scattered radio waves, $F(\psi)$, provided that the relation between t and ψ has the form (4.98). The coefficient of proportionality depends on instrumental parameters; actually, however, it is insignificant because $F(\psi)$ is normalized to unity at $\psi = 90°$. The dependencies $F(\psi)$ were obtained experimentally for the Moon and some planets of the solar system.

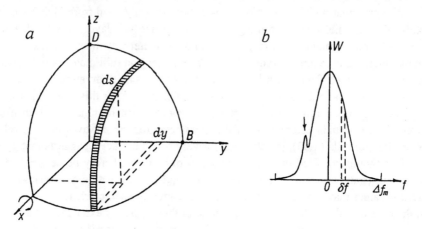

Fig. 4.6. Effect of the rotation of a planet on the power spectrum of reflected radio waves.

Let us consider the energy spectrum of the radio waves reflected from a rotating planet, $W(f - f_0)$ [13, 14]. In Fig. 4.6a, the planet rotates around the x-axis; the z-axis coincides with the direction towards the ground-based radar. The Doppler frequency shift of the radio waves scattered from a small area ds is given by

$$\Delta f = 2 f y \Omega c^{-1} = y f_m a^{-1} , \tag{4.101}$$

where Ω is the angular velocity of the rotating planet, and $f_m = 2 f a \Omega c^{-1}$ is the maximal frequency shift corresponding to point B.

From expression (4.101) it follows that the Doppler frequency shift isolines correspond to the sphere intersection with the plane $y = $ const. In the power spectrum of scattered radio waves (Fig. 4.6b), the narrow frequency band δf corresponds to the radio waves scattered from a narrow hatched annular region on the sphere (Fig. 4.6a). Then we have

$$d(\Delta f) = f_m a^{-1} dy . \tag{4.102}$$

The wave power dW that is backscattered from the area ds and received by the ground-based radar is given by

$$dW = A_1 F(\psi) ds = \frac{A_1 F(\psi) a\, dz\, dy\, d(\Delta f)}{\sqrt{a^2 - z^2 - y^2}},$$ (4.103)

where A_1 is a constant dependent on instrumental parameters. From (4.101)–(4.103), we can derive the following expression for the spectral density:

$$\frac{dW}{d(\Delta f)} = \frac{\dfrac{A_1}{a}\left(\dfrac{a}{f_m}\right) F(\psi)\, dz}{\sqrt{1 - \left(\dfrac{z}{a}\right)^2 - \left(\dfrac{\Delta f}{\Delta f_m}\right)^2}}.$$ (4.104)

Taking into account that $za^{-1} = \sin\psi$ and $dz = a\cos\psi\, d\psi$ and integrating over the hemisphere, we obtain the following expression for the energy spectrum of radio waves:

$$W(\Delta f) = \int_{\psi_1}^{0} \frac{F(\psi)\cos\psi\, d\psi}{\sqrt{\cos^2\psi - \dfrac{\Delta f^2}{\Delta f_m^2}}}.$$ (4.105)

where

$$\psi_1 = \arccos\left(\frac{\Delta f}{\Delta f_m}\right).$$

If $F(\psi)$ is given by (4.43), then

$$W(\Delta f) = \left[1 - \left(\frac{2\Delta f}{\Delta f_m}\right)^2\right]^{\frac{m}{2}}.$$ (4.106)

Thus, we showed that the time and frequency discrimination of reflected signals allow two transections of the planetary surface, which correspond to the hatched regions in Figs. 4.5 and 4.6. The intersection of these transections gives a small element of the planetary surface. This makes it possible to estimate the intensity of the radio waves scattered from a small area with particular coordinates. The distribution of this intensity in coordinates gives a visual image of a radar mapped object. This mapping technique is applicable not only to planets but also to celestial bodies of an irregular form, such as asteroids.

Let us now turn our attention to experimental radar data. The first celestial body explored by a radar technique was the Moon. The ground-based remote sensing of the Moon made it possible to determine the effective cross sectional area of scattered radio waves, σ, and then, based on formulae (4.35) and (4.96), to estimate the Fresnel reflection coefficient M. According to (4.17), the Fresnel coefficient M depends solely on the permittivity ε of surface soil layer. It is this approach that enabled the estimation of ε for lunar regolith. In [15–24], the results of the remote sensing of the Moon's surface with radar wavelengths ranging from $\lambda = 0.86$ mm to $\lambda = 19.2$ m are presented. Table 4.1 compiled from those literature data shows experimental values for the effective radar cross sectional areas normalized to the visible lunar disk area and corresponding values of the lunar soil permittivity. In each column of Table 4.1, the value for ε is the mean of four or five replicate measurements in the indicated wavelength band. It should also be noted that the data presented refer to the central part of the lunar disk. As is evident from the table, the dependence of the effective permittivity ε on wavelength in the interval $\lambda = 10–200$ cm is very weak. At the same time, the permittivity shows a strong dependence on wavelength in the interval $\lambda = 3–10$ m. As is shown in [7–9], the wavelength dependence of ε is due to a deeper penetration of longer wavelengths into lunar soils, where their density is higher. In the interval $\lambda = 20–200$ cm, the dielectric permittivity ε can be taken as constant and equal to 2.9.

Table 4.1. Mean values of $\sigma/\pi a^2$ and ε for the Moon

λ, cm	0.8–3.6	10–23	33–75	100–200	250–780
$\sigma/\pi a^2$	0.063	0.052	0.078	0.068	0.087
ε	2.7	2.6	3.1	2.9	3.4

More detailed radar data on the permittivity of lunar soil are presented in Fig. 4.7, from which it is evident that ε increases with wavelength. It should be noted that, at $\lambda > 5$ m, the accuracy of radar measurements decreases. On the other hand, centimeter radio waves are significantly scattered, which is difficult to take into account during the determination of ε. In [15–23], the authors obtained the experimental dependencies $F(\psi)$ for the Moon in 8.6-mm, 3.8-cm, 23-cm, 68-cm, 6-m, and 11.3-m wavelength bands (four of these dependencies are shown in Fig. 4.8, where numerals along the curves indicate wavelength in cm). As is shown in [15], the functions $F(\psi)$ for meter and decimeter radio waves are well described by formula (4.91) provided that the value of parameter C_0 has been appropriately chosen (the mean values of C_0 for the lunar surface are listed in Table 4.5). From Fig. 4.8 it is evident that the dependencies $F(\psi)$ in the meter and decimeter wavelength bands are similar, but have a different diffuse character in the centimeter and millimeter wavelength bands. Such behavior of the functions $F(\psi)$ is due to different contributions from the diffuse scattering and quasi-specular reflection of radio waves of different wavelengths.

Let us now discuss experimental data on the differential scattering cross sectional area σ_0. This quantity is given by the approximate formulae (4.87) or (4.90) with the parameters γ and C_0, respectively. Calculations in terms of these formulae yield only rough estimates for σ_0. Moreover, the formulae are not applicable for the case of millimeter and centimeter radio waves.

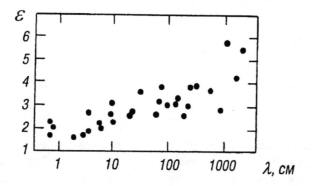

Fig. 4.7. Dielectric permittivity of lunar soil versus the radar wavelength.

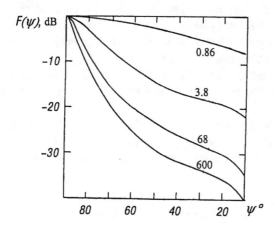

Fig. 4.8. The indicatrix of the lunar surface backscattering function for four different radio waves.

Thus, the remote sensing of the Moon and planets provided information about $F(\psi)$ and σ, from which the average value for σ_0 can be derived using formula (4.94):

$$\sigma_0 = \frac{\sigma}{2\pi a^2 \int\limits_{\pi/2}^{0} F(\psi)\cos\psi\, d\psi} \tag{4.107}$$

This relation shows how σ_0 can be obtained from the experimentally measured σ and $F(\psi)$. It was found that $\sigma_0 = 0.1 \div 0.15$ for $\lambda = 0.8$ cm and is as high as 5 to 9 for $\lambda = 3$ m [24]. The average values of σ_0 are listed in Table 4.2.

Table 4.2. Mean values of σ_0 for the lunar surface [24]

σ_0	0.15	0.6	1.5	3	5	8
λ, cm	1	3	10	30	100	300

Radar altimeters installed on board lunar landers enabled direct measurements of σ_0 [25–29]. Two *Luna* landers gave a value $\sigma_0 = 0.5$–1 in the 3-cm wavelength band, whereas five *Serveyors* landers gave a value $\sigma_0 = 0.2$–0.4 in the 2.3-cm wavelength band. Direct measurements in various lunar regions showed that σ_0 varies from -1 dB to -6 dB in the wavelength band $\lambda = 3.2$–2.3 cm and that $\gamma = 9$–14° for centimeter radio waves. It should be noted that the estimation of γ from the results of measurements of σ_0 in the centimeter wavelength band is not reasonable, since centimeter radio waves undergo diffuse scattering rather than quasi-specular reflection and, hence, the application of formulae (4.86) and (4.87) or (4.90) and (4.91) may give incorrect results.

The upgrading of ground-based radars allowed radar images of the lunar surface to be obtained. The first radio image of the central part of the Moon in the 70-cm wavelength band [30] demonstrated the feasibility of the radiovision technique for investigating the solar system objects. The first radar observations of the polar lunar regions are described in [147].

Extensive radar observations of Mars made it possible to obtain the scattering indicatrix $F(\psi)$, permittivity ε, soil density ρ, and parameter C_0 [31–42] (the ratios $\sigma/\pi a^2$ and permittivities of the Martian surface obtained at different wavelengths are presented in Table 4.3).

Table 4.3. Radar data on $\sigma/\pi a^2$ and permittivity of the Martian surface soil

λ, cm	3.8	12.5	23	40	70
$\sigma/\pi a^2$	0.11	0.086	0.14	0.07	0.07
ε	4.0	3.3	4.7	3.0	3.0

As is evident from this table, the mean permittivity of the Martian surface soil is 3.6. At the same time, light and dark surface regions display different values of ε (2.2 and 4.3, respectively). The mean values of the parameter C_0 for the Martian relief irregularities obtained in three wavelength bands are listed in Table 4.5. At $\lambda = 5$ cm, the mean value of γ was found to be about 5°.

The high-resolution radar observations of Mars in the 3.5-cm wavelength band described in [41] were performed using a radar system whose receiving block included 27 antennae arranged in an interferometer that had a 36-km baseline. These observations made it possible to obtain the σ_0-maps of large subequatorial regions of Mars. The values of σ_0 were derived for the matched and inverse circular polarization of radio waves. The experimental dependencies $F(\psi)$, parameter σ_0, and its diffuse component σ_α greatly varied for particular Martian regions. Thus, an extended subequatorial region between 160 and 135° longitude exhibited low values of σ_0, whereas the region between 108 and 120° longitude exhibited an increased value of the diffuse component. The radar measurements of ε allowed the density ρ of the surface soil to be determined. The author of [42] found that the mean density of a 0.5-m layer of the Martian soil is equal to 1.4 g cm^{-3}, varying from 1 to 2 g cm^{-3} for different regions of this planet. Soil density increases with depth. Thus, the spacecraft-borne radar studies of Martian subsurface soils are possible in a decameter wavelength band [43].

Table 4.4. Radar estimations of $\sigma/\pi a^2$ and permittivity ε for Venus

λ, cm	23 ÷ 41	68 ÷ 70	784
$\sigma/\pi a^2$	0.16	0.14	0.14
ε	5.4	4.7	4.7

Table 4.5. Mean values of parameter C_0 for some celestial bodies

λ, cm	3.6	23	68
Moon	25	80	100
Venus	100	230	300
Mars	300	600	900

Radar observations provided valuable information on the relief of the subequatorial Martian region. The major portion of the returning wave energy is due to the reflection of radio waves from the subradar point D (Fig. 4.1); this allows a precise timing of the radio waves traveling from the transmitter to point D. As the planet rotates, this point shifts, thereby enabling the deviation of large-scale relief features from a sphere to be determined. For the Martian surface, these deviations reach ± 6 km; therefore, height differences on the surface of this planet may be as high as 12 km [34, 35, 37, 39, 60].

Most radar observations are concerned with Venus. The first stage of such investigations involved the determination of the radar cross-section σ and the power spectrum of reflected radio waves, $W(\Delta f)$, for several wavelengths. The data obtained were then used to evaluate the permittivity of the surface soil, ε, and the backscattering indicatrix $F(\psi)$ [44–49]. It was found that the experimental indicatrix $F(\psi)$ is well described by formula (4.91) and, therefore, the parameter C_0 can be deduced from the comparison of experimental and calculated data. Detailed studies [50–60] provided more information on the dependencies $W(\Delta f)$ and $F(\psi)$ and the parameters σ, ε, and C_0. Unlike C_0, parameters σ and ε do not depend on the wavelength λ. Table 4.4 summarizes the values of $\sigma/\pi a^2$ and ε calculated from experimental data by formulae (4.96) and (4.17). For meter and decimeter radio waves, $\sigma/\pi a^2 = 0.15$ and $\varepsilon = 4.9$. The experimental values of the parameter C_0 for three wavelengths are given in Table 4.5 (it should be noted that the data presented in Tables 4.4 and 4.5 refer to the subequatorial Venusian region). Radar measurements in the 3.6-cm wavelength band gave lower values for σ, which can be due to a stronger absorption of centimeter radio waves in the Venusian atmosphere.

Further progress in the ground-based radar technique made it possible to obtain detailed power spectra of reflected radio waves and to improve the accuracy of timing of the radio wave trip to the subradar point D. The resolution of reflected signals in the delay time–frequency coordinates led to the discovery of a bright region and the observation of the dynamics of an anomaly in the power spectrum of the reflected signals. The precise timing of the radio wave trip to the subradar point made it possible to investigate the planetary relief in the equatorial region, where height differences reach 4 km. The discovery of the bright region allowed researchers to introduce a zero Venusian meridian, i.e., to establish the latitude–longitude coordinate system of this planet. Analysis of the power spectra of the reflected radio waves showed that these spectra have an anomaly (shown in Fig. 4.6 b by the arrow) induced by an increased reflectance of a particular region of the planet. Due to the planet rotation, the anomaly shifts along the spectrum, thereby enabling the rotation period of the planet to be determined. The rotation period of Venus was found to be equal to 243 Earth days. The determination of the travel time of radio signals performed in various years led to a precise estimation of the astronomical unit, a measure of distances in the solar system, and planetary ephemerides [59, 60]. The development of technique for the digital processing of reflected signals enhanced the radar resolution along the planetary surface and made it possible to obtain radar images of the large-scale Venusian relief [61–66].

The principle of radiovision lies in signal discrimination in the time delay–frequency coordinate system. In Figs. 4.5 and 4.6, the hatched areas correspond to the planetary regions for which, respectively, the delay of radio signals and the Doppler frequency shift are constant. The intersection of these areas defines the coordinates of an element of the planetary surface. In these figures, the z-axis points towards the Earth, the x-axis coincides with the planet rotation axis, and the line DB corresponds to the equator. The resolution of the first radar images of Venus, which was of about 50 km, was then improved to

≈ 2 km. The Venusian relief was found to be complex, exhibiting circular structures, ridges, volcano craters, and large flats.

Radar altimeters installed on board the artificial Venusian satellites *Pioneer-Venus*, *Venera 15*, and *Venera 16* made possible the profiling of the Venusian relief and its hypsometric imaging [67–75]. In these experiments, the accuracy of the height estimation was about 200 m, and the horizontal resolution was about 100 m. The high-resolution radar observations of Venus provided detailed information on the characteristics of radio wave scattering from different Venusian terrain forms. The parameters γ and C_0 for these terrain forms can be found in [63–75].

The reflectance characteristics of Mercury's surface are nearly the same as those of the Moon: $\gamma = 6°\div9°$ and $\mu = 0.06$ [41, 76, 77]. The power spectrum width of 13-cm radio waves due to the planet's rotation is 110 Hz.

High-performance planetary radars provided valuable information on Jupiter's satellites Ganymede, Callisto, and Europa [78, 88]. The characteristics of reflectivity of these bodies greatly differed from those of the Moon and Mercury, exhibiting, in particular, increased μ and $\sigma/\pi a^2$ values and widened power spectra of the radio signals reflected from these Jupiter satellites. The backscattering indicatrix $F(\psi)$ derived from the power spectra of reflected signals was close to that predicted by the Lambert law (see formula (4.43)) with the index m less than 2 (implying that radio waves are scattered from a very rough surface). Together with other observations, radar-derived data suggest that the rough surfaces of Ganymede, Callisto, and Europa are icy and covered with frost. Presumably, the surfaces of these satellites are composed of broken ice with numerous fractures, faces, and inclusions of silicate materials. Some radar characteristics of the three satellites obtained in the 12.6-cm wavelength band are listed in Table 4.6 [78, 88]. The satellite's radii are given according to optical data.

Table 4.6. Radar characteristics of Jupiter's satellites

Satellite	$\sigma/\pi a^2$	μ	m	a, km
Ganymede	1.6	1.55	1.46	2631
Callisto	0.7	1.19	1.43	2400
Europa	2.5	1.56	1.73	1569

Conversely, the radar characteristics of Io, the fourth of the Jupiter satellites, are very close to those of the Moon.

The remote sensing of the largest Saturnian satellite Titan in the 3.6-cm wavelength band revealed variations in the parameter $\sigma/\pi a^2$ from 0.07 to 0.3, with an average value of 0.12 [41]. Analysis of the power spectra of the returning radio signals showed that they have undergone diffuse scattering.

High-resolution planetary radars were also employed for the remote sensing of asteroids and comets. The earliest radar observation of minor celestial bodies was performed back in 1968 when asteroid 1566 Icarus approached the Earth within 6.3 million km.

Since that time 47 asteroids in the asteroid belt, 74 asteroids approaching the Earth, and 5 comets have been investigated by the radar technique [79–87, 148, 149].

The remote sensing of asteroids involves the estimation of the effective scattering cross section for matched polarization, σ, an arbitrary radius a, and the power spectrum width of reflected signals, Δf_m. To estimate surface roughness, one has to determine the ratio of the powers of reflected signals for unmatched and matched polarization, μ. If the asteroid's radius a is known from optical measurements, the reflectivity M is sought from the experimental value of σ using formula (4.96). Given M, the permittivity of the asteroid's matter, ε, can be found by formula (4.17). Knowing the radius a and the spectral width of the reflected signals, one can estimate the asteroid's rotation period. The employment of modulated signals allows the discrimination of the radio waves reflected from various regions of the asteroid's surface, i.e., the obtaining of a radar image of the asteroid. To that end, the reflected signal should be analyzed in two coordinates: the modulated signal can be resolved to yield the range, while the resolution in frequency due to body rotation gives the second coordinate. The precise radar-aided estimations of the distance to asteroids and their radial (with respect to the Earth) velocity allow the ephemerides of asteroids to be refined in terms of celestial mechanics.

The authors of [79–81] carried out the radar observations of 20 asteroids in a 13-cm wavelength band and, for each of them, estimated the effective scattering cross section for matched polarization, σ, the ratio of the powers of reflected signals for unmatched and matched polarization, μ, and the power spectrum width of reflected signals, Δf_m. These parameters, together with asteroids' radii derived from optical measurements, are summarized in Table 4.7. The large values of $\sigma/\pi a^2$ (from 0.2 to 0.6) found for 5 Astraea, 16 Psyche, and 554 Peraga, are indicative of a high metal content of these asteroids.

Possible cataclysmic encounters of large asteroids with the Earth have stimulated investigations of regularly approaching asteroids, such as Toutates, which approaches the Earth every four years. Detailed radar observations of this asteroid were performed in 1992, when Toutates was 3.6 million km from the Earth [82–85]. In particular, the radar image of Toutates was obtained with a linear resolution of 200 m, or an angular resolution of 10^{-3}" (it should be noted that the resolving power of advanced ground-based radars exceeds that of spacecraft-borne optical telescopes [84, 85]). Toutates was found to have an irregular shape and a size of $1.9 \times 2.4 \times 4.6$ km (the accuracy of these measurements was 100 m). Analysis of the Doppler spectra showed that the rotation period of this asteroid around one axis was 5.4 Earth days, and 7.3 days around the other axis. The remote sensing parameters were found to be $\mu = 0.25$, $\sigma \approx 1$ km^2, and $\sigma/\pi a^2 \approx 0.08$. Other information on the characteristics of 4179 Toutates, 216 Cleopatra, and 1998 KY26 can be found in [157–159].

Radar observations of five comets showed that their remote sensing parameters σ and μ are close to those of asteroids [86, 87]. At the same time, the power spectra of the radar signals reflected from two comets differed from those of asteroids in having a narrow spectral component corresponding to the reflection of radio waves from the cometary nuclei and a broad component corresponding to their scattering from the hazy cometary heads [88].

The results of the remote sensing of planets, their satellites, asteroids, and comets are comprehensively reviewed in [88].

Table 4.7. The remote sensing parameters of asteroids [79–81, 88]

Asteroid	$2\Delta f_m$, Hz	σ, km^2	μ	$\sigma/\pi a^2$	$2a$, km
1 Ceres	1500	33000	0.04	0.047	950
2 Pallas	590	21000	0.05	0.092	538
4 Vesta	900	12000	0.40	0.054	530
5 Astraea	120	2400	0.20	0.21	121
6 Hebe	600	4300	–	0.13	205
7 Iris	280	5900	0.08	0.17	208
8 Flora	220	1500	0.16	0.073	162
9 Metis	180	3500	0.18	0.11	203
12 Victoria	190	2100	0.14	0.15	135
16 Psyche	540	14000	0.14	0.29	250
19 Fortuna	550	3200	0.04	0.083	221
41 Daphnia	500	2900	0.13	0.090	203
46 Hestia	110	900	–	0.067	131
80 Sappho	77	650	0.25	0.12	83
97 Klotho	45	1100	–	0.12	108
139 Juewa	75	1300	0.10	0.056	172
144 Vibilia	140	1800	0.18	0.13	131
356 Liguria	72	1800	0.12	0.095	155
554 Peraga	150	1600	0.06	0.20	101
694 Ekard	170	610	–	0.076	101
4179 Toutates	3.5	0.7–1.3	0.25	0.15	1.5–4
6489 Golevka	0.8–3.7	0.05	–	0.2	0.5

4.5 Reflection of radio waves in bistatic remote sensing

In the bistatic remote sensing of planets and the Moon (Fig. 4.9), radio signals transmitted from a satellite (point A) are reflected from the hatched region around point D, which corresponds to the condition of equality between the angles of incidence of radio waves and their reflection, and received at a ground station (point B). During the bistatic radar observations of a planet, the planet-to-Earth distance $BO = R_2$ is much greater than the planet's center-to-satellite distance $AO = R_1$ and the angle ABO is small. Therefore, the ground-based station receives the direct radio waves propagating along the path AB and reflected radio waves propagating along the path ADB. The station can discriminate between these waves because of the different Doppler frequency shifts (due to the satellite motion) for the direct and reflected radio waves propagating along the paths AB and ADB, respectively.

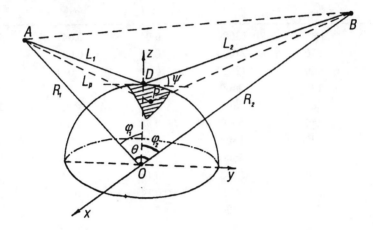

Fig. 4.9. Graphical representation of the bistatic radar observations of a planetary surface.

Bistatic radar observations can yield the reflection coefficient η (and, hence, the permittivity ε) and the shape of the power spectrum of radio waves, $W(f - f_0)$ (and, hence, the parameter γ). As the satellite moves, the reflecting region D shifts along the planet's surface, thereby enabling the estimation of ε and γ for various regions of the investigated planet or the Moon. The results of the theoretical and experimental bistatic radar investigations of the Moon, Mars, and Venus are described in [89–119].

Let us consider the theoretical expressions relating the reflectivity of radio waves, η^2, and the effective cross sectional area σ. In the case of a sufficiently smooth spherical surface irradiated with meter radio waves, we can use formulae (4.37) and (4.38), which are exact if the rigorous condition (4.14) is satisfied. Actually, planetary reliefs are always rough and have irregularities whose dimensions are comparable with the wavelength of radio signals; in this case formulae (4.75) and (4.76), which account for relief roughness, are more appropriate. Using (4.75) and (4.76) and taking into account (4.70) and (4.72), we have

$$\eta^2 = \frac{M^2}{16\pi\gamma^2 L_1^2} \int_S \exp\left(-\frac{q_x^2 + q_y^2}{2\gamma^2 q_z^2}\right) ds \, , \qquad (4.108)$$

and

$$\sigma = \frac{M^2}{4\gamma^2} \int_S \exp\left(-\frac{q_x^2 + q_y^2}{2\gamma^2 q_z^2}\right) ds, \tag{4.109}$$

These expressions were derived by removing the slowly changing terms from the integrand and taking into account the following relations:

$$L_1 \ll L_2,$$

$$q_z = -2\sin\psi,$$

$$\bar{\alpha} \cdot \bar{\beta} = \cos 2\psi, \tag{4.110}$$

$$\gamma^2 = \frac{2z_0^2}{l_1 l_2},$$

$$\int_0^\infty I_0\{k|q_\rho|\rho\}\rho \cdot \exp\left(-\frac{k^2 q_z^2 \gamma^2 \rho^2}{2}\right) d\rho = \frac{1}{k^2 \gamma^2 q_z^2} \cdot \exp\left(-\frac{q_x^2 + q_y^2}{2\gamma^2 q_z^2}\right).$$

In (4.108) and (4.109), the Fresnel reflection coefficient M is given by (4.16) for the glancing angle ψ found for the specular reflection point D. The effect of the radio wave polarization on the scattered electromagnetic field is described by the dependence $M(\psi)$. If a satellite situated at point A bears an omnidirectional antenna, the integration region S is determined by the condition of its observability from both the satellite and the ground-based station. With the directional antenna pointing to point D, S is determined by the beamwidth of this antenna. Let us introduce the relative scattering surface

$$A_1 = \frac{\sigma}{\pi a^2 M^2} = \frac{1}{4\pi a^2 \gamma^2} \int_S \exp\left(-\frac{q_x^2 + q_y^2}{2\gamma^2 q_z^2}\right) ds. \tag{4.111}$$

From (4.108) and (4.111) it follows that the radio wave reflectance η^2 is related to the relative surface A_1 as

$$\eta^2 = \frac{A_1 M^2 a^2}{4 L_1^2}. \tag{4.112}$$

The representation of the scattering coefficient in the form (4.112) allows a separate analysis of the effects of the surface roughness and sphericity (these effects are taken into account by the parameter A_1) and of the electrodynamic properties of the surface (these are taken into account by the Fresnel coefficient M). Let us introduce a spherical coordinate system with the z-axis passing through the specular reflection point D (Fig. 4.10).

The position of the element ds on the scattering surface at point P is fixed by coordinates φ and χ. Figure 4.10 shows the unit vectors of the incident and scattered radio waves ($\overline{\alpha}$ and $\overline{\beta}$, respectively); in this case the vector $\overline{\alpha}$ is in the direction from the satellite located at point A towards an arbitrary point P on the planet's surface, whereas the vector $\overline{\beta}$ points towards the Earth.

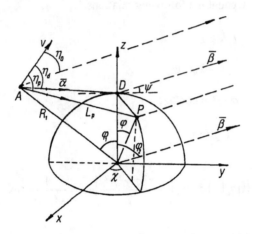

Fig. 4.10. Spherical coordinate system used for the determination of the scattering coefficient and the Doppler frequency shift.

Let us express the projections of the scattering vector $\overline{q} = \overline{\alpha} - \overline{\beta}$ through the coordinates of the spherical system:

$$q_z = \left(\frac{a}{R_1} + \sin\chi \cdot \sin\varphi \cdot \sin\varphi_1 - \cos\varphi \cdot \cos\varphi_1\right) \cdot I_2^{-1} -$$
$$- \sin\chi \cdot \sin\varphi \cdot \sin\varphi_2 - \cos\varphi \cdot \cos\varphi_2 \ ,$$
$$q_x = \left(\sin\chi \cdot \cos\varphi \cdot \sin\varphi_1 + \sin\varphi \cdot \cos\varphi_1\right) \cdot I_2^{-1} +$$
$$+ \sin\varphi \cdot \cos\varphi_2 - \sin\chi \cdot \cos\varphi \cdot \sin\varphi_2 \ , \tag{4.113}$$
$$q_y = \sin\varphi_1 \cos\chi \cdot I_2^{-1} - \sin\varphi_2 \cos\chi \ ,$$
$$I_2 = \left[\left(\frac{a}{R_1}\right)^2 + 1 + 2\left(\frac{a}{R_1}\right)\left(\sin\chi \cdot \sin\varphi \cdot \sin\varphi_1 - \cos\varphi \cdot \cos\varphi_1\right)\right]^{1/2} \ .$$

Here α is the radius of the planet or the Moon, φ_2 and φ_1 are the angles between the z-axis and directions towards the Earth or the satellite, respectively. The meaning of the other parameters is clear from Figs. 4.9 and 4.10.

Formulae (4.111) or (4.112) and (4.113) can be used to analyze energy relationships during the bistatic radar observations of relief irregularities [90–92]. Figures 4.11a and

4.11b present the results of the computation of the relative surface $A_1(\psi)$ (solid lines) for $\gamma = 0.15$ and 0.3 (these values of the parameter γ are typical of weakly and very rough surfaces, respectively). For comparison, the dashed lines in these figures show the angular dependence of the relative surface for a smooth sphere. Calculations were performed for three values of the parameter a/R_1 equal to 0.685, 0.813, and 0.897 (curves *1*, *2*, and *3*, respectively). For the Moon, these curves correspond to observation from a satellite occurring at altitudes of 800, 400, and 200 km, respectively. For Mars, curves *1*, *2*, and *3* correspond to observations from altitudes of 1560, 780, and 390 km.

As is evident from Fig. 4.11, at the glancing angles $\psi > 30°$, the relative surface depends only slightly on the surface roughness; this dependence becomes, however, essential at $\psi < 20°$. If $\gamma < 0.08$, a rough spherical surface reflects radio waves like a smooth sphere. At $\psi < 10°$, A_1 exhibits a steep decline. At small glancing angles, the above expressions for A_1 are approximate because they do not take into account the shading of a part of the scattering surface by its irregularities. The comparison of the A_1 values for rough and smooth spherical surfaces shows that the effect of the surface irregularities is not great at $\psi > 40°$. It should be emphasized that, in the case under consideration, the heights of irregularities are considerably greater than the wavelength, and, therefore, condition (4.14) is not satisfied. Nevertheless, the parameters A_1 and, hence, η^2 and σ turned out to be close to those for a smooth sphere. As is evident from Fig. 4.11, at $\psi < 10°$, the A_1 values for rough and smooth spherical surfaces differ appreciably; the theory is the least accurate just at these glancing angles.

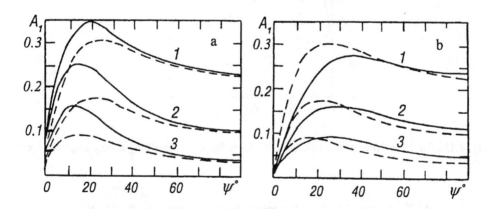

Fig. 4.11. Dependence of the scattering cross section on the glancing angle of radio waves for various satellite altitudes (the solid and dashed lines correspond to rough and smooth spheres, respectively).

In calculating the scattering of radio waves emitted from the satellites of Venus, Mars, and the Moon, we shall use the radar-derived values of ε and γ presented in Section 4.4.

Knowing ε, γ, and A_1, we can determine the dependence of the reflection coefficient η on the glancing angle ψ and the satellite altitude H.

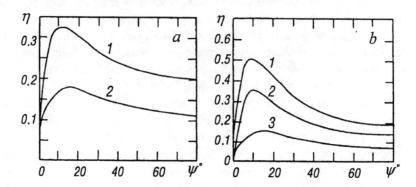

Fig. 4.12. Field strength reflection coefficient as a function of the glancing angle for (*a*) Venusian and (*b*) Martian satellites.

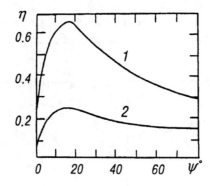

Fig. 4.13. The dependence of the field strength reflection coefficient on glancing angle for lunar satellite.

Figure 4.12b presents the dependencies of the reflectance η of the Martian surface on the glancing angle ψ for horizontally polarized radio waves (curves *1*, *2*, and *3* correspond to $\gamma = 0.05$ and satellite altitudes of 750, 1500, and 6000 km, respectively) and Fig. 4.12a presents the same dependencies for the Venusian surface (curves 1 and 2 correspond to $\gamma = 0.09$ and satellite altitudes of 2800 and 11000 km, respectively). Calculations were performed without accounting for the effect of the Venusian atmosphere. A more close consideration of reflectivity with allowance for the effect of the Venusian atmosphere showed that the refractive attenuation of radio waves in radar observation should be accounted only if $\psi < 15°$.

Let us consider the results of the calculation of the lunar surface reflectance for horizontally polarized radio waves. Figure 4.13 illustrates the dependencies $\eta(\psi)$ for $\varepsilon = 3$ and $\gamma = 0.15$. Curves *1* and *2* are for satellite elevations of 100 and 800 km, respectively. It should be noted that the dependencies of the reflectance on the glancing angle of radio waves, $\eta(\psi)$, presented in Figs. 4.12 and 4.13 correspond to linear wave polarization. In the case of vertical polarization, the Fresnel coefficient M has a deep minimum when ψ is equal to the Brewster angle; therefore, the dependence $\eta(\psi)$ must exhibit a steep decline to zero at this angle. The polarization of radio waves in bistatic radar observations is primarily determined relative to the plane tangent to the sphere at point D.

The radio waves scattered from different surface areas show different Doppler frequency shifts because of the motion of the satellite, which leads to spectrum broadening and the midband frequency shift. The wave power spectra and the Doppler frequency shift during radar observations were analyzed in a number of publications [91–94, 99, 100]. Let us use the data presented therein to analyze the frequency difference of the incident and reflected waves, Δf, in relation to the geometry of observations and the trajectory of the satellite. This problem is of great importance since the knowledge of this frequency difference makes it possible to perform the spatial separation of radio signals corresponding to the rays AB and ADB through their frequency discrimination (Fig. 4.9). To estimate Δf, we must find the Doppler frequency shift for the incident wave:

$$\Delta f_0 = fc^{-1}\left(\bar{V}\cdot\bar{\beta}\right) \tag{4.114}$$

and for the wave reflected from the vicinity of point D:

$$\Delta f_d = fc^{-1}\left(\bar{V}\cdot\bar{\alpha}\right). \tag{4.115}$$

In expressions (4.114) and (4.115), $\bar{V}$ is, in accordance with Fig. 4.10, the satellite velocity vector, vector $\bar{\beta}$ points toward the Earth, and vector $\bar{\alpha}$ is directed from the satellite located at point A to the specular-reflection point D on the surface. From (4.114) and (4.115) it follows that

$$\Delta f = \Delta f_d - \Delta f_0 = fc^{-1}\left[\left(\bar{V}\cdot\bar{\alpha}\right)-\left(\bar{V}\cdot\bar{\beta}\right)\right]. \tag{4.116}$$

If vectors $\bar{\beta}$ and $\bar{\alpha}$ lie in the orbital plane, then from (4.116) we have

$$\Delta f = fc^{-1}V\left(\cos\eta_d - \cos\eta_0\right), \tag{4.117}$$

where η_d is the angle between vectors $\overline{V}$ and $\overline{\alpha}$ and η_0 is the angle between vectors $\overline{V}$ and $\overline{\beta}$. From (4.117) we can obtain

$$\Delta f = 2Vfc^{-1}\sin\left(\frac{\eta_0 + \eta_d}{2}\right)\sin\left(\frac{\eta_0 - \eta_d}{2}\right). \qquad (4.118)$$

It follows from (4.118) that when a satellite immerses behind the planet or emerges from behind it, the frequency difference Δf tends to 0, since in this case $\psi \to 0$ and, consequently, $\eta_d \to \eta_0$. For the backscattered radio waves, when vectors $\overline{\alpha}$ and $\overline{\beta}$ point in opposite directions and $\psi = 90°$, Δf is also zero, since $\eta_0 - \eta_d = 180°$. At $\psi \approx 40°$, Δf is at a maximum. When the satellite reappears from behind the planet, Δf is negative; conversely, Δf is positive when the satellite disappears behind the planet.

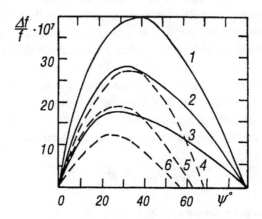

Fig. 4.14. Dependence of the differential frequency magnitude of incident and reflected radio waves on the glancing angle for Martian satellites.

Figure 4.14 shows typical dependencies of the differential frequency difference, $\Delta f/f$, on the glancing angle ψ for Martian satellites occurring at different altitudes and differently tilted orbits. Curves 1, 2, and 3 are for satellite altitudes of 1600, 800, and 400 km, respectively; the earthward vector $\overline{\beta}$ lies in the orbital plane. Dependencies 4, 5, and 6 are for the same altitudes, but vector $\overline{\beta}$ makes an angle of 45° with the orbital plane. It can be inferred from Fig. 4.14 that, in bistatic radar observations, the incident and reflected waves can reliably be distinguished by the frequency discrimination method. It should be noted that the data presented in Fig. 4.14 were obtained under the assumption that radio waves are reflected from the vicinity of point D. Actually, however, radio waves are reflected from a larger area; this causes a broadening of the power spectrum of the reflected radio signals. In this case, Δf corresponds to the maximum of this spectrum.

Analysis of the power spectra of the scattered radio waves and their relationship with the surface roughness requires the knowledge of the distribution of the Doppler shifts of the radio signals scattered from arbitrary points on the planetary surface. The Doppler frequency shift of the radio waves scattered earthward from a small surface area with its center at point P (Fig. 4.10) is given by

$$\Delta f_p = f c^{-1} V \cdot \cos \eta_p , \qquad (4.119)$$

where η_p is the angle between the velocity vector $\overline{V}$ and the direction to point P. The simple expression (4.119) allows a graphical representation of the lines of equal Doppler frequencies on a scattering sphere since $\Delta f_p = $ const if η_p are constant. These lines represent the intersections of the scattering sphere with the cones that have a vertex angle equal to η_p, cone element AP, and axis coinciding with the velocity vector of the satellite, $\overline{V}$ (Fig. 4.10).

From the diagram presented in Fig. 4.10 follows the expression for the frequency shift Δf_p:

$$\Delta f_p = -\frac{a R_1 f \dfrac{d\varphi_1}{dt} \left(\sin \varphi \cdot \sin \chi \cdot \cos \varphi_1 - \cos \varphi \cdot \sin \varphi_1 \right)}{c \left[a^2 + R_1^2 - 2 a R_1 \left(\sin \varphi \cdot \sin \chi \cdot \sin \varphi_1 + \cos \varphi \cdot \cos \varphi_1 \right) \right]^{1/2}} . \qquad (4.120)$$

Here R_1 and φ_1 are the coordinates of the satellite (i.e. of point A); φ and χ are the coordinates of the arbitrary point P on the sphere of radius a.

To estimate the resolving power of the frequency discrimination technique, let us consider the Doppler frequency shift for the radio waves scattering from the lunar surface. It is convenient to consider the frequency difference

$$\Delta f_1 = \Delta f_p - \Delta f_d , \qquad (4.121)$$

which shows the extent to which the frequency of the wave reflected from the surface area with its center at point P differs from the frequency of the radio wave reflected from the vicinity of the specular-reflection point D.

Figure 4.15 illustrates the dependence of the differential frequency $\Delta f_1 / f$ on coordinate φ for several values of χ. The curves are for the lunar satellite flying at a height $H = 200$ km and $\psi = 45°$. Curves *1*, *2*, *3*, *4*, and *5* are for χ equal to 180, 130, 90, 50, and 0°, respectively. From 4.15 it follows that, as P shifts from point D in the direction for which $\chi = 100°$, Δf_1 changes insignificantly. However, Δf_1 exhibits a steep rise with increasing φ in the directions for which $\chi = 180$ and 0°, thereby indicating a high resolution of the frequency discrimination technique for these directions.

Let us now analyze the power spectrum of scattered radio waves. For a statistically rough surface, the scattered power flux density is mainly determined by the behavior of the exponential factor in (4.108). At point D, this factor is equal to unity and shows a steep decline as the scattering surface element shifts from this point. Figure 4.16 presents the dependencies of the exponential factor on the angle φ for the lunar satellite flying at a height of 200 km and $\psi = 45°$; γ was taken to be 0.15. Curve *1* corresponds to the shift of point P relative to point D in the direction for which $\chi = \pm 90°$, and curve *2* corresponds to the shift for which $\chi = 0°$. From the figure it follows that the scattered power flux density decreases five- to sevenfold as the scattering surface element shifts from point D by an angle of 3°.

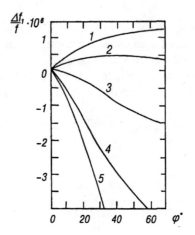

Fig. 4.15. Differential frequency $\Delta f_1/f$ versus the coordinate φ of point P in the bistatic radar observations of the Moon.

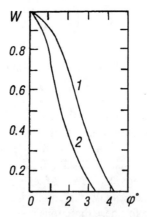

Fig. 4.16. Decline in the energy of the scattered radio waves when point P shifts relative to the specular-reflection point D.

A comparison of Figs. 4.14 and 4.16 gives an idea of the formation of the power spectra of scattered radio waves. To calculate the power spectrum, one has to evaluate the power of the radio waves scattered from various regions of a rough surface in a unit frequency interval [93, 94]. According to (4.108), the power flux density dP of the radio waves scattered from the surface area ds is given by

$$dP = \frac{1}{\gamma^2 L_p^2} \exp\left(-\frac{q_x^2 + q_y^2}{2\gamma^2 q_z^2}\right) ds \ . \tag{4.122}$$

Formula (4.122) does not contain the term $M^2/16\pi$, since the power spectrum of the scattered radio waves, W, is normalized to unity at $\Delta f_1 = 0$. Let us introduce a new parameter,

$$\tau = \frac{(\bar{a} \cdot \bar{q})}{a \cdot |\bar{q}|} \ , \tag{4.123}$$

where $\bar{a}$ is a radius-vector in the direction from the center of the scattering sphere toward point P and $\bar{q}$ is the scattering vector. Then expression (4.122) takes the form

$$dP = \frac{a^2}{\gamma^2 L_p^2} \exp\left(\frac{1 - \tau^2}{2\gamma^2 \tau^2}\right) \sin \varphi \, d\varphi \, d\chi \ . \tag{4.124}$$

To calculate the power spectrum, one has to integrate (4.124) over the areas on the spherical surface that correspond to the fixed frequencies of scattered radio waves.

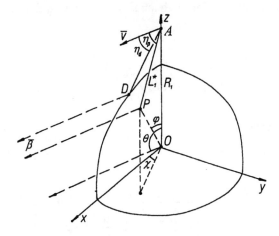

Fig. 4.17. Diagram illustrating the formation of the power spectra of scattered radio waves.

Let us consider the geometry of the problem presented in Fig. 4.17, where the z-axis passes through the source of radio signals (a satellite at point A), and the orbital plane coincides with the xz-plane. Make use of the following parameters: a is the planet's radius, $R_1 = AO$ is the planet center-to-satellite distance, $L_1 = AD$ is the distance between the satellite and the specular-reflection point D, $L_p = AP$ is the distance between the satellite and an arbitrary point P on the planetary surface, V is the velocity vector of the satellite, η_p is the angle between the vector V and the line AT. Point P has coordinates φ and χ; point D has coordinates φ_1 and χ_1. The angles φ and φ_1 are measured from the z-axis, and χ and χ_1 are measured relative to the x-axis. Figure 4.17 represents a particular case, when the vector $\bar{\beta}$ lies in the orbital plane. However, analytical expressions are given for the general case, when $\chi_1 \neq 0$ and vector $\bar{\beta}$ pointing towards the Earth does not lie in the orbital plane.

The differential frequency of the radio waves scattered from small surface areas around points P and D is given by

$$\Delta f_1 = fVc^{-1}\left(\cos\eta_p - \cos\eta_d\right). \tag{4.125}$$

For a circular orbit, from expression (4.125) and the diagram presented in Fig. 4.17 it follows that

$$\frac{\Delta f_1 R_1 c}{faV} = \frac{\cos\chi \cdot \sin\varphi}{\left[1 + \left(\dfrac{a}{R_1}\right)^2 - 2\left(\dfrac{a}{R_1}\right)\cos\varphi\right]^{1/2}} - \frac{\cos\chi_1 \cdot \sin\varphi_1}{\left[1 + \left(\dfrac{a}{R_1}\right)^2 - 2\left(\dfrac{a}{R_1}\right)\cos\varphi_1\right]^{1/2}}. \tag{4.126}$$

This equation defines the geometrical locus of points $P(\varphi, \chi)$ on the sphere that corresponds to a fixed value of Δf_1. For the boundary of the vision field of the satellite-borne radar, we have

$$\sin\varphi\left[1 + \left(\frac{a}{R_1}\right)^2 - 2\left(\frac{a}{R_1}\right)\cos\varphi\right]^{-1/2} = 1 \tag{4.127}$$

and the expression for χ:

$$\chi_3 = \arccos\left[\frac{R_1 c\Delta f_1}{aVf} + \frac{\cos\chi_1 \cdot \sin\varphi_1}{\left[1 + \left(\dfrac{a}{R_1}\right)^2 - 2\left(\dfrac{a}{R_1}\right)\cos\varphi_1\right]^{1/2}}\right]. \tag{4.128}$$

Designating $v = c\Delta f_1 V^{-1} f^{-1}$ and $e = a R_1^{-1}$, we can rewrite the equation (4.126) in the form:

$$v = e \left[\frac{\cos \chi \cdot \sin \varphi}{(1 + e^2 - 2e \cdot \cos \varphi)^{1/2}} - \frac{\cos \chi_1 \cdot \sin \varphi_1}{(1 + e^2 - 2e \cdot \cos \varphi_1)^{1/2}} \right]. \tag{4.129}$$

or in the explicit form:

$$\varphi = \arccos \left[be + [(b - 1)(be^2 - 1)]^{1/2} \right],$$

where

$$b = \left[\frac{v(1 + e^2 - 2e \cdot \cos \varphi_1)^{1/2} + e \cdot \cos \chi_1 \sin \varphi_1}{e \cdot \cos \chi (1 + e^2 - 2e \cdot \cos \varphi_1)^{1/2}} \right]^2. \tag{4.130}$$

Let us go over to a new variable v in expression (4.124), using the Jacobian transform. Then we have

$$I_3 = \frac{(1 + e^2 - 2e \cdot \cos \varphi)^{3/2}}{e \cdot \cos \chi [(1 + e^2 - 2e \cos \varphi) \cos \varphi - e \cdot \sin^2 \varphi]}. \tag{4.131}$$

After appropriate transformations in (4.124), we have

$$dP = \frac{a^2 |I_3| \sin \varphi}{\gamma^2 L_p^2} \cdot \exp \left(\frac{1 - \tau^2}{2\gamma^2 \tau^2} \right) d\chi dv. \tag{4.132}$$

To calculate the power spectrum of scattered radio waves, one has to know the spectral power density W, for which one has to integrate the expression (4.132) with respect to the variable v. After integration and transformations we can obtain the following expressions for the power spectrum of scattered radio waves:

$$W(v) = \gamma^{-2} \int_{-\chi_3}^{\chi_3} \frac{|I_3| e^2}{L_p^2} \cdot \exp \left(\frac{1 - \tau^2}{2\gamma^2 \tau^2} \right) \sin \varphi d\chi \tag{4.133}$$

when

$$1 - L_1 R_1^{-1} \cos \chi_1 \sin \varphi_1 \geq v \cdot e^{-1} \geq -L_1 R_1^{-1} \cos \chi_1 \sin \varphi_1 \quad ,$$

and

$$W(v) = \gamma^{-2} \int_{-\chi_3}^{2\pi-\chi_3} \frac{|I_3|e^2}{L_p^2} \cdot \exp\left(\frac{1-\tau^2}{2\gamma^2\tau^2}\right) \sin\varphi d\chi \qquad (4.134)$$

when

$$-\left(1 + L_1 R_1^{-1} \cos\chi_1 \sin\varphi_1\right) \le v \cdot e^{-1} \le -L_1 R_1^{-1} \cos\chi_1 \sin\varphi_1 .$$

In expressions (4.133) and (4.134), integration limits are determined by the values of the angle χ at the boundary of the area on the spherical surface that is observable from both the Earth and the satellite:

$$\chi_3 = \arccos\left(\frac{v L_1 R_1^{-1} - e \cdot \cos\chi_1 \sin\varphi_1}{L_1 R_1^{-1}e}\right). \qquad (4.135)$$

Since this expression is derived under the condition of the irradiated surface area observability from the satellite, it must be valid for rather great glancing angles ψ. For small glancing angles ψ, the condition of the surface area observability from the Earth is not essential, because of a steep decline of the exponential factor in expressions (4.133) and (4.134).

The numerical analysis of the complex expressions (4.133) and (4.134) for the case of a lunar satellite moving in a circular orbit such that its orbital plane passes through the Earth shows that the half-power spectral width depends weakly on the satellite altitude. Indeed, the increasing altitude of the satellite results in a decrease in its velocity (this factor leads to spectrum narrowing) and in an increase in the surface area that is visible from both the Earth and the satellite (this factor causes spectrum broadening). Acting together, these factors weaken one another. The spectrum is asymmetric: for a satellite entering behind the Moon, the positive wing of the spectrum has a more gentle slope than its negative wing. The spectrum shape is strongly dependent on the glancing angle ψ. Figure 4.18 presents theoretically derived spectra for the lunar satellite with $H = 200$ km and $\gamma = 0.14$. Curves *1*, *2*, and *3* correspond to ψ equal to 14, 29, and 45°, respectively. It can be seen that the spectrum narrows as the angle ψ decreases and widens as ψ rises. The dependence of the spectrum width on γ is especially strong at large glancing angles. It should be emphasized that the shape of the power spectrum does not depend on the permittivity of the scattering surface.

Numerical analysis shows that the half-power spectral width of the scattered radio waves, ΔF, can be given by the following approximate formula:

$$\Delta F = 4.7 f V_4 c^{-1} \gamma \cdot \sin\psi . \qquad (4.136)$$

Here V_4 is the velocity of the motion of the specular-reflection point along the planet's surface. Formula (4.136) can be used to estimate γ from the spectral line bandwidth of scattered radio waves.

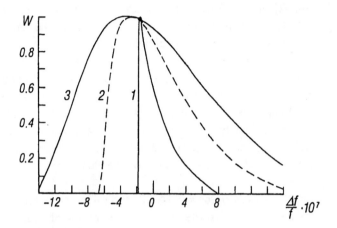

Fig. 4.18. Theoretically derived power spectra of scattered radio waves for different glancing angles ψ.

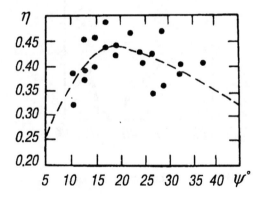

Fig. 4.19. Experimental (dots) and theoretically derived (dashed line) reflection coefficients during the bistatic radar observations of the Moon [92].

Let us now consider the results of the bistatic radar observations of the Moon, which included the measurements of the reflection coefficient η and the power spectrum of radio signals, $W(\Delta f)$. These parameters can be used to determine the permittivity ε, soil density ρ, and mean slope roughness γ for various surface areas corresponding to the varying positions of point D on the surface [89–102].

Such measurements were carried out during the missions of the Soviet satellites *Luna 11, 12,* and *14* [89, 90, 92], and the American satellites *Orbiter 1, Explorer 35, Apollo 14,* and *Apollo 15* [96–98, 102]. Figure 4.19 shows the dependence of the reflection coefficient η on ψ for $H = 250$ km and the horizontal polarization of radio waves ($\lambda = 1.7$ m) derived from *Luna 14* data. The scattering coefficient reached a maximum value of 0.44 at $\psi = 18\div22°$. With decreasing ψ, the scattering coefficient diminished to make up 0.35 at $\psi = 10°$. At $\psi < 5°$, η values were at the sensitivity threshold. These data suggest that the reflectance of the Moon tends to zero with $\psi \to 0$. Experiments during the mission of *Explorer 35* were concentrated on the study of linearly polarized radio waves ($\lambda = 2.2$ m). The scattering coefficients for vertically polarized radio waves and the angles ψ close to the Brewster angle are presented in Fig. 4.20. It can be seen that η has a distinct minimum at $\psi = 30°$.

The *Luna 14*-derived values of η for $\psi = 14–22°$ and different satellite altitudes were used to plot the dependencies of the scattering coefficient on the altitude H (Fig. 4.21). This figure shows that, at $H = 500\div1000$ km, the dependence of the reflection coefficient η on H is weak; but at $H < 300$ km, η rises steeply with decreasing altitude of the satellite.

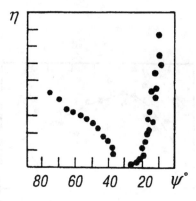

Fig. 4.20. Reflection coefficient in relative units versus the glancing angle for vertically polarized radio signals [97].

Let us compare the results of the bistatic radar observations of the Moon with the relevant theoretical data or, to be more exact, compare the experimental dependencies $\eta(\psi,H)$ with the theoretical curves plotted for different values of ε and γ. The theoretical curves in Figs. 4.19 and 4.21 (dashed lines) calculated for $\varepsilon = 2.8$ and $\gamma = 0.14$ fit the respective experimental dependencies well. Parameter γ weakly affects the theoretical estimations of reflectivity, so that the theory and experimental observations agree well at $\gamma = 0.1$–0.2. The reflectivity of vertically polarized meter radio waves given in Fig. 4.20 are in agreement with the theory at $\varepsilon = 3$. The fact that η has a distinct minimum suggests that the effect of the surface roughness was insignificant in this experiment.

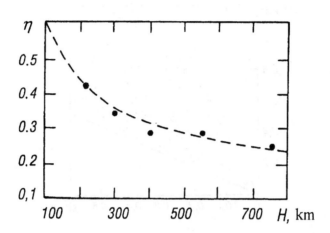

Fig. 4.21. Dependence of scattering coefficient on the lunar satellite altitude for the glancing angles $\psi = 14$–$22°$ [92].

Detailed analysis of the reflected radio signal spectra and their dependence on the lunar relief were performed in [92–95, 99–101]. Experiments were carried out in the meter wavelength band using the *Luna 14* and *Explorer 35* spacecraft and in the decimeter wavelength band using the *Luna 19* and *Luna 20* spacecraft. The meter-band experiments showed a high variability of the spectra of scattered radio waves and their dependence on the relief of the scattering surface. Comparison between the spectra and the lunar relief made in [93, 94] showed that the reflection of 1.7-m radio waves from relatively flat regions of the Sea of Calmness gave narrow spectra with a half-power width of 10÷12 Hz. Measurements of the radio signals scattered from the Carpathian Mountains provided a good illustration of a dependence of the spectra of scattered radio waves on the surface relief. In particular, the spectra of the radio signals scattered from the hilly regions located southwest of the Kunovsky crater were rather narrow, with a half-power width of 40÷50 Hz. At the same time, the spectra of the radio waves scattered from the Carpathian piedmont were rather wide (their half-power width was 200÷300 Hz) and possessed a few maxima. The complex relief of the Lame, Balmer, and Humboldt craters gave wide and greatly varied spectra of the reflected signals. As can be seen from these data, the power spectra of reflected radio signals are very sensitive to the surface relief in the vicinity of the specular-reflection point. The theoretical analysis of experimental spectra allowed the values of γ to be determined. Thus, the reflection of radio waves from the Sea of Calmness is characterized by $\gamma = 0.03$, whereas $\gamma = 0.15$ for the regions of the Siccard, Humboldt, and Wendelin craters.

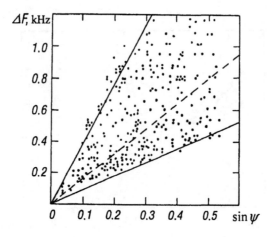

Fig. 4.22. The half-power spectrum width of scattered radio signals versus $\sin\psi$ [94].

Luna 19 and *Luna 22* provided valuable information on the power spectra of scattered decimeter-band radio waves [94, 95]. The satellites moved in near-circular orbits at a height of about 100 km. In this case the area irradiated with 32-cm radio waves moved at a rate of about 1 km s^{-1} along the lunar surface. These experiments confirmed the fact of the sensitivity of the power spectra of scattered radio signals to the scattering relief. As is evident from (4.136), the half-power spectral line width ΔF is proportional to $\sin\psi$; that is why Fig. 4.22 gives the dependence of ΔF on $\sin\psi$. The large spread of data points is due to the effect of the relief irregularities rather than to experimental errors. The solid straight lines in this figure correspond to maximal and minimal values of γ ($\gamma_{min} = 0.04$ and $\gamma_{max} = 0.23$), whereas the dashed line represents the average dependence of ΔF on $\sin\psi$, for which $\gamma = 0.1$. Figure 4.23 illustrates changes in the energy spectrum of scattered radio waves for a 300-km shift of the scattering area when the glancing angle varied from 21 to 12°. In this figure, the zero value on the abscissa corresponds to the frequency of the incident wave and the arrows show how the spectral anomaly, which corresponds to the reflection of radio waves from a sharp feature of the surface relief, shifts along the spectrum. The role of the frequency discrimination in relief studies was considered in more detail in the publication [95], whose authors analyzed spectral changes in the reflected modulated signals induced by the motion of the reflecting area along the Rokka crater and succeeded in obtaining its height profile.

The first bistatic radar observations of Mars from *Mariner 7* flying past the planet included the recording of reflected radio signals and the estimation of the reflectivity of the Mars surface [103, 104]. More detailed bistatic radar investigations of Mars in the 13-cm wave band were carried out during the *Viking 2* mission [105, 106]. The polar orbit of this planetary probe allowed the observation of various regions of the planet; for eleven of them, the authors presented γ values and photographs with the indicated trajectories of the specular-reflection point. The parameter γ differed for different regions, being maximum (7.6–9°) in the region between 26–34° N and 134–142° Martian longitude and minimum (1–1.5°) in the northern polar-cap region between 85–88° N and 128–170° longitude. In [106], the authors compared γ values obtained from the data of bistatic satellite-based and monostatic ground-based radar observations and found their good agreement. They also determined the reflectivity and permittivity ε of the surface rocks for the equatorial ($\varepsilon = 2.8$–3.2) and polar ($\varepsilon = 1.8$–2.3) regions. The lower values of ε and γ in the polar regions of Mars led the authors to the suggestion of the presence of snow in these Martian regions.

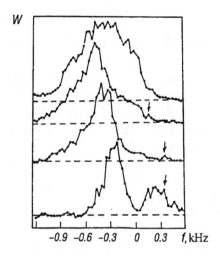

Fig. 4.23. Typical changes in the power spectra of decimeter radio waves produced by the motion of scattering area along the lunar surface [94].

The bistatic radar observations of Venus from *Veneras 9, 10, 15,* and *16* in the 32-cm wave band involved the measurements of surface parameters in five large equatorial regions and two regions near the North Pole of the planet [107–117]. Parameter γ determined from the spectral width of scattered radio signals along the trajectory of the specular-reflection point was found to be small, varying from 0.6 to 1° for particular regions. For some large planetary planes, the value of γ was as low as $\approx 0.2°$. At glancing angles $\psi < 15°$, bistatic radar measurements were considerably influenced by the dense Venusian atmosphere. This influence manifested itself as a decrease in the reflectivity of the planet and a change in the frequency difference between the incident and reflected radio waves because of their strong refraction at small angles ψ. The frequency difference data can be used to elucidate the relationship between the refraction and glancing angles of radio signals. The effect of the dense Venusian atmosphere on bistatic radar measurements was analyzed in detail in [111, 112, 118].

The employment of two satellites, one emitting and the other receiving radio waves, makes possible the bistatic radar observations of the Earth. The first experiment of this kind was described in [119].

All the above bistatic radar experiments used common space communication channels with monochromatic signals, which allowed the Doppler effect–based discrimination of reflected radio signals only in one coordinate. The use of modulated signals makes it possible to find the second coordinate of the surface element ds from the radio wave time delay data and thus accomplish a bistatic radar vision.

4.6 Side-looking radar technique for studying planetary surfaces

Radio images of planets can be obtained by side-looking radars operated at modulated signals. The principle of this method is shown in Fig. 4.24. The directional antenna of a radar installed on board a satellite located at point A is oriented in such a way that its major lobe is normal to the satellite's velocity vector v and makes an angle η with the normal to the planet's surface (Fig. 4.24). The antenna irradiates a small area of the planet's surface bound in the figure by the dashed line. The major lobe of the antenna is directed along the line AB; $AA_1 = H$ is the satellite's height. The xy-plane is tangent to the planet's surface at the subsatellite point A_1; the satellite's orbit lies in the xz-plane. Let us assume for simplicity that the orbit is circular, so that the velocity vector v is parallel to the xy-plane. If the tilt angle η is small, the coordinates of an arbitrary point P in the xy-plane and its projection onto the planetary surface, P_1, are close. Assume for simplicity that points P and P_1 coincide. The relief element ds at point P retransmits radio waves back to the radar located at point A. To obtain a radio image of the surface, it is necessary to find the dependence of the intensity of radio waves reflected from the surface element ds on the coordinates of point P. The modulation of the radar signals allows one to determine the first coordinate $\rho = A_1 P$ through the time delay of the radio signals traveling along the path AP. The lines of equal time delay t represent, in the xy-plane, a set of circles of radius

$$\rho = \left(4c^2t^2 - H^2\right)^{1/2} \ . \tag{4.137}$$

The second coordinate of point P can be determined from the Doppler frequency shift of radio signals scattered from the surface element ds. The frequency shift for point P is given by

$$\Delta f = 2\lambda^{-1}(\overline{V} \cdot \overline{\alpha}) = 2\lambda^{-1}V \cdot \cos\zeta , \qquad (4.138)$$

where $\overline{V}$ is the satellite velocity vector, $\overline{\alpha}$ is the unity vector in the direction AP, and ζ is the angle between $\overline{V}$ and $\overline{\alpha}$. For points on the line A_1B, angle $\zeta = 90°$ and, therefore, $\Delta f = 0$; for point B_1, $\zeta > 90°$ and, consequently, Δf is negative; and for the surface element located at point B_2, $\zeta < 90°$ and Δf is positive. The dot-and-dash frequency isolines in the xy-plane (Fig. 4.24) correspond to constant values of the angle ζ; the isolines of the Doppler frequency shift Δf are formed by the xy-plane intersection of a cone with an vertex angle ζ. Thus, the modulated highly stable coherent radar signals allow of coordinates ρ and ζ to be determined. The coordinates of point P on the planet's surface can then be found using geometrical relations.

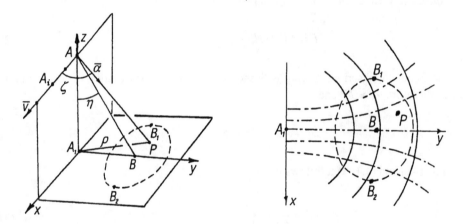

Fig. 4.24. Graphical representation of the side-looking radar observations of a planetary surface.

The resolving power of a side-looking radar in the range AP, i.e., along coordinate ρ, is determined by the modulated signal bandwidth ΔF. The resolution in the time delay of signals $\delta t \approx \Delta F^{-1}$; therefore, according to (4.137), the resolution in coordinate ρ is given by the expression

$$\delta\rho = 2c\Delta F^{-1}\left[1 + \left(\frac{H}{\rho}\right)^2\right]^{1/2} . \qquad (4.139)$$

For the line A_1B, coordinates ρ and y coincide. Since the radar antenna is tilted at an angle η to the normal, then, for the center of the irradiated area, $y_{mean} = H \tan\eta$ and hence

$$\delta y = 2c\Delta F^{-1}(1 + \mathrm{ctg}^2\,\eta)^{1/2}. \tag{4.140}$$

The resolution in the angular coordinate $\delta\zeta$ depends on the difference δf in the Doppler frequency shifts of the radio signals reflected from the two surface elements ds_1 and ds_2 located at the same distance AP. Then, allowing for expression (4.138), we have for the frequency difference

$$\delta f = 2\lambda^{-1}V\delta\zeta. \tag{4.141}$$

Frequency can be measured only within a time interval T, until point P leaves the irradiated area bound in Fig. 4.24 by the dashed line. Let the half-power beamwidth be symbolized Δ; then the time interval T is given by

$$T = \Delta V^{-1}(H^2 + \rho^2)^{1/2}. \tag{4.142}$$

The frequency discrimination df is limited by the time of observation of the harmonic signal; therefore, $df \approx T^{-1}$ and

$$\Delta f = V\Delta^{-1}(H^2 + \rho^2)^{-1/2}. \tag{4.143}$$

From (4.141) and (4.143) it follows that

$$\delta\zeta = \frac{\lambda}{2\Delta(H^2 + \rho^2)^{1/2}}. \tag{4.144}$$

The resolution in coordinate x in the xy-plane is

$$\delta x = \delta\zeta(H^2 + \rho^2)^{1/2}. \tag{4.145}$$

Taking into account that the half-power beamwidth of the directional antenna, Δ, is related to its aperture D as $\Delta = \lambda D^{-1}$, we obtain from (4.144) and (4.145) the maximum theoretical resolution in x:

$$\delta x = \frac{D}{2}. \tag{4.146}$$

Relations (4.140) and (4.146) are valid if the transmitter power is sufficiently high to neglect the effects of noise and other instrumental factors. Actually, the resolving power of side-looking radars is lower by a factor of about 2.

The operation of the side-looking radar can be outlined in terms of a synthetic aperture antenna. As the satellite moves, its radar antenna with aperture D is sequentially located at points A_i (Fig. 4.24). The summation of returning signals with allowance for their delay is equivalent to the operation of a synthetic linear coherent antenna of length $L = VT$, where T is the time of the signal summation determined by the above condition of the occurrence of point P in the irradiated area. The synthetic linear coherent antenna of length L has a beamwidth $\delta\zeta = (VT)^{-1}\lambda$; therefore, again, the resolving power in the direction of motion, δx, is equal to $D/2$.

Generally, the determination of coordinate x in the direction of motion can be rationalized in terms of the Doppler frequency discrimination, synthetic aperture, or frequency filtration. These three approaches to the explanation of the principle of side-looking radars are completely equivalent [126–128].

The radar antenna must have a relatively narrow beam with weak sidelobes. The backscattering pattern, which depends on the surface relief, may have different indicatrixes $F(\psi)$. However, the reflected signals are always maximal at $\psi \approx 90°$, i.e., when the direct signals are vertically incident. In this case a strong reflection of radar signals from the vicinity of the subsatellite point produces severe interfering noise; this imposes stringent requirements on the intensity of sidelobes.

An essential parameter of radars is the desired contrast of images, which depends on the wavelength, the backscattering indicatrix $F(\psi)$ of the surface to be investigated, and the tilt angle of the antenna, η. Since various types of surfaces (deserts, sea, land) are characterized by very different dependencies $F(\psi)$, the contrasts of their radar images may also greatly differ. If surfaces are relatively smooth, such as Venusian flats or sea swell, the backscattering of radio waves is weak and radar images are low-contrast. If a relief contains small-scale features, such as stony irregular hills or sea ripples, radar images are also be low-contrast because of the diffuse scattering of radio waves. It can be anticipated that the radar images of mountains with steep slopes extending parallel to the satellite's velocity vector are the most contrasting.

The side-looking radar observations of the planetary surface from a satellite are associated with the necessity to transfer a large volume of information to a ground-based station, with a high velocity and altitude of the satellite varying due to the orbit ellipticity, and with the displacement of the orbital plane relative to the planet because of its rotation. The satellite-borne side-looking radar must perform the digital processing of reflected signals to determine their intensity and the coordinates of the respective surface element. All of this substantially complicates practical implementation of the imaging system composed of a side-looking radar and blocks for signal telemetry, processing, and image orientation [128, 134].

The first side-looking radar observations of a planetary relief were performed from the satellites *Venera 15* and *Venera 16* [72, 120, 123]. The satellite-borne side-looking radar with a 6×1 m antenna operating at 8-cm wavelengths was tilted at an angle $\eta = 10°$ to the normal. It should be noted that the choice of an optimal value of the angle η is of crucial importance for the contrasting imaging of planetary surfaces. The dependence of the average backscattering indicatrix on the angle η for Venus was derived earlier using a ground-based radar. The data thus obtained showed that the mean power of reflected signals decreases by about 10 dB if the tilt angle varies from 0 to 10°. In such a case, the radar images of the Venusian relief must be contrasting. Indeed, mountain slopes of $\pm 8°$ produce the attenuation of reflected signals ranging from −2 to −18 dB; therefore, such slopes must be in contrast to the radar images of the Venusian surface. *Veneras 15* and *16* were placed in elliptical polar orbits with the minimum pericenter height $H = 1000$ km. During the period of one satellite revolution, the observation session lasted for 16 min, with an accessibility swath of 140 × 7500 km. Due to the rotation of Venus, the next satellite circuit allowed the imaging of the neighboring region of the Venusian surface. The output signals of the radar were retranslated to a ground-based processing center, where they were subjected to high-speed matched Fourier-filtration with respect to 127 delay times t and 31 Doppler frequency shifts Δf to give 3920 pixels of the radar image. One radar swath gave about 2 million pixels with estimated intensities and coordinates. The visual image of the Venusian surface was generated using a special photodevice. Dark elements of the image corresponded to strong reflected signals, while light elements corresponded to weak signals. The image resolution was $\delta x \approx \delta y \approx 1500$ m, which corresponded to relief features visible to the naked eye from a height of about 1000 km. That was the first image of the Venusian surface 115 million km^2 in area. Radar altimeters installed on board *Venera 15* and *16* ensured relief profiling along the trajectory of the subsatellite point (point A_1 in Fig. 4.24). Eventually, these remarkable studies culminated in the creation of the first Venusian atlas [72].

The planetary probe *Magellan*, launched 6 years later, carried a side-looking radar operated at 12.6-cm radio waves and equipped with a parabolic antenna 3.7 m in diameter [124, 125]. The vehicle had a pericenter height of 297 km and the apocenter height $H = 2100$ km. Such a low-height orbit allowed radar imaging over a greater part of the satellite circuit, with a survey swath of 25 × 16 000 km. The angle of view, η, varied from 18° in polar regions to 50° in the pericenter. The image resolution was $\delta x \approx \delta y \approx 140$ m. The radar altimeter installed on board the *Magellan* probe accomplished the height profiling of the Venusian surface irregularities.

Thus, the flights of *Venera 15*, *Venera 16*, and *Magellan* made it possible to obtain detailed high-quality images disclosing the pristine Venusian surface without any indications of erosion. Analysis of the images revealed impact craters with clearly seen rock outbursts, volcanic craters, faults and mountain ridges, large flats, and recent volcanic formations, such as huge lava domes [125].

Side-looking radars also proved their potency for the investigation and ecological monitoring of the Earth [131–146]. The side-looking radar observations of the Earth from satellites are anticipated to ensure the contrast imaging of different types of the Earth's surface, including sea, mountains, tundra, deserts, and ice fields, since these relief features have diverse scattering indicatrixes $F(\psi)$. It should be noted that the scattering characteristics of planetary features vary with time. Thus, the scattering of radio waves from woodlands depends on humidity and, therefore, on the season; on the other hand, sea surface irregularities depend on meteorological conditions, especially on winds. The backscattering of radio signals from various terrestrial reliefs and a rough sea has been extensively studied. The relevant experimental indicatrixes $F(\psi)$ for various terrestrial reliefs and sea roughnesses can be found in [129, 130]. Analysis of published data shows that it is impossible to choose such values of η and λ that would be optimal for the radar observations of both terrain and sea. It can be stated with some exaggeration that one must first create a side-looking system and then look for potential benefits that could be derived from its implication in geology, agriculture, ocean monitoring, the control of ice conditions, etc. The major parameters of different side-looking radar systems taken from [131–136] are summarized in Table 4.8.

Table 4.8. Parameters of different satellite-borne side-looking radars [131–135].

Satellite	Year	Wave-length, cm	Resolu-tion, m	Survey width, km	Height, km	Antenna area, m × m	Angle η, deg.	Owner
Venera 15,16	1983	8	1500	140	1000–1600	6×1	10	Russia
Magellan	1990	12.6	140	30	300–2000	3.7	18–50	United States
Kosmos-1870	1987	9.6	25	40	300			Russia
Almaz-1	1991	9.6	15–20	40	330	15×1.5	30–50	Russia
Seasat	1978	23	25	100				United States
SIR-A	1981	23	40	50				United States
SIR-B	1984	23	40	20–50				United States
SIR-C	1994	3.1	30	15–45				United States
ERS-1	1991	5.6	25	100	780	9.8×1	23	European Space Agency
ERS-2	1995	5.6	25	100	780	9.8×1	23	European Space Agency
IERS-1	1993	23	18	75	568	12×2.4	35	Japan
RADAR-SAT	1995	5.6	10–60	50–500				Canada

In Russia, the first side-looking radar observations of the Earth were accomplished from the Earth-orbiting satellite *Kosmos-1870* [133, 138]. These observations demonstrated the potentiality of side-looking radars for oceanography, in particular, for investigating wind-induced waves, swell, dynamics of streams, surface manifestations of subsea waves, as well as for detecting oil-polluted regions. The first Russian radar, operated at 9.6-cm radio waves, ensured a resolution of $\delta x \approx \delta y \approx 25 \div 30$ m.

An upgraded side-looking radar installed on board *Almaz-1* had a better resolution of $\delta x \approx \delta y \approx 15 \div 20$ m and ensured the simultaneous survey of two Earth's swaths $30 \div 40$ km in width to the right and to the left of the satellite's trajectory. This radar has demonstrated the feasibility of the side-looking radar technique for the investigation of geological structures, as well as for the ecological monitoring of tundra and crop control in agriculture [143–145]. Furthermore, these experiments showed the feasibility of precise relief profiling by the interferometric method, which uses data on the phase difference of radio waves reflected from the same area during neighboring satellite circuits [146].

Radar observations appear to be an efficient tool for geological survey. In particular, characteristic rectilinear regions, or lineaments, on radar-derived images can provide information about the geological conditions of a given region. Differences in the density of radar images estimated on a gray scale reflect respective lithological differences and thus can also yield useful geological information [143].

The side-looking radar technique is also applicable for investigating the vegetation cover of the Earth, since plants occurring at different growth stages greatly differ in their reflectivity. Reliable data on the state of vegetation cover can be obtained by implementing a multi-frequency radar survey with radio waves linearly polarized in perpendicular directions.

Of great importance are the side-looking radar observations of ice conditions in the northern and arctic regions, where aerial survey is difficult because of heavy cloudiness and poor illumination. Here radar images may help to determine the position of large water openings, icebergs, and even to distinguish between young, hummocked, and deep ices.

The side-looking radar observations of the ocean from the *Almaz* satellite greatly contributed to the investigation of surface and gravitational capillary waves and the extended fronts of high-amplitude subsurface waves [135, 139–142].

The operation of the Russian *Almaz*-borne side-looking radar coincided in time with the operation of the side-looking radar on board the European remote sensing satellite ERS-1; this allowed the comparison of these two systems. The side-looking radar carried by ERS-1 operated at 5.6-cm radio waves and had a resolution of $\delta \approx 25$ m. Analysis of the characteristics of a rough sea obtained by the two radars led to the conclusion that "both satellite systems are not optimal for obtaining sea surface images as being primarily intended for terrain survey" [139].

Experience shows that the efficiency of side-looking radars for ecological purposes and for investigating the terrain and ocean can be further improved by using both decimeter and centimeter radio waves at a time, as was done in the latest version of the American side-looking radar SIR-C operated at 3.1-cm and 23-cm radio waves.

Satellite-borne side-looking radars are commonly linked to a radar altimeter or scatterometer. Thus, the ERS satellite is equipped with a radar altimeter that ensures height measurements with an accuracy of 5–10 cm. This makes possible the investigation of terrain and sea surfaces, so that even the height of large waves can be determined. The ERS-borne radar scatterometer generates three radio-beams – leading, middle, and back, which allows one to estimate the backscattering of radio waves, i.e., the values of function $F(\psi)$ for different ψ. Knowing $F(\psi)$, one can estimate, for instance, the velocity and direction of surface winds based on the relationship between the backscattering functions $F(\psi)$ and the parameters of capillary waves, or shallow ripples.

Thus, there is convincing experimental evidence that the combined application of side-looking radars, radar altimeters, and scatterometers is very effective for oceanographic studies. Recent advances in the radar observations of the Earth from satellites are described in [150–156].

References

1. Feinberg, E.L. (1961) *Propagation of Radio Waves along Earth's Surface*, Moscow: AN SSSR (in Russian).
2. Krotikov, W.D. (1962) *Radiofizika*, **5**, 6: 1057 (in Russian).
3. Olhoefft, G.R. and Strangwag, D.W. (1975) *Earth and Planetary Science Letters*, **24**: 394.
4. Fok, W.A. (1970) *Diffraction of Electromagnetic Waves*, Moscow: Sovetskoe Radio (in Russian).
5. Kerr, D.E. (1951) *Propagation of Short Radio Waves*, N.Y.: Mc Grow-Hill.
6. Brehowskih, L.M. (1973) *Wave Propagation in Inhomogeneous Media*, Moscow: Nauka (in Russian).
7. Matveev, J.G. (1968) *Radiofizika*, **11**, 1: 147, (in Russian).
8. Ward, S.H., Jiracek, G.R., and Linlor, W.J. (1968) *J. Geoph. Res.*, **73**, 4: 1355.
9. Jiracek, G.R. and Ward, S.H. (1971) *J. Geoph. Res.* **76**, 26: 6237.
10. Rytov, S.M., Kravtsov, Yu.A., and Tatarskii, V.I. (1989) Principles of Statistical Radiophysics, in: *Wave Propagation through Random Media*, Berlin: Springer-Verlag, vol. **4.**
11. Beckman, P. and Spizzichino, A. (1963) *The Scattering of Electromagnetic Waves from Rough Surfaces*, London–New York: Pergamon–Macmilln.
12. Fung, A.K. and Moore, R.K. (1966) *J. Geoph. Res.*, **71**, 12: 2939.
13. Karpenter, R.L. (1964) *Astronom. J.*, **69**, 1: 2.
14. Goldstein, R.M. (1964) *IEEE Trans. on Antenn. and Prop.* **AP-12**, 7: 865.
15. Beckmann, P. and Klemperer, W.K. (1965) *Radio Sci.*, **69D**, 12: 1669.
16. Davis, I.R., Rohlfs, D.C., Skaggs, G.A., and Joss, J.W. (1965) *Radio Sci.*, **69D**, 12: 1659.
17. Evans, J.V. (1965) *Radio Sci.*, **69D**, 12: 1637.
18. Evans, J.V. and Hagfors, T. (1966) *J. Geoph. Res.*, **71**, 20: 4871.
19. Beckmann, P. (1968) *J. Geoph. Res.*, **73**, 2: 649.
20. Burns, A.A. (1970) *J. Geoph. Res.*, **75**, 8: 1467.
21. McCue, J.J. and Crocher, E.A. (1972) *J. Geoph. Res.*, **77**, 22: 4069.
22. Evans, J.V. and Hagfors, T. (Eds.) (1968) *Radar Astronomy*, N.Y.: Mc.Grow-Hill Book Co.
23. Simpson, R.A. and Tyler, G.L. (1982) *IEEE Trans. on Ant. and Prop.*, **AP-30**: 438.

24. Krupenio, N.N. (1971) *Radar Investigation of the Lunar Surface*, Moscow: Nauka (in Russian).
25. Brown, W.E. (1967) *J. Geoph. Res.*, **72**, 2: 791.
26. Jaff, L.D., Batterson, S.A., Brown, W.E., *et al.* (1968) *J. Geoph. Res.*, **73**, 12: 3983.
27. Krupenio, N.N., Ruzskii, E.G., and Cherkasov, V.V. (1976) *Kosmicheskie Issledovaniya* (Cosmic Res.), **14**, 3: 460 (in Russian)*.
28. Krupenio, N.N. (1979) *Kosmicheskie Issledovaniya* (Cosmic Res.), **17**, 2: 291 (in Russian)*.
29. Krupenio, N.N. (1972) Kosmicheskie Issledovaniya, **10**, 4: 569 (in Russian).
30. Thompson, T.W. (1987) *Earth, Moon, and Planets*, **37**, 1: 59.
31. Rzhiga, O.N. (1967) *Astronomicheskii Zhurnal*, **44**, 1: 147 (in Russian).
32. Zachs, A. and Fung, A.K. (1969) *Space Sci. Rev.*, **10**, 3: 442.
33. Rogers, A.E., Ash, M.E., Counselman, C.C., et. al. (1970) *Radio Sci.*, **5**, 2: 465.
34. Downs, G.S., Goldstein, R.M., Green, H.R., *et al.* (1973) *Icarus*, **18**, 1: 81.
35. Downs, G.S., Reinchlen, P.E., and Green, R.R. (1975) *Icarus*, **26**, 3: 273.
36. Lipa, B. and Tyler, G.L. (1976) *Icarus*, **28**, 2: 301.
37. Simpson, R., Tyler G., and Lipa, B. (1977) *Icarus*, **32**, 2: 147.
38. Krupenio, N.N. (1977) *Kosmicheskie Issledovaniya* (Cosmic Res.), **15**, 3: 470 (in Russian)*.
39. Downs, G.S., Green, R.R., and Reichley, P.E. (1978) *Icarus*, **33**, 3: 441.
40. Calvin, W.M., Jakosky, B.M., and Christiansen, B. (1988) *Icarus*, **76**, 3: 513.
41. Muhleman, D.O., Grossman, A.W., and Butler, B.I. (1995) *Annual Rev. Earth Planet.*,**23**: 337.
42. Krupenio, N.N. (1978) *Kosmicheskie Issledovaniya* (Cosmic Res.), **16**, 3: 443 (in Russian)*.
43. Andrianov, V.A. (1992) *Radiotekhnika i Elektronika* (J. Communications Technology and Electronics), **37**, 11: 1937 (in Russian)*.
44. Kotelnikov, W.A., Aleksandrov, J.N., and Apraksin, L.W. (1965) *Doklady Akademii Nauk SSSR*, **163**, 1: 50 (in Russian).
45. Aleksandrov, J.N., Zjatickii, W.A., and Rzhiga, O.N. (1967) *Astronomicheskii Zhurnal*, **44**, 5: 1060 (in Russian).
46. Aleksandrov, J.N., Kuznezov, B.I, and Rzhiga, O.N. (1968) *Astronomicheskii Zhurnal*, **45**, 2: 371 (in Russian).
47. Evans, J.V., Ingalls, R.P., Rainville, L.P., and Silva, R.R. (1966) *Astronom. J.*, **71**, 9: 902.
48. James, J.C., Ingals, R.P., and Rainville, L.P. (1967) *Astronom. J.*, **72**, 8: 1047.
49. Evans, J.V. (1968) *Astronom. J.*, **73**, 3: 125.
50. Ingals, R.P. and Evans, J.V. (1969) *Astronom. J.*, **74**, 2: 258.
51. Jurgens, R.F. (1970) *Radio Sci.*, **5**, 2: 435.
52. Campbell, D.B., Dyce, R.B., Ingalls, R.P., *et al.* (1972) *Science*, **175**, 4021: 514.
53. Hagfors, T. and Campbell, D.B. (1974) *Astronom. J.*, **79**, 4: 493.
54. Rumsey, H.C., Morris, G.A., Green, R.R., and Goldstein, R.M. (1974) *Icarus*, **23**, 1: 1.
55. Rogers, A.E., Ingalls, R.P., and Pettengill, G.H. (1974) *Icarus*, **21**, 3: 237.
56. Campbell, D.B., Dyce, R.B., and Pettengill, G.H. (1976) *Science*, **193**, 4258: 1123.
57. Golovkov, V.K., Kuznetsov, B.I., Petrov, G.M., and Khasyanov, A.F. (1976) *Radiotekhnika i Elektronika* (Radio Engineering and Electronic Physics), **21**, 9: 1801 (in Russian)*.
58. Aleksandrov, Y.N., Golovkov, V.K., Dubrovin, V.M., *et al.* (1980) *Astronomicheskii Zhurnal* (Soviet Astronomy) , **24**, 2: 139 (in Russian)*.

59. Kislik, M.D., Koluka, J.F., Kotelnikov, W.A., *et al.* (1978) *Doklady Akademii Nauk SSSR*, **241**, 5: 1046 (in Russian).

60. Kotelnikov, V.A., Rzhiga, O.N., Aleksandrov, Y.N., *et al.* (1985) Radar Observations of Planets in the USSR, in: *Problems of Modern Radio Engineering and Electronics*, Kotelnikov, V.A. (Ed.), Moscow: Nauka.

61. Rogers, A.E. and Ingalls, R.P. (1970) *Radio Sci.*, **5**, 2: 425.

62. Goldstein, R.M. and Rumsey, H.C. (1970) *Science*, **169**, 3949: 974.

63. Goldstein, R.M., Green, R.R., and Rumsey, H.C. (1976) *J. Geoph. Res.*, **81**, 26: 4807.

64. Goldstein, R.M., Green, R.R., and Rumsey, H.C. (1978) *Icarus*, **36**, 3: 334.

65. Campbell, D.B. and Burns, B.A. (1980) *J. Geoph. Res.*, **85**, A13: 8271.

66. Jurgens, R.F., Goldstein, R.M., Rumsey, H.C., *et al.* (1980) *J. Geoph. Res.*, **85**, A13: 8282.

67. Masursky, H., Eliason, E., Ford, P.G., *et al.* (1980) *J. Geoph. Res.*, **85**, A13: 8232.

68. Pettengill, G.H., Eliason, E., Ford, P.G., *et al.* (1980) *J. Geoph. Res.*, **85**, A13: 8261.

69. Garvin, J.B., Head, J.W., Pettengill, G.H., and Zisk, S.H. (1985) *J. Geoph. Res.*, **90**, B8: 6859.

70. Head, J.W., Peterfreund, J.B., Garvin, J.B., and Zisk, S.H. (1985) *J. Geoph. Res.*, **90**, B8: 6873.

71. Davis, P.A., Kozak, R.G., and Schaber, G.G. (1986) *J. Geoph. Res.*, **91**, B5: 4979.

72. Kotelnikov, V.A. (Ed.) (1989) *Atlas of the Surface of Venus*, Moscow: Nauka (in Russian).

73. Ford, P.G. and Senske, D.A. (1990) *J. Geoph. Res.*, **17**, 9: 1361.

74. Tyler, G.L., Simpson, R., Maurer, M.J., and Holman,E. (1992) *J. Geoph. Res.*, **98**, E8: 13115.

75. Ford, P.G. and Pettengill, G.H. (1992) *J. Geoph. Res.*, **97**, E8: 13103.

76. Ingalls, R.P. and Rainvill, L.P. (1972) *Astronom. J.*, **77**, 2: 185.

77. Harmon, J.K., Campbell, D.B., Bindschadler, D.L., *et al.* (1986) *J. Geoph. Res.*, **91**: 385.

78. Ostro, S.J. (1982) in: *Satellites of Jupiter*, Morrison D. (Ed.), The University of Arizona Press.

79. Ostro, S.J. (1984) *Meteoritics*, **19**, 4: 286.

80. Ostro, S.J., Connelly, B., and Belkora, L. (1988) *Icarus*, **73**, 1: 15.

81. Ostro, S.J., Campbell, D.B., and Shapiro, J.J. (1985) *Science*, **229**, 4712: 442.

82. Zaitsev, A.L., Sokol'skii, A.G., Rzhiga, O.N., *et al.* (1993) *Radiotekhnika i Elektronika* (J. Communications Technology and Electronics), **38**, 10: 1842 (in Russian)*.

83. Zaitsev, A.L., Al'tenkhof, V., Vilebinskii, P., *et al.* (1995) *Doklady Rossiiskoi Akademii Nauk*, **342**, 4: 480 (in Russian).

84. Osto, S.J., Hudson, R.S., Jurgens, R.F., *et al.* (1995) *Science*, **270**: 80.

85. Hudson, R.S. and Osto, S.J. (1995) *Science*, **270**: 84.

86. Harmon, J.K., Campbell, D.B., Hine, A.A., *et al.* (1989) *Astrophys. J.*, **338**, 2: 1071.

87. Campbell, D.B., Harmon, J.K., and Shapiro, J.J. (1989) *Astrophys. J*, **338**, 2: 1094.

88. Ostro, S.J. (1999) *Encyclopedia of the Solar System*, p. 773.

89. Yakovlev, O.I. and Efimov, A.I. (1967) *Doklady Akademii Nauk SSSR*, **174**, 3: 583 (in Russian).

90. Yakovlev, O.I., Efimov, A.I., and Matugov, S.S. (1968) *Kosmicheskie Issledovaniya*, **6**, 3: 432 (in Russian).

91. Pavelev, A.G. (1969) *Radiotekhnika i Elektronika*, **14**, 11: 1923 (in Russian).

92. Yakovlev, O.I., Matugov, S.S., and Schwatchkin, C.M. (1970) *Radiotekhnika i Elektronika*, **15**, 7: 1339 (in Russian).

93. Matugov, S.S. and Yakovlev, O.I. (1971) *Radiotekhnika i Elektronika*, **16**, 9: 1545 (in Russian).

94. Kaevitser, V.I., Matyugov, S.S., Pavelyev, A.G., *et al.* (1974) *Radiotekhnika i Elektronika* (Radio Engineering and Electronic Physics), **19**, 5: 936 (in Russian)*.
95. Zaitsev, A.L., Kaevitser, V.I., Kucheryavenkov, A.I., *et al.* (1977) *Radiotekhnika i Elektronika* (Radio Engineering and Electronic Physics), **22**, 10: 67 (in Russian)*.
96. Tyler, G.L., Eshleman, V.R., Fjeldbo, J., *et al.* (1967) *Science*, **157**, 3785: 193.
97. Tyler, G.L. (1968) *Nature*, **219**, 5160: 1243.
98. Tyler, G.L. (1968) *J. Geophys. Res.*, **73**, 24: 7609.
99. Tyler, G.L. and Simpson, R.A. (1970) *Radio Sci.*, **5**, 2: 263.
100. Tyler, G.L. and Ingalls, D.H. (1971) *J. Geoph. Res.*, **76**, 20: 4775.
101. Parker, M.N. and Tyler, G.L. (1973) *Radio Sci.*, **8**, 3: 117.
102. Tyler, G.L. and Howard, H.T. (1973) *J. Geophys. Res.*, **78**, 23: 4852.
103. Fjeldbo, G., Kliore, A., and Seidel, B. (1972) *Icarus*, **16**, 3: 502.
104. Tang, G., Boak, T., and Grossi, M. (1977) *J. Geoph. Res.*, **82**, 28: 4305.
105. Simpson, R.A. and Tyler, G.L. (1981) *Icarus*, **46**, 3: 361.
106. Simpson, R.A., Tyler, G.L., and Schaber G.G. (1984) *J. Geoph. Res.*, **89**, B12: 10385.
107. Pavelyev, A.G., Yakovlev, O.I., Matyugov, S.S., *et al.* (1975) *Radiofizika* (Radiophysics and Quantum Electronics), **18**, 6: 816 (in Russian)*.
108. Kolosov, M.A., Yakovlev, O.I., Pavelyev, A.G., *et al.* (1976) *Radiotekhnika i Elektronika* (Radio Engineering and Electronic Physics), **21**, 9: 6 (in Russian)*.
109. Pavelyev, A.G., Kolosov, M.A., Yakovlev, O.I., *et al.* (1978) *Radiotekhnika i Elektronika* (Radio Engineering and Electronic Physics), **23**, 10: 1 (in Russian)*.
110. Pavelyev, A.G., Kaevitser, V.I., and Kucheryavenkov, A.I. (1978) *Kosmicheskie Issledovaniya* (Cosmic Res.), **16**, 5: 582 (in Russian)*.
111. Pavelyev, A.G. and Kucheryavenkov, A.I. (1978) *Radiotekhnika i Elektronika* (Radio Engineering and Electronic Physics), **23**, 7: 13 (in Russian)*.
112. Kucheryavenkov, A.I., Yakovlev, O.I., Pavelyev, A.G., *et al.* (1979) *Radiofizika* (Radiophysics and Quantum Electronics), **22**, 6: 467 (in Russian)*.
113. Pavelyev, A.G., Yakovlev, O.I., and Kucheryavenkov, A.I. (1981) *Radiofizika* (Radiophysics and Quantum Electronics), **24**, 1: 3 (in Russian)*.
114. Kolosov, M.A., Yakovlev, O.I., Pavelyev, A.G., *et al.* (1981) *Icarus*, **48**, 2: 188.
115. Pavelyev, A.G., Yakovlev, O.I., Milekhin, O.E., and Kucherjvenkov A.I. (1986) *Acta Astronautica*, **13**, 1: 39.
116. Kucheryavenkov, A.I., Milekhin, O.I., Pavelyev, A.G., and Yakovlev, O.I., (1986) in: *Electromagnetic Waves in the Atmosphere and Space*, Moscow: Nauka (in Russian).
117. Pavelyev, A.G., Yakovlev, O.I., Rzhiga, O.N., *et al.* (1990) *Kosmicheskie Issledovaniya* (Cosmic Res.), **28**, 1: 125 (in Russian)*.
118. Pavelyev, A.G. and Kucheryavenkov, A.I. (1994) in: *Itogi Nauki i Tekhniki, Radiotekhnika*, **44**: 81 (in Russian).
119. Rubashkin, S.G., Pavelyev, A.G., and Yakovlev, O.I. (1993) *Radiotekhnika i Elektronika* (J. Communications Technology and Electronics), **38**, 9: 447 (in Russian)*.
120. Kotelnikov, W.A., Bogomolov, A.F., and Rzhiga, O.N. (1985) *Adv. Space Res.*, **5**, 8: 5.
121. Bogomolov, A.F., Zherikhin, N.V., and Sokolov, G.A. (1985) *Radiofizika* (Radiophysics and Quantum Electronics), **28**, 3: 259 (in Russian)*.
122. Aleksandrov, Yu.N., Dubrovin, V.M., Zakharov, A.I., *et al.* (1987) in: *Problems of Modern Radioengeenering and Electronics*, Kotelnikov, V.A. (Ed.), Moscow: Nauka (in Russian).
123. Aleksandrov, Yu.N., Bazilevsky, A.I., Petrov, G.M., *et al.* (1987) *Itogi Nauki i Tekhniki, Astronomiya*, Moscow: VINITI, **32**: 201 (in Russian).
124. Tyler, G.L., Simpson, R.A., Maurer, M.I., and Holdmann, E. (1992) *J. Geoph. Res.*, **97**, E8: 13115.

125. Magellan at Venus (1992) *J. Geoph. Res.*, **97**, E8 and E10.
126. Cutrona, L.J. (1990) Synthetic Aperture Radar, Chapt. 21, in: *Radar Handbook*, 2nd ed., Skolnik, M.J. (Ed.), N.Y.: Mc Grow-Hill.
127. Wehner, D.R. (1994) High-Resolution Radar, Chapt. 6, in: *Synthetic Aperture Radar*, Boston – London: Artech House.
128. Elachi, C. (1988) *Spaceborne Radar Sensing: Applications and Techniques*, IEEE Press.
129. Skolnik, M.I. (Ed.) *Radar Handbook*, Mc Grow-Hill Co.
130. Ulaby, F.T. and Dobson M.C. (1989) *Handbook of Radar Scattering Statistics for Terrain*, Nor Wood MA: Artech House.
131. Sorokin, I.V. (1994) *Space Systems for Remote Sensing of the Earth*. Review–Handbook. Series Rocket-Space Technique, Moscow: Energiya (in Russian).
132. Attema, E.P. (1991) *Proc. IEEE*, **79**, 6: 791.
133. Salganik, P.O., Efremov, G.A., Neronskii, L.B., *et al.* (1990) *Issledovaniya Zemli iz Kosmosa*, 2: 70 (in Russian).
134. Neronskii, L.B., Koberchenko, V.G., and Zraenko, S.M. (1993) *Issledovaniya Zemli iz Kosmosa*, 4: 33 (in Russian).
135. Viter,V.V., Efremov,G.A., Ivanov, A.Yu., *et al.* (1993) *Issledovaniya Zemli iz Kosmosa*, 6: 63 (in Russian).
136. Hasselman, K., Raney, R., Plant, W., *et al.* (1985) *J. Geoph. Res.*, **90**, C3: 4659.
137. Hasselman, K. and Hasselman, S. (1991) *J. Geoph. Res.*, **96**, C6: 10713.
138. Chelomei, V.N., Efremov, G.A., Litovchenko, K.F., *et al.* (1990) *Issledovaniya Zemli iz Kosmosa*, 2: 80 (in Russian).
139. Al'pers, V., Bryuning, K., Vil'de, A., *et al.* (1994) *Issledovaniya Zemli iz Kosmosa*, 6: 83 (in Russian).
140. Pereslegin, S.V., Korolev, A.M., and Marov, M.N. (1994) *Issledovaniya Zemli iz Kosmosa*, 2: 84 (in Russian).
141. Mal'tsev, I.G., Marov, M.N., Ramm, N.S., *et al.* (1995) *Issledovaniya Zemli iz Kosmosa*, 3: 56 (in Russian).
142. Litovchenko, K.Ts., Raev, M.D., Semenov, S.S., *et al.* (1995) *Issledovaniya Zemli iz Kosmosa*, 3: 47 (in Russian).
143. Rundkvist, I.K., Devisov, A.N., Zakharov, V.I., *et al.* (1994) *Issledovaniya Zemli iz Kosmosa*, 2: 94 (in Russian).
144. Bel'chanskii, G.I., Ovchinnikov, G.K., Shebchenko, B.I., and Duglas, D. (1994) *Issledovaniya Zemli iz Kosmosa*, 3: 44 (in Russian).
145. Zabolotskii, V.R. (1996) *Issledovaniya Zemli iz Kosmosa*, 2: 106 (in Russian).
146. Elizavetin, I.V. and Ksenofontov, E.A. (1996) *Issledovaniya Zemli iz Kosmosa*, 1: 75 (in Russian).
147. Stacy, N.J., Campbell, D.B., and Ford, P.G. (1997) *Science*, **276:** 1527.
148. Ostro, S.J., Jurgens, R.F., Rosema, K.D., *et al.* (1996) *Icarus*, **121:** 46.
149. Ostro, S.J. (1997) *Celestial Mechanics and Dynamical Astronomy*, **66:** 87.
150. Gens, R. and Vangenderen, J.L. (1996) *Int. J. Remote Sensing*, **17:** 1803.
151. Zribi, M., Taconet, O., Vidal, D., *et al.* (1996) *Remote Sensing of Environment*, **59**, 2: 256.
152. Schmullius, C.C. and Evans, D.L. (1997) *Int. J. Remote Sensing*, **18**, 13: 2713.
153. Kasischke, E., Melack, J., and Dobson, M. (1997) *Remote Sensing of Environment*, **59**, 2: 141.
154. Shi, J. and Dozier, J. (1997) *Remote Sensing of Environment*, **59**, 2: 294.
155. Soares, J., Formaggio, R., and Yanasse, C. (1997) *Remote Sensing of Environment*, **59**, 2: 234.
156. Wang, J., Hsu, J., Shi, C., *et al.* (1997) *Remote Sensing of Environment*, **59**, 2: 308.

157. Ostro, S.J., Hudson, R.S., Rosema, K.D., *et al.* (1999), *Icarus*, **137**, 122.
158. Ostro, S.J., Hudson, R.S., Nolan, M.C., *et al.* (2000), *Science*, **288**, 836.
159. Ostro, S.J., Pravec, P., Benner, L.A., *et al.* (1999), *Science*, **285**, 557.

Chapter 5

Conditions of radio communications in space

5.1 Conditions of radio communications in the solar system

Progress in the investigation of the solar system using radiophysical facilities is largely determined by the efficiency of space radio communications. The main problem with this is how to separate weak radio signals from the unwanted noise of a receiving system.

The quality of space radio communications depends on the wavelength, antenna sizes, receiver sensitivity, transmitter power, the effect of the environment on propagating radio waves, and some other factors. The effect of the Earth's atmosphere on space radio communications, which was considered in Chapter 1 of this book, is weak for wavelengths $\lambda = 3 \div 50$ cm.

The power flux density of radio waves propagating in free space over a distance L is given by

$$P = \frac{W_1 G}{4\pi L^2} , \tag{5.1}$$

where W_1 is the power of a transmitter and G is the gain of the spacecraft antenna.

The input power W_2 of the signal received at a ground-based station is the product of the power flux density P and the effective cross section A of the receiving antenna: $W_2 = PA$. Apart from useful signals, there is ever-present noise, whose power W_3 can be expressed through the effective noise temperature T and the bandwidth Δf of the receiving system:

$$W_3 = KT\Delta f = K(T_1 + T_2)\Delta f , \tag{5.2}$$

where $K = 1.38 \cdot 10^{-23}$ W K^{-1} Hz^{-1}, T_1 is the noise temperature of the receiving antenna, and T_2 is the equivalent noise temperature of the receiver. In radio communications, the signal-to-noise ratio $\alpha = W_2 W_3^{-1}$ is of great significance. From (5.1) and (5.2) follows the expression for the minimum necessary transmitter power W_1 that ensures radio communications in the bandwidth Δf:

$$W_1 = 4\pi L^2 K\Delta f \alpha (T_1 + T_2)(AG)^{-1} . \tag{5.3}$$

This expression yields the following formula for the admissible signal bandwidth in deep-space radio communications:

$$\Delta f = \frac{W_1 A G}{4\pi \alpha L^2 K(T_1 + T_2)} \; . \tag{5.4}$$

Expressions (5.3) and (5.4) allow the conditions of radio communications in space to be analyzed [1–4]. The minimum necessary power W_1 and bandwidth Δf of antennae depend on the wavelength of radio signals , since this parameter influences the noise temperature, the gain of the transmitting antenna, and the effective area of the receiving antenna.

Let us consider the effect of antenna parameters on radio communications given by a factor AG. Deep-space radio communications commonly employ parabolic antennae. The gain of the parabolic transmitting antenna, which depends on its diameter, can be given by the approximate formula

$$G = 4 D_1^2 \lambda^{-2} \; , \tag{5.5}$$

where D_1 is the aperture diameter of the antenna.

The diameter D_1 must not be very large; this is not for only engineering reasons, but also on account of the necessary accuracy of the antenna's orientation toward the Earth. Indeed, the half-power beamwidth of a directional antenna can be approximated by $\Delta\varphi = \lambda D^{-1}$; therefore, big transmitting antennae should be oriented more accurately than small ones.

The effective cross section of the receiving antenna, A, is determined by its aperture. For big parabolic antennae, the effective cross section A depends on the wavelength, which is due to random deviations in the antenna's shape from an ideal paraboloid. Thus, the present-day big ground-based parabolic antenna 100 m in diameter described in [5] has nearly an ideal parabolic shape. The random deviations of its surface from the ideal paraboloid are within ± 0.5 cm, and so this antenna can operate at a minimum wavelength of $\lambda \approx 3$ cm. Irregularities in the antenna surface bring about random phase inaccuracies and, consequently, a decrease in its effective cross section given by the approximate formula

$$A = \frac{\pi D_2^2}{8} \exp\left[-\left(\frac{4\pi\mu}{\lambda}\right)^2\right] \; , \tag{5.6}$$

where D_2 is the aperture diameter of the ground receiving antenna and μ is the mean deviation of the antenna surface from an ideal paraboloid [6]. From (5.6) it follows that, if $\mu\lambda^{-1} < 5\cdot10^{-2}$, the effective area of the antenna is equal to half of its aperture area and does not depend on the wavelength. At $\mu\lambda^{-1} > 10^{-1}$, however, A shows a steep decline with decreasing wavelength.

From (5.5) and (5.6) we can derive the following wavelength dependence for the antenna factor AG:

$$AG = \frac{\pi}{2}\left(\frac{D_1 D_2}{\lambda}\right)^2 \exp\left[-\left(\frac{4\pi\mu}{\lambda}\right)^2\right]. \tag{5.7}$$

The orientation of spacecraft strongly affects the conditions of its communications. If spacecraft is not stabilized in space, its antenna should have a wider beamwidth to ensure good communications. In this case, G does not depend on the wavelength and the antenna factor is given by

$$AG_0 = \frac{\pi D_2^2 G_0}{8} \exp\left[-\left(\frac{4\pi\mu}{\lambda}\right)^2\right], \tag{5.8}$$

where G_0 is the gain of a low-directional antenna.

Let us consider now the effect of noise on radio communications. From (5.2) it follows that the total noise power is the sum of the noise powers of the transmitting and receiving antennae. The noise temperature of a receiving system depends on how its input stages are designed. Thus, masers operated on centimeter radio waves have a noise temperature $T_2 = 10 \div 20$ K; cooled parametric amplifiers operated in a decimeter or meter bandwidth have a $T_2 = 40 \div 90$ K. Basically, the noise temperature T_2 can be lowered to about 10 K. The noise temperature of receiving antennae is determined by background radiation coming from space, as well as from the Earth's atmosphere and surface. A more detailed analysis of the noise of big antennae is given in [7–9].

Figure 5.1 presents the wavelength dependencies of the noise temperature T of an antenna tilted at an angle greater than 20° to the horizon. In a meter bandwidth, antenna noise is determined by space radiation; in this case T depends on the orientation of the antenna in the sky. Curves *1* and *3* in Fig. 5.1 correspond to antenna orientations toward sky regions with maximal and minimal levels of space radiation, respectively. In the wavelength band $\lambda = 5 \div 20$ cm, temperature T is determined by space radiation and the Earth's radiation received by the antenna's sidelobes. At $\lambda < 3$ cm, the noise temperature rises due to radiation from the Earth's troposphere. Curve *2* shows the average dependence $T(\lambda)$ for big antennae. As can be seen, the minimal noise temperature corresponds to the wavelength range $\lambda = 3 \div 20$ cm. The total noise temperature of the whole receiving system is somewhat higher, due to a contribution from the receiver.

Formulas (5.3), (5.4), (5.7), and (5.8), together with the dependences $T(\lambda)$ presented in Fig. 5.1, allow the conditions of radio communications in space to be analyzed.

Let a ground-based parabolic antenna have a diameter $D_2 = 100$ m and let the inaccuracy of manufacture of its surface be characterized by a parameter $\mu = 5$ mm (the manufacture of big full-revolving parabolic antennae with a lower value of μ is difficult because of their possible deformation). Assume for definiteness that the transmitting antenna of a spacecraft has a diameter $D_1 = 3$ m; path length $L = 3 \cdot 10^8$ km; signal-to-noise ratio $\alpha = 10$; and frequency band $\Delta f = 100$ Hz. The wavelength dependence of the minimum necessary power of the transmitter, W_1, can be determined using formulas (5.3) and (5.7). In Fig. 5.2, curve *2* shows the dependence $W_1(\lambda)$ for a spacecraft oriented towards the Earth. As is evident from this curve, the range of optimal operating wavelengths of oriented antennae is narrow: $\lambda = 3$–8 cm. This range is determined solely by the wavelength dependence of the antenna factor AG and the noise temperature T.

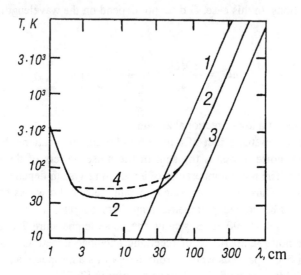

Fig. 5.1. The noise temperature of antenna versus its operating wavelength.

If a spacecraft is not oriented towards the Earth, its transmitting antenna must be omnidirectional. The gain G of such an antenna is equal to about 1 and is independent of the wavelength. In Fig. 5.2, curve 1 represents the wavelength dependence of the minimal necessary power of the transmitter for this case. When plotting this curve, we used the same values for the above parameters, i.e., the antenna diameter $D_2 = 100$ m, $\mu = 5$ mm, $L = 3 \cdot 10^8$ km, the signal-to-noise ratio $\alpha = 10$, except that Δf was taken to be equal to 10 Hz. As is evident from curve 1, the optimal range of operating wavelengths for omnidirectional antennae is from 10 to 30 cm.

The conditions of radio communications in the solar system can be described by the channel capacity, which is characterized by the transmitted frequency band Δf. According to formula (5.4), the admissible transmitted bandwidth decreases with increasing path length as $\Delta f \sim L^{-2}$. The optimal operating wavelength for both directional and omnidirectional spacecraft antennae is about $\lambda = 10$ cm. If the spacecraft antenna is directional and its diameter $D_1 = 3$ m, this would ensure radio communications with $\Delta f = 100$ Hz even over distances as great as 10^{10} km (i.e., within the entire solar system), provided that the ground-based receiving antenna has a power $W_1 = 100$ W and diameter $D_2 = 100$ m. Since spacecraft must always have a communication channel employing an omnidirectional antenna with $G \approx 1$, the admissible frequency band is actually narrower than the aforementioned value. Thus, at $L = 3 \cdot 10^8$ km, $W_1 = 100$ W, and $D_2 = 100$ m, the admissible frequency band is equal to 30 Hz.

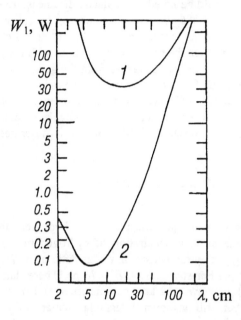

Fig. 5.2. Minimum necessary transmitter power versus the wavelength of radio signals.

The feasibility of radio communications to spacecraft and radio observations of planets strongly depends on their distance from the Earth. In view of this, let us now illustrate some distances in the solar system.

Every four months, Mercury approaches the Earth to a distance of about 90 million km, the maximum Mercury-to-Earth distance being around 210 million km. The minimum distance to which Venus approaches the Earth is 45 million km (every 19 months), the maximum Venus-to-Earth distance being about 250 million km. The minimum Mars-to-Earth distance observed biennially comprises 75 million km, the maximum separation being 380 million km. In 1974–1975, the minimum Jupiter-to-Earth distance was about 600 million km, the maximum separation being 900 million km. In spite of these large distances, current radio facilities allow reliable radio communications and radiophysical studies throughout the solar system.

5.2 Radio communications and propagation of radio waves in the Galaxy

Let us consider the propagation of radio waves in the Galaxy based on experimental data presented in [10–13]. It should be noted that galactic distances are commonly expressed in units of parsecs equal to $3.085 \cdot 10^{13}$ km (to compare, the astronomical unit, i.e., the mean distance from the Sun to the Earth, is equal to $1.5 \cdot 10^8$ km). Let us introduce a cylindrical coordinate system in which the plane $z = 0$ coincides with the galactic plane and the system origin is at the galactic center. Distances from the galactic center will be given by coordinate r. For instance, the Sun's distance r from the galactic center is about 10 kiloparsec (kpc). The diameter of our Galaxy is about 30 kpc. Interstellar matter in the Galaxy is distributed so that most of it is in or very close to the galactic plane [10–13]. The densities of stars, hydrogen, and electrons in the Galaxy is described by the exponential function

$$n = n_0 \exp\left(-|z|H^{-1}\right), \qquad (5.9)$$

where z is the distance from the galactic plane. For hydrogen distribution, parameter $H = 150 \div 300$ pc, and the mean concentration of hydrogen in the galactic plane $n_0 \approx 0.7$ cm^{-3}. Correspondingly, electron density in the galactic plane $N_0 = 0.03$ cm^{-3} and parameter $H = 300 \div 1000$ pc. For star density, $H = 76$ pc. These data refer to the galactic region with $4 < r < 12$ kpc. For $r > 12$ kpc, parameters n_0 and N_0 show a decline; the dependence $n(z)$ is, however, still unknown. Interstellar matter in the Galaxy is distributed nonuniformly, especially in the galactic plane. The aforementioned information gives only a rough idea of our Galaxy.

Let us consider the absorption of radio waves by interstellar plasma. As shown in [14], the wave attenuation Y and absorption coefficient γ in a rarefied plasma are given by

$$Y = e^{-\tau} = \exp\left[-\int_0^L \gamma(l)dl\right], \qquad (5.10)$$

$$\gamma = 9.75 \cdot 10^{-3} N^2 T_e^{-3/2} f^{-2} \ln\left(4.97 \cdot 10^7 T_e^{3/2} f^{-1}\right).$$

where τ is the "optical" thickness, L is the path length, dl is the length element, T_e is the electron temperature, N is the electron density expressed in cm^{-3}, γ is expressed in cm^{-1}, and f has dimensions Hz. As is evident from (5.10), the absorption of radio signals must be higher in regions with a more dense and severely ionized plasma, which has a relatively low temperature. To calculate the absorption of radio waves, one has to know the electron density and temperature, which makes such calculations unreliable. Relevant information on the attenuation of radio waves can be more reliably derived from the analysis of the low-frequency spectra of radiation emitted from natural sources. Such spectra have a strong experimental and theoretical underpinning; in particular, the low-frequency falloff of the radio signal spectrum is interpreted as indicating the absorption of radio waves by interstellar plasma [10–12, 15]. The absorption of radio waves can also be studied by measuring the low-frequency background galactic radiation. Experiments employing omnidirectional antennae showed that brightness temperature is maximal at a frequency of 3 MHz provided that the antenna is oriented in the direction opposite to that toward the galactic center. At $f < 3$ MHz, the noise temperature of antenna decreases obviously because of the absorption of radio waves in the Galaxy. At high galactic latitudes, the absorption of the extragalactic background radiation at a frequency $f = 1$ MHz corresponds to $\tau \approx 1.5$. This gives $N \approx 0.03$–0.05 cm^{-3} and $T_e \approx 4000$–1000 K in formula (5.10). Measurements of the galactic background radiation at frequencies of 9 and 6 MHz with the aid of directional antennae show that its intensity decreases in the direction towards the galactic plane [15]. This effect was theoretically rationalized by assuming that radio waves are adsorbed by plasma clouds, which possess an increased electron density of about 0.5 cm^{-3} and occupy 0.1–0.6% of space. The intensity of nonthermal background radiation in the galactic plane exhibits a lower-frequency falloff at 50–100 MHz. The absorption of radio waves increases in the direction towards the galactic center, that is why the lower-frequency falloff in this direction occurs already at $f \approx 200$ MHz. As is evident from formula (5.10), absorption decreases with frequency by the law $\gamma \sim f^{-2}$; therefore, it can be anticipated that the absorption of radio waves at $f > 1000$ MHz must be low. Analysis of the distribution of brightness temperature near the strong source of radiation, Sagittarius A, provided direct evidence in favor of the weak absorption of centimeter radio waves in the direction toward the galactic center. Indeed, Sagittarius A is located very close to the center of the Galaxy. Since the absorption of radio waves and the respective heat emission of the absorbing medium are related, the occurrence of a distributed absorbing medium must lead to a significant thermal radiation with a characteristic spectrum. Analysis of the distribution of centimeter radiation showed that thermal radiation is distributed in the direction of the galactic center nonuniformly, i.e., there exist low-angular-sized discrete sources of radiation [13]. The brightest source of radio waves, Sagittarius A, has an angular size of 3.5′; its spectrum is, however, nonthermal. Other sources that occur near the galactic center and emit thermal radiation have low angular sizes. Thus, except for an angular range of ±2° in the direction toward the galactic center, the absorption of centimeter radio waves, even if they propagate in the galactic plane throughout the Galaxy, can be neglected. Interstellar hydrogen absorbs radio waves at a resonant frequency of 1420 MHz. The discovery of pulsars stimulated relevant investigations, since the pulse radiation of pulsars with frequencies close to 1420 MHz made possible the determination of the effective thickness τ by measuring the absorption of radio waves by hydrogen along the pulsar–Earth path. Measurements of the radiation of pulsar PSR 0329+54 at the maximum of hydrogen absorption gave $\tau \approx 1$–2, and the respective measurements for pulsar PSR 1933+16 gave a lower value $\tau \approx 0.6$–0.8. The distances to these pulsars are 2.5 and about 6 kpc, respectively; therefore, for the distances 2–6 kpc, $\tau \approx$ 1–2, even if this parameter is measured at the maximum of the hydrogen absorption line.

Let us now consider the dispersion of radio waves in interstellar plasma. For high-frequency radio signals, the group velocity, as was shown by the authors of [14], can be given by

$$V_g = c\left[1 - Ne^2\left(2\pi m f^2\right)^{-1}\right], \tag{5.11}$$

where e and m are the charge and the mass of electron, respectively. For two radio frequencies, f_1 and f_2, it takes different times, t_1 and t_2, to cover the distance L. From (5.11) it follows that the time delay $\Delta t = t_1 - t_2$ is proportional to the integral electron concentration:

$$\Delta t = e^2\left(2\pi mc\right)^{-1}\left(f_2^{-2} - f_1^{-2}\right)\int_0^L N\,dl \ . \tag{5.12}$$

The pulse radiation of pulsars allows the high-accuracy determination of the pulse delay at the lower frequency f_2 relative to the time of the pulse arrival measured at the higher frequency f_1. Using this method, the integral electron density, which is a measure of the radio wave dispersion, was determined for 300 pulsars [11]. As the distances L to some pulsars have already been known, this allowed the mean electron density to be determined: $N = 0.03$ cm^{-3}. At present, data on the integral electron density are used to estimate the pulsar distances L by comparing the radio wave dispersion with the above mean value of N. The distance to pulsar PSR 0950+0.8, which is nearest to the Earth, was found to make 0.1 kpc, while those of the outermost pulsars range from 10 to 20 kpc. This method yields more reliable data for the distances to the pulsars that occur at high galactic latitudes, since the electron density changes little in this direction. On the other hand, the measurement of the radio wave dispersion gives only rough estimates of distances in the galactic plane, since interstellar plasma is fairly inhomogeneous in this direction.

A large body of experimental evidence on the integral electron density in different galactic directions unambiguously indicates that the modulated signals covering distances of up to 10 kpc must undergo dispersion-induced disturbances in interstellar plasma. These disturbances can be estimated by the criterion

$$\Delta f \cdot \Delta t \ll 1 \ , \tag{5.13}$$

where Δf is the transmitted signal bandwidth, and Δt is the time delay of arrival of the border spectral frequencies. From (5.12) and (5.13) we can get an expression for the bandwidth of a signal that can be transmitted without considerable dispersion-induced disturbances:

$$\Delta f \le (2\pi m c)^{\frac{1}{2}} e^{-1} f_0 \left[\int_0^L N\, dl \right]^{-\frac{1}{2}}, \qquad (5.14)$$

where f_0 is the carrier frequency.

Experimental studies of short meter-band subpulses of pulsars showed that the subpulses, indeed, undergo severe dispersion-induced disturbances to become totally diffused when propagating for long distances. On the other hand, formula (5.14) and relevant experimental data suggest that the effect of interstellar plasma on centimeter-band radio waves must be weak.

The motion of interstellar plasma inhomogeneities or the radio source must lead to the fading of radio waves. It is known that radio waves strongly fade if their source is pointlike (i.e., has low angular sizes). Pulsars have low angular sizes, as they represent neutron stars of about 10 km in diameter. Observations have demonstrated that, indeed, the envelope of pulsar pulses undergoes fluctuations. Analysis showed that fast changes in the amplitude with the characteristic time of several seconds are inherent to the pulsar itself, whereas slow fades are related to interstellar plasma inhomogeneities. Slow fades with the characteristic time of tens of minutes correlate with the dispersion measure of the pulsar, thus indicating that it is the interstellar plasma inhomogeneities that cause radio wave fading. The fading of meter radio waves is deep (the scintillation index $S \approx 1$), while centimeter radio waves undergo a more shallow fading. If fading is not saturated, one can estimate the parameters of plasma inhomogeneities. Such estimations showed that $\Delta N \approx 5 \cdot 10^{-5}$ cm^{-3} with a plasma inhomogeneity scale of 10^{10}–10^{12} cm. Analysis of pulsar scintillations showed that plasma inhomogeneities are described by the power-law spectrum with a power index of 4 [16–18]. The spatial spectrum of electron density fluctuations has a complex pattern with two characteristic scales, $3 \cdot 10^{10}$ and $3 \cdot 10^{13}$ cm. It should be noted that the presently available experimental data are insufficient to allow a comprehensive discussion of the origin and spectrum of electron density inhomogeneities. Analysis of pulsar scintillation patterns recorded at two ground-based widely spaced stations showed that they are correlated and shifted in time relative to each other. Based on these data one can estimate the velocity of plasma irregularities normal to the line-of-sight. This velocity, which is due to the fast motion of pulsars, varies from 40 to 200 km s^{-1}.

Interstellar plasma inhomogeneities lead to the pulsar pulse broadening. This effect is sizeable for meter radio waves, but shows a decline at shorter wavelengths.

Let us consider, following the publications [19–22], the conditions of radio communications in the Galaxy in relation to other technical civilizations. As was mentioned above, centimeter and millimeter radio waves, in contrast to meter radio waves, almost do not undergo the effect of interstellar plasma. On the other hand, millimeter radio waves are absorbed by planetary atmospheres and galactic dust clouds. Therefore, galactic radio communications must be optimum in the centimeter waveband. Using formula (5.4) and assuming that the powers of a desired signal and noise are equal, we can derive the following expression for the maximum distance of radio communications:

$$L = (W_1 AG)^{1/2} (4\pi KT\Delta f)^{-1/2} .$$ (5.15)

The noise temperature depends primarily on galactic and atmospheric radiation (see Section 5.1). Space radiation shows a steep decline with its wavelength, so that in the centimeter waveband the contribution from cosmic noise is negligible. Noise in this waveband is determined by the absorption and emission of radio waves in the atmosphere of a planet where a technical civilization may exist. Therefore, the dependence $T(\lambda)$ also indicates that centimeter radio waves are preferable for interstellar communications.

Let us estimate the maximum distance of radio communications with an advanced civilization whose technical competence corresponds to our level. Narrow-band radio communications, such as the transmission of signal letters, need a bandwidth Δf as small as 10 Hz. Assume that a parabolic antenna 100 m in diameter is used for the signal transmission and reception, and that a transmitter has a power of 10^5 W. Such an antenna has an effective cross section $A \approx 4000$ m^2 and gain $G = 1.4 \cdot 10^7$ for $\lambda = 6$ cm. The minimum noise temperature T for an Earth-like planet in the centimeter waveband must be about 20 K. Having this in mind, we can find that $L_{max} = 4 \cdot 10^{17}$ m, or 13 pc. Recall that this value of L_{max} corresponds to the current technical level of our civilization. Actually, the term $W_1 GA$ can relatively easily be increased 1000-fold by increasing the transmitter power to 10^6 W and the diameter of the parabolic antenna to 300 m. In this case narrow-band radio communications will be possible throughout our Galaxy. In other words, once a sufficiently stable monochromatic signal is transmitted by an extraterrestrial civilization, we shall receive it whatever the distance to this civilization is.

Table 5.1. Natural wavelengths that may be appropriate for interstellar communications with other technical civilizations

Parameter	λ_1 of hydroxyl, cm	λ_2 of hydrogen, cm
λ	18	21
λ/e	6.64	7.72
λ/π	5.73	6.70
$\lambda/2e$	3.32	3.86
$\lambda/2\pi$	2.86	3.35

The question now arises: at what wavelengths are radio communications with extraterrestrial civilizations plausible? We may assume that these are the basic natural frequencies of the emission lines of hydroxyl and hydrogen (λ_1 and λ_2, respectively), since they lie in the short-wave region of the decimeter waveband and must be known for technically advanced civilizations. The derivative wavelengths, such as λ/π or λ/e must be even more appropriate, since their absorption in interstellar space is not so strong when compared with the absorption of the λ_1 and λ_2 wavelengths.

Table 5.1 summarizes the wavelengths that are hypothetically appropriate for interstellar communications. Of great interest is the closeness of the derivative wavelengths λ_1/e and λ_2/π (6.64 and 6.70, respectively), which allows the suggestion that the most likely wavelengths for radio communications with extraterrestrial civilizations are between 6.6 and 6.8 cm. On the other hand, if a planet on which an extraterrestrial civilization resides has a rather thin atmosphere, such as the Earth's atmosphere, which does not absorb centimeter radio waves, radio signals from that civilization may be expected around the derivative wavelengths $\lambda_1/2e$ or $\lambda_2/2\pi$, that is, within the wavelength band 3.3–3.4 cm.

Thus, narrow-band radio communications with another technical civilization within our Galaxy are possible if that civilization is at least at our current technical level. The main problem with this is that we are not sure of the direction from which the signals from an extraterrestrial civilization should be expected. Nor do we know the communication frequency and the rate of the Doppler frequency shift. The use of big receiving antennae, though expected to increase the signal-to-noise ratio and thereby the probability of recording the signals from another technical civilization, will narrow the beamwidth and thus diminish the probability of the proper choice of the direction toward this civilization. The narrowing of the receiver bandwidth should increase the signal-to-noise ratio and thus make the reception of signals from the extraterrestrial civilization more likely, but will diminish the probability of the proper choice of the operation frequency of the receiver, since the actual frequency transmitted by the extraterrestrial civilization is unknown. The foregoing demonstrates that the problem of communication with extraterrestrial civilizations is very complex and, if resolved, will be a breakthrough in space radio science.

5.3 Propagation of radio waves in gravitational fields

Like other media, gravitational fields can cause a refraction of radio waves, change their frequency and amplitude, and increase their travel time.

The theory of general relativity suggests that the propagation of electromagnetic waves in a gravitational field can be considered as a problem of the wave propagation in Euclidean space, provided that the effect of the gravitational field is taken into account by the refractive index of a vacuum, n, which is related to the gravitational potential U as [23–26]

$$n = 1 + N = 1 + \frac{2U}{c^2} \quad . \tag{5.16}$$

Here N is the refractivity of a vacuum and c is the speed of light. Formula (5.16) is approximate since it does not take into account the rotation of a gravitational field. In particular, this formula is not valid for high gravitational potentials. For the gravitational field of a stellar or planetary mass M, $U = \gamma_1 M r^{-1}$ and, therefore,

$$N = \frac{2\gamma_1 M}{c^2 r} = \frac{R_g}{r}. \tag{5.17}$$

Here $R_g = 2\gamma_1 M c^{-2}$ is the gravitational radius, γ_1 is the gravitational constant, and r is the distance from the center of the star or planet to an arbitrary point P. From (5.17) it follows that the stellar gravitational field is equivalent, from the standpoint of wave propagation, to a spherically symmetric medium whose refractive index decreases with distance by the law $N \sim r^{-1}$. Therefore, the gravitational field must cause ray bending toward the gravitational center, bring about refractive changes in the energy flux density, and induce a radio wave time delay and frequency shift.

The problem of the radio wave propagation in a gravitational field can be treated as follows (the graphical representation of this problem is shown in Fig. 5.3): point O is at the star's center, TM is the ray path, ξ is the angle of refraction, $TK = p$ is the impact parameter, a is the star's radius. Let the source of radio waves be situated at point T, while signals are received at the station located at point M. To determine angle ξ, we shall avail ourselves to expression (2.2) for refraction in a spherically symmetric medium and take into account (5.17). Then we obtain

$$\xi = 2pR_g \int_{r_1}^{\infty} \frac{dr}{r^2\left(1 + \dfrac{R_g}{r}\right)\left[r^2\left(1 + \dfrac{R_g}{r}\right)^2 - p^2\right]^{1/2}}, \tag{5.18}$$

where r_1 is the minimal distance from the ray to the star's center. Taking into account that $R_g \ll r$ and $r_1 \approx p$, we can easily calculate the integral (5.18):

$$\xi = \frac{2R_g}{p}. \tag{5.19}$$

Angle ξ attains its maximal value, ξ_m, when the ray grazes the star's surface:

$$\xi_m = \frac{2R_g}{a}. \tag{5.20}$$

Here a is the planetary or stellar radius. This expression was derived in [27]. For the Sun, $R_g = 3$ km, and, therefore, $\xi_m = 1.75''$.

Optical measurements of the angle ξ_m confirmed the validity of formula (5.20) to an accuracy of ± 30%. A better accuracy was achieved in 1970 by two independent groups of radio astronomers, who employed radio interferometers to measure ξ_m for two radio sources of small angular sizes, 3C273 and 3C279 [28, 29]. Due to the orbital motion of the Earth around the Sun, the lines of sight of these sources approach the Sun at various times. Because of the refraction of radio waves in the gravitational field of the Sun, the angle between the lines of sight toward these sources changes; this can reliably be recorded by radio interferometers. The authors of [28] used a radio interferometer with a 24-km baseline operated at a frequency f = 2388 MHz, while the authors of [29] used an interferometer operated at f = 9602 MHz. With allowance for the effect of the circumsolar plasma, these measurements yielded close values, ξ_m = 1.82±0.2″ [28] and ξ_m = 1.77±0.2″ [29]. Independent measurements by the same method gave a much lower value, ξ_m = 1.57±0.08″ [30]. The authors of [31] achieved a greater accuracy, as they used three-base interferometers operated at two frequencies, 2695 and 8085 MHz, and observed three radio sources. This complex experiment yielded ξ_m = 1.77±0.02″, i.e., the value which is very close to that predicted theoretically by formula (5.20). Even better accuracy of measurements was achieved by the authors of [32], who used an interferometer with a 3100-km baseline and observed 14 radio sources of small angular sizes. The experiments described confirmed the validity of formula (5.20) to an accuracy of ±0.01″. Some additional information on the refraction of radio waves in the gravitational field of the Sun can be found in [33].

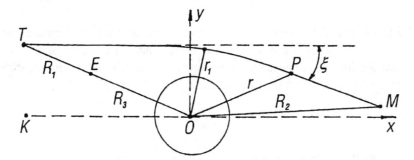

Fig. 5.3. Graphical representation of the bending effect of gravitational fields on propagating radio waves.

Ray bending in the gravitational field of a star must produce a focusing effect; in other words, the gravitational field must behave as a focusing lens [23, 35]. As follows from formula (5.20) and the geometry presented in Fig. 5.3, the minimum distance x_m at which the parallel beam of radio waves crosses the x-axis and, therefore, produces the maximum focusing effect is given by

$$x_{\mathrm{m}} = \frac{\alpha}{\xi_{\mathrm{m}}} = \frac{\alpha^2}{2R_{\mathrm{g}}}. \tag{5.21}$$

At $x > x_{\mathrm{m}}$, the rays that were bent by the angle ξ will also meet on the x-axis; therefore, the gravitational lens has a focal axis $x \geq x_{\mathrm{m}}$ rather than a single focal point. The magnification factor of the gravitational lens can be found in the same manner as the refractive field changes during radio occultation studies (see Section 2.1). Using (2.21) or the approximate formula (2.22) and taking into account (5.19), we obtain the following expression for the refractive magnification of the gravitational lens:

$$X = \frac{\sqrt{2R_{\mathrm{g}}x}\left(\dfrac{y^2}{4R_{\mathrm{g}}x} + 1\right)}{y\sqrt{\dfrac{y^2}{8R_{\mathrm{g}}x} + 1}}. \tag{5.22}$$

Here x and y are the coordinates of the point M of gravitational focusing. For $y \ll x$, where the focusing effect is strong, the following formula is valid:

$$X = \frac{\sqrt{2R_{\mathrm{g}}x}}{y}. \tag{5.23}$$

Formulae (5.22) and (5.23) were derived in terms of the ray concept; therefore, they are inapplicable at $y = 0$, where the focusing effect is maximum. The authors of [36–38] took into account the diffraction phenomenon and found that in this case the magnification of the gravitational field for $y = 0$ can be described by the following formula

$$X_{\mathrm{m}} = 4\pi R_{\mathrm{g}}\lambda^{-1}. \tag{5.24}$$

The authors of [39, 40] analyzed the magnification of gravitational lenses for radio waves and found that plasma inhomogeneities restrain the range of wavelengths that can be focused by such lenses. The focusing effect of gravitational fields seems to be essential for only shorter-wavelength centimeter and millimeter radio waves. In the case of the gravitational field of the Sun, the magnification for 1-cm radio waves may reach 10^7. As follows from formula (5.21), such a strong focusing effect can be observed only at distances as great as 10^{11} km from the Sun. It should be noted that formulas (5.22)–(5.24) are valid for point sources. For extended sources, which are typical in radio astronomy, the magnification of gravitational fields must be lower. Thus, the magnification of the gravitational field of the Sun for centimeter and millimeter radio waves is $X \approx 10^5$ [40,

41]. The image of a point source observed through a gravitational lens was extensively analyzed by the author of [44], who showed that, if the viewing point is situated on the focal x-axis, the source will be perceived as a bright ring embracing the gravitational lens of the star. But if the viewing point lies beyond the x-axis, the viewer will observe two imaginary images of the radio source occurring symmetrically relative to the star. Such symmetrical images were actually observed with a long-base interferometer [41]. The authors of that work believe that it was the focusing effect of a gravitational lens that they dealt with. A more detailed consideration of the focusing effect of gravitational fields was given in [42, 43] and monograph [44].

Let us consider now the time delay of radio waves, ΔT, in a gravitational field. With allowance for (5.17), the delay is given by the formula

$$\Delta T = \frac{1}{c} \int N(r) dl = \frac{R_g}{c} \int \frac{dl}{r}, \tag{5.25}$$

where integration is carried out along the ray path TM and dl is the length element of the ray. From (5.25) follows

$$\Delta T = \frac{2R_g}{c} \ln\left(\frac{R_1 + R_2 + L}{R_1 + R_2 - L}\right), \tag{5.26}$$

where R_1 and R_2 are the distances from the star's center to the transmitter and receiver, respectively; and L is the length of the path TM (Fig. 5.3). The delay must be maximal when the ray passes near the Sun's photosphere; for this case the theory gives $\Delta T = 200 \, \mu s$. Experimentally, the gravitational delay of radio waves was estimated from the radar measurements of distances to Venus and Mercury [45]. In this case radio waves pass through the gravitational field of the Sun twice, when they travel to those planets and back to the Earth; therefore, the time delay of radio waves must double. The errors of ΔT determination might be due to the inaccuracy of measurements of the planet's orbit, as well as to the effects of planetary relief and circumsolar plasma. The time delay found in [45] agreed with the theory, i.e., it corresponded to formula (5.26), within an accuracy of 20%. More accurate measurements of ΔT were made during the missions of the *Mariners 6* and *7* and *Viking* spacecraft [46–51]. In those experiments, again, ΔT was measured under conditions of the double passing of radio waves through the gravitational field of the Sun. The accuracy of measurements, determined primarily by the effect of the circumsolar plasma, was essentially improved. Measurements gave ΔT values that corresponded to formula (5.26) with an error not exceeding ±1%. More information on the gravitational delay of electromagnetic waves can be found in [52, 53].

Now consider the effect of a gravitational field on the frequency of radio waves. The relative positions of the transmitter and receiver of radio waves define two extreme cases. The first case corresponds to the motion of a spacecraft in a circular orbit of radius R_2, where gravitational potential is constant, and the ray *TM* undergoes the refraction by angle ξ in the gravitational field (Fig. 5.3). In the second case, the ray *TE* is only slightly bent, and the source of radio waves (point T) and the receiver (point *E*) are located in the regions with essentially different gravitational potentials. Clearly, there may also be situations intermediate between the above two cases.

Let us consider the first case, when a spacecraft moves, at a velocity V_1, so that the ray passes nearby the Sun at a minimum distance r_1 (Fig. 5.3). The frequency shift ΔF induced by the gravitational field can be expressed in terms of either the time delay ΔT or angle ξ. In terms of the time delay ΔT, ΔF is given by the following expression

$$\Delta F = \frac{1}{2\pi} \cdot \frac{d\varphi}{dt} = f \frac{d\Delta T}{dt} = f \frac{dp}{dt} \cdot \frac{d\Delta T}{dp} ,\qquad(5.27)$$

where we took into account that $r_1 \approx p$. From (5.26) and (5.27) follows

$$\Delta F = f U_1 \frac{2R_g}{c} \cdot \frac{d}{dp}\left(\ln \frac{R_1 + R_2 + L}{R_1 + R_2 - L} \right),\qquad(5.28)$$

where $U_1 = dp/dt$ is the rate of impact parameter changing. When differentiating (5.28), it should be taken into account that R_1 and R_2 are constant and L is related to p as

$$L = \sqrt{R_1^2 - p^2} + \sqrt{R_2^2 - p^2} .\qquad(5.29)$$

After respective rearrangements, we obtain the following approximate formula for the gravitational shift in frequency:

$$\Delta F = 4 f U_1 R_g L p c^{-1} \left(R_1 + R_2\right)\left(R_1^2 - p^2\right)^{-\frac{1}{2}}\left(R_2^2 - p^2\right)^{-\frac{1}{2}}\left[\left(R_1 + R_2\right)^2 - L^2\right]^{-1}.\qquad(5.30)$$

The problem of the gravitational shift in frequency can be simplified by expressing ΔF through the angle ξ [54]. In Section 2.1, we considered the analogous problem of frequency shift during the remote sensing of a spherically symmetric medium and derived expression (2.46), which relates the frequency shift ΔF and refraction angle ξ. This expression is valid for the remote sensing of the gravitational field of the Sun as well. Then from (2.46) and (5.19) immediately follows

$$\Delta F = \frac{2V_1 R_g}{\lambda p},\qquad(5.31)$$

where V_1 is the component of the spacecraft velocity normal to the ray *TM*. The coherent mode of space communications, often used in practice, implies that a spacecraft transponder receives radio signals of highly stable frequency emitted from a ground-based station and retransmits them back. In this case radio waves travel the path *TM* twice, so that the frequency shift ΔF doubles.

Let us estimate the expected value of ΔF assuming that $\lambda = 8$ mm, $V_1 = 10$ km s^{-1}, and the system operates in the coherent mode. We chose millimeter wave signals since the interfering effect of the circumsolar plasma on such wavelengths is weak. Figure 5.4 shows the dependence of the gravitational shift in frequency on p/a for the spacecraft moving in such a way that the impact parameter p varies. For the ray approaching the Sun and receding from it, ΔF has different signs. If so, the maximum difference in frequencies measured during the approach and receding stages at $p = a$ is described by the formula

$$\Delta F_m = \frac{8V_1 R_g}{\lambda \alpha} .$$

(5.32)

For the above values of the relevant parameters, $\Delta F_m \approx 40$ Hz.

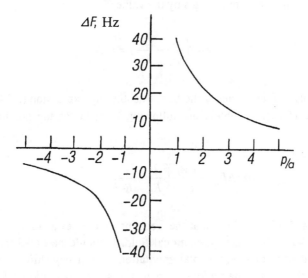

Fig. 5.4. Gravitational shift in frequency for 8-mm radio waves.

Let us now consider the second case, when radio waves travel between two points with different values of gravitational potential U. The gravitational shift in frequency, ΔF, in this case does not depend on the motion of the radio source and is determined only by the difference in the gravitational potentials at these points. If a radio wave travels from point T (Fig. 5.3) with a gravitational potential U_1 to point E with a gravitational potential U_2, the quantum energy changes by the value $\Delta U = U_1 - U_2$. The change in the energy of a quantum is equivalent to the change in its frequency, so that $U = h^* \Delta F$, where h^* is Planck's constant. The gravitational shift in frequency is given by

$$\Delta F = \frac{U_1 - U_2}{h^*} \quad . \tag{5.33}$$

In terms of quantum mechanics, a radio wave can be characterized by the quantum mass m given by

$$m = \frac{h^* f}{c^2} \quad . \tag{5.34}$$

If a quantum with the mass m travels from a point located at a distance R_1 from the center of a planet or the Sun to a point occurring at a distance $R_3 = OE$ from this body, the potential energy of the quantum changes by the value

$$\Delta U = \gamma_1 mM \left(\frac{1}{R_1} - \frac{1}{R_3} \right) = h^* \Delta F \quad , \tag{5.35}$$

where M is the mass of the planet or the Sun. Substituting expression (5.34) for the quantum mass m into (5.35), we obtain the following formula for the gravitational shift in frequency:

$$\Delta F = \frac{\gamma_1 M f}{c^2} \left(\frac{1}{R_1} - \frac{1}{R_3} \right) \quad . \tag{5.36}$$

From formula (5.36) it follows that the frequency increases as radio waves propagate toward a region with a stronger gravitational field. It should be noted that for a system operating in the coherent mode, the total gravitational frequency shift is zero (this is also the case during the radar observations of planets), since gravitational frequency shifts have opposite signs for direct and back radio signals. The authors of [55] described the experiment when a spacecraft-borne highly stable frequency standard was launched to an altitude of 10^4 km. To improve the accuracy of measurements and eliminate the interfering effect of the ionospheric plasma, two coherent radio waves were used. The results of the experiment unequivocally demonstrated the validity of formula (5.36). More experimental evidence on the effect of gravitational fields on radio waves can be found in [34, 56–59].

In the above analysis of gravitational effects, we postulated the validity of formula (5.16) and used the important relation (5.34). A more stringent consideration is possible in terms of the theory of general relativity. The influence of gravitational fields on radio waves is insignificant; however, the precise measurements of the respective gravitational effects is of crucial importance for the experimental verification of the theory of general relativity [56, 57].

The effects emerging during the propagation of radio signals in space can also be used for the detection of gravitational waves, since they influence the distance between two spacecraft located far away from each other. The induced change in the distance can be detected by measuring precisely the Doppler frequency shift in the radio signals used for communications between these spacecraft [60, 61].

References

1. Perelman, S., Russel, W., and Diksar, F. (1960) *IRE Trans. Milit. Electron.*, **4**, 2/3: 184.
2. Pratt, H.I. (1960) *IRE Trans. Communic. Syst.*, **8**, 4: 214.
3. Mueller, G.E. (1960) *Proc. IRE*, **48**, 1: 557.
4. Perelman, S., Kelly, L., and Stuart, W. (1965) *IRE Trans.*, **CS-7**, 3: 167.
5. Sky and Telescope (1970), **40**, 6: 338.
6. Shifrin, S.Ya. (1970) *Questions of the Statistical Theory of Antennae*, Moscow: Soviet Radio (in Russian).
7. Smith, A.G. (1960) *Proc. IRE*, **48**, 4: 593.
8. Forward, A.G. (1960) *Microwave J.*, **3**, 12: 73.
9. Giddis, A.R. (1963) *IEEE Trans. and Aerospace*, **AS-1**, 2: 887.
10. Verschuur, G.L. and Kellerman, K.I. (1974) *Galactic and Extragalactic Radio Astronomy*, Springer-Verlag.
11. Manchester, R.N. and Taylor, I.H. (1977) *Pulsars*, San Francisco: FREEMAN and Company.
12. Smith, F.G. (1977) *Pulsars*, N.Y.: Cambridge University Press.
13. Riegler, G.R. and Blanford, R.D. (1982) *The Galactic Center*, American Inst. of Physics.
14. Ginsburg, W.L. (1970) *Propagation of Electromagnetic Waves in Plasma*, Pergamon Press.
15. Benediktov, E.A., Efimova, T.V., and Skrebkova, L.A. (1969) *Astronomicheskii Zhurnal*, **46**, 2: 286 (in Russian).
16. Altunin, V.I. (1980) *Astronomicheskii Zhurnal* (Soviet Astronomy J.), **24**, 6: 677 (in Russian)*.
17. Shishov, V.I. (1980) *Astronomicheskii Zhurnal* (Soviet Astronomy J.), **24**, 2: 321 (in Russian)*.
18. Pynzar, A.V. and Shishov, V.I. (1980) *Astronomicheskii Zhurnal* (Soviet Astronomy J.), **24**, 6: 685 (in Russian)*.
19. Kaplan, S.A., (Ed.), (1969) *Extraterrestrial Civilizations*, Moscow: Nauka (in Russian).
20. Eruchimov, L.M. (1986) Effect of the Radio Wave Propagation in the Interstellar Medium on the Signal of Extraterrestrial Civilizations, in: *Problems of the Search of the Life in the Universe*, Moscow: Nauka (in Russian).
21. Tarter, D. (1986) The Space Haystack and Current Programms SETI in USA, in: *Problems of the Search for Life in the Universe*, Moscow: Nauka (in Russian).
22. Tarter, D. (1986) SETI Observations Worldwide, in: *The Search for Extraterrestrial Life*, Papagiannis, M.D. (Ed.), JAU, p. 271.
23. Lodge, O. (1919) *Nature*, **104**, 2614: 354.
24. Skrozky, G.V. (1957) *Doklady Akademii Nauk SSSR*, **114**, 1: 73 (in Russian).

25. Plebanski, I. (1960) *Phys. Rev.*, **118**, 5: 1396.
26. Volkov, A.M., Izmestjev, A.A., and Skrozky, G.V. (1970) *Zhurnal Experimentalnoi i Teoreticheskoi Fiziki*, **59**, 10: 1254 (in Russian).
27. Einstein, A. (1911) *Annalen Phys.*, **35**: 898.
28. Muhleman, D.O., Ekers, R.D., and Fomalont, E.B. (1970) *Phys. Rev. Lett.*, **24**, 24: 1377.
29. Seilstad, G.A., Sramek, R.A., and Weiler, K.W. (1970) *Phys. Rev. Lett.*, **24**, 24: 1373.
30. Sramek, R. (1971) *Astrophys. J.*, **167**, 2: 155.
31. Fomalont, E.B. and Sramek, R.A. (1976) *Phys. Rev. Lett.*, **36**, 25: 1475.
32. Robertson, D.C. and Carter, W.E. (1984) *Nature*, **310**, 6012: 572.
33. Cowling, S.A. (1984) *Mon. Notic. Roy. Astron. Soc.*, **209**, 2: 415.
34. Krisher, T.P., Anderson, I.D., and Taylor, A.H. (1991) *Astrophys. J.*, **373**: 665.
35. Einstein, A. (1936) *Science*, **84**: 506.
36. Bjalko, A.V. (1969) *Astronomicheskii Zhurnal*, **46**, 5: 998 (in Russian).
37. Ohanian, H. (1973) *Phys. Rev.*, **D8**, 8: 2734.
38. Ingel, L.K. and Rubakha, N.R. (1978) *Radiofizika* (Radiophysics and Quantum Electronics), **21**, 1: 58 (in Russian)*.
39. Bliokh, P. and Minakov, A. (1975) *Astrophys. and Space Sci.*, **34**, 2: 17.
40. Eshleman, R. (1979) *Science*, **205**, 4411: 1133.
41. Young, P., Gunn, I.E., Kristian, W.I., *et al.* (1981) *Astrophys. J.*, **244**, 3: 736.
42. Deguchi, S. and Watson, W.D. (1986) *Astrophys. J.*, **307**, 1: 30.
43. Kayser, R., Surdej, I., Condon, I., *et al.* (1990) *Astrophys. J.*, **364**, 1: 15.
44. Bliokh, P.V. and Minakov, A.A. (1989) *Gravitational Lenses*, Kiev: Naukova Dumka (in Russian).
45. Shapiro, I.I., Ash, M.E., Ingalls, R.P., *et al.* (1971) *Phys. Rev. Lett.*, **26**, 18: 1132.
46. Anderson, I.D., Exposito, P.B., Martin, W., and Muhleman, D.O. (1972) *Space Res.*, **12**: 1623.
47. Anderson, I.D., Esposito, P.B., Martin, W., and Muhleman, D.O. (1975) *Astrophys. J.*, **200**, 1: 221.
48. Fomalont, E.B. and Sramek, R.A. (1975) *Astrophys. J.*, **119**, 3: 749.
49. Reasenberg, R.D., Shapiro, I.I., and Goldstein, R.B. (1979) *Astroph. J. Letters*, **234**: 219.
50. Shapiro, I.I. (1980) in: *General Relativity and Gravitation*, A. Held (Ed.), vol. 2, p. 469.
51. Reasenberg, R.O., Shapiro, I.I., Goldstein, R.B., and Macneil, P.E. (1982) *Acta Astronautica*, **9**, 2: 91.
52. Press, W.H., Rybicki, G.B., and Hewitt, I.N. (1992) *Astrophys. J.*, **385**, 2: 416.
53. Van Ommen, T.D., Jones, D.L., Preston, R.A., and Iauncey, D.L. (1995) *Astrophys. J.*, **44**, 2: 561.
54. Yakovlev, O.I. (1974) *Radio Wave Propagation in the Solar System*, Washington: NASA (Technical translation).
55. Vessot, R.F. and Levine, M.W. (1979) *Gen. Relat. and Gravit.*, **10**, 3: 181.
56. Konopleva, N.P. (1977) *Uspekhy Fizicheskikh Nauk*, **123**, 4: 537 (in Russian).
57. Will, C.M. (1981) *Theory and Experiment in Gravitational Physics*, London–N.Y.: Cambrige University Press.
58. Krisher, T.P., Anderson, I.D., and Campbell, I.K. (1990) *Phys. Rev. Lett.*, **64**, 12: 812.
59. Krisher, T.P., Morabito, D.D., and Anderson, I.D. (1993) *Phys. Rev. Lett.*, **70**, 15: 218.
60. Bertott, B. *et al.* (1993) *Astron. Astrophys.*, **269**, 608.
61. Tinto, M. and Armstrong, J. (1998) *J. Phys. Rev.*, **D58**, 2002.

Subject Index

I-MONTH